Advances in Industrial Control

For further volumes:
http://www.springer.com/series/1412

German Ardul Munoz-Hernandez
Sa'ad Petrous Mansoor • Dewi Ieuan Jones

Modelling and Controlling Hydropower Plants

Springer

German Ardul Munoz-Hernandez
Instituto Tecnologico de Puebla
Av. Tecnologico, Col. Maravillas 420
Puebla
Mexico

Sa'ad Petrous Mansoor
School of Computer Science
Bangor University
Dean Street
LL57 1UT Bangor
United Kingdom

Dewi Ieuan Jones
GWEFR Cyf
Pant Hywel
Penisarwaun
Caernarfon
Gwynedd LL55 3PG
United Kingdom

ISSN 1430-9491 ISSN 2193-1577 (electronic)
ISBN 978-1-4471-2290-6 ISBN 978-1-4471-2291-3 (eBook)
DOI 10.1007/978-1-4471-2291-3
Springer London Dordrecht Heidelberg New York

Library of Congress Control Number: 2012940343

Printed on acid-free paper

Springer is part of Springer Science+Business Media (www.springer.com)

Series Editors' Foreword

The series *Advances in Industrial Control* aims to report and encourage technology transfer in control engineering. The rapid development of control technology has an impact on all areas of the control discipline. New theory, new controllers, actuators, sensors, new industrial processes, computer methods, new applications, new philosophies . . . , new challenges. Much of this development work resides in industrial reports, feasibility study papers and the reports of advanced collaborative projects. The series offers an opportunity for researchers to present an extended exposition of such new work in all aspects of industrial control for wider and rapid dissemination.

Hydroelectric power has an important contribution to make to the supply of electricity in many countries. In the developed nations, those countries that have suitable topological features and plentiful rainfall have long exploited their hydroelectric potential. Norway, for example, derives 98% of its electric supply from hydroelectric power resources. In the so-called BRIC countries (Brazil, Russia, India and China) and the developing nations, some very large hydroelectric power plants have been constructed, are being constructed or being planned, thus there is a significant growth in hydroelectric power resources worldwide.

Hydroelectric power stations have many advantages in the context of climate change and the desired global reduction in CO_2 emissions. Yet, the focus of attention in the renewable energy field is often on solar power, wind energy and biomass, while hydroelectric resources are often overlooked. Nevertheless, the exploitation of hydropower is a long-standing, well-proven, reliable technology.

Pumped-storage hydroelectric facilities are one particular configuration within the field of hydroelectric power station technologies. Unlike many renewable energy technologies, hydroelectric power is not a recent development; schemes have been operating from the beginning of the last century. Pumped storage is a recirculant hydroelectric system that is able to use cheaper nighttime electricity tariffs to store energy (the pumped storage) ready for use in daytime electricity generation. However in modern times, the fast response times of hydroelectric power stations have been put to strategic use in the "peak-lopping" and network frequency control roles within large-scale electrical power networks. All this and more can be found in this *Advances in Industrial Control* monograph by German A. Munoz-Hernandez, Sa'ad P. Mansoor and Dewi I. Jones.

The book pursues three main themes for hydroelectric power plants, and pumped storage in particular:

1. Hydropower stations per se, comprising presentations on historical perspectives, a survey of the technology and its classification, industrial standards and the outlook for the future development and installation of hydroelectric power technology.
2. Control for pumped-storage facilities, comprising full presentations on plant models, model identification, control mechanisms, simulations (software and "hardware-in-the-loop"), culminating in detailed design and assessment material on classical controllers (PID) and model predictive control for these systems.
3. Case study material on Dinorwig Hydroelectric Power Station. This is a pumped-storage facility in North Wales, UK. Modelling activities showed this plant to contain non-minimum phase process behaviour, and the identification experiments for the plant controller revealed signal conditioning components in the PID controller.

The monograph shows how all the skills of the control discipline are used in a real-world control engineering study. As such, the volume is a very appropriate addition to the *Advances in Industrial Control* series. Two chapters in the monograph examine the potential benefits that are obtained by using methods from the model predictive control paradigm in place of PID control. For the reader interested in learning more about model predictive control, the *Advances in Industrial Control* series contains the monographs:

Predictive Functional Control, Jacques Richalet and Donal O'Donovan, ISBN 978-1-84882-492-8, 2009

Model Predictive Control System Design and Implementation with MATLAB®, Liuping Wang, ISBN 978-1-84882-330-3, 2009

Moreover, from the *Advanced Textbooks in Control and Signal Processing* series there is the now classic textbook:

Model Predictive Control (Second Edition), Eduardo F. Camacho and Carlos Bordons, ISBN 978-1-85233-694-3, 2004.

Industrial Control Centre
Glasgow, Scotland, UK

M.J. Grimble
M.A. Johnson

Preface

Hydraulic turbines have been used as prime movers from the earliest days of electricity generation. Drawing its fuel from natural processes, hydroelectricity is widely regarded as a sustainable technology that makes a substantial contribution to low-carbon energy production. Because of its maturity, hydropower is sometimes associated with sluggish, even moribund, technology development. The truth is very different because, as for any competitive business, survival has required the hydropower industry to take full advantage of advances in science and engineering. There exists today a very active niche area of hydro oriented research, encompassing a whole range of topics – from dams and water management to turbines and power electronics – and one of these topics is control engineering. This monograph describes some of the techniques used for controlling hydro-electric plant, with particular emphasis on pumped storage schemes, whose modes of operation give rise to the most demanding set of control problems.

Effective control of hydroelectric plant is important because small improvements in generating efficiency and reduced operating costs can make the difference between profit and loss in a de-regulated trading environment that is very different to that of just 20 years ago. However, this desire for performance has to be tempered with the absolute need for safety and reliability, at the generating station itself and in a wider sense for the power system as a whole on which society is so dependent. The role of pumped storage schemes in particular is changing as variable energy sources, such as wind and solar, are progressively introduced onto power networks. The need to counteract the poor predictability of these sources and regulate the supply has changed the role of pumped storage schemes from traditional load-shifting to more active intervention on a shorter time scale. This trend is set to continue and advanced control methods will play a vital part in achieving future performance goals.

For many years, the authors have been privileged to work alongside one of Europe's largest pumped storage schemes, located at Dinorwig in north-west Wales. Dinorwig power station has six 330 MW rated hydro-turbines which, when operating in 'spinning in air' mode, can ramp from zero to full load in 12–15 s. It is a vital component of control for the United Kingdom power network. Much of this monograph draws on our experience with this plant in the form of an extended case study. However, the material is mostly generic in nature and applies equally well to many other stations around the world which are presently being refurbished, extended or built as new.

We have attempted to collect together in one place the essential material on modelling, simulation and control of hydroelectric plant and to place it in its proper industrial context, its history, current status and potential future role. In that respect, the first two parts of the book will hopefully be attractive to practising control engineers in the hydropower industry, both as an introduction to the topic and as a reference. Further, it is hoped that the final part of the book, which demonstrates the promise of advanced control methods, will stimulate the industry to move them forward from the realm of theory and simulation to being implemented on real plant – the ultimate test! We also believe that the mixture of classical and modern control, discussed in the book in terms of a specific and real application, will be useful reading for graduate students. Much of the material should also be accessible to final year undergraduate students studying control systems.

The book is divided into three parts:

- The first part includes a brief historical introduction to hydropower and how it developed to its present state. This is followed by a review of different types of hydroelectric schemes with a particular focus on the role played by pumped storage schemes in regulating grid supply. Finally, an overview of control methods is given, from early mechanical governors to today's electronic implementation. This part ends by outlining the primary industrial standards which apply to controlling hydroelectric plant.
- The second part deals with system modelling, covering all the major components required to set up a computer simulation of a hydroelectric plant connected synchronously to a power network. A typical hydroelectric station has a single tunnel drawing water from an upper reservoir into a manifold, which splits the main flow into several penstocks, each supplying a pump/turbine that drives a motor/generator on a common shaft. The power is modulated by means of a guide vane that regulates the flow of water, normally under feedback control of an electronic governor. In frequency control mode, the reference input to the power loop is the frequency deviation of the power network from its set point. The book reviews linear and nonlinear models of hydroelectric power stations, with both single-input, single-output (SISO) and multivariable characteristics. The dynamics of all the main subsystems are discussed, including conduit hydraulics, turbine, generator, guide vanes and the power network. A chapter that explains the benefits of hardware-in-the-loop simulation is included.
- The final part of the book focuses on methods of control. Starting with the classical approach that is prevalent in the industry at present, it progresses to the use of feed-forward as a means of achieving closer tracking to demanded power. The possibilities offered by using more advanced techniques such as model predictive control (MPC) and mixed logical dynamical generalised predictive control (MLD-GPC) are then considered. Simulation shows that MPC offers significantly better performance than a conventional governor, across the plant's

operating range. Constrained GPC produces a faster response when the station is operating with a single unit while preserving stability as the operating conditions change when multiple units are on-line. MLD-GPC control is faster and less sensitive than constrained GPC.

The book ends by briefly reviewing the status of hydroelectricity in the current global market, concluding that ample opportunities remain for large-scale base load development. For the moment, pumped storage schemes remain the only industry-tested and economically viable means of rapidly storing and releasing energy in bulk. However, steady improvements in other methods of energy storage, such as flywheels and chemical batteries, may soon pose a realistic challenge to this hitherto unique capability. These technologies may well meet and clash in competition for the role of maintaining power system stability and quality of supply. To compete effectively, pumped storage must seek to improve in three key areas:

- Accuracy of power delivery
- Speed of response
- Flexibility of operation

It is our hope that this book will help in this aspiration.

Acknowledgements

This book, and the research programme from which it largely derives, would not have been possible without the assistance of Capel Aris (Consultant) and Glyn Jones, Arwel Jones and Toni Jones (all at the Control & Instrumentation Group, First Hydro Company). We also wish to thank all our colleagues who have been part of the research effort, especially David Bradley and David King (both of Abertay University) and Carlos Arturo Gracios-Marin (at Instituto Tecnologico de Puebla). Our thanks go to the series editors and the staff at Springer for their assistance. Finally, the authors would like to thank their families for all their support and consideration during the writing of the book.

Contents

Part III Controlling the Power Plant

Part I
Hydropower Plants

Chapter 1
Hydropower: A Historical Perspective

1.1 Introduction

The principles of hydropower are simple. Heat from the sun causes large amounts of water vapour to rise, thus raising its potential energy. Clouds serve to concentrate the vapour into droplets which fall as rain on high ground. This is further concentrated by natural channelling into rivers, providing a convenient source of 'free' renewable fuel. Extracting energy from the water consists of:

- Diverting some of the water into an artificial channel (leat)
- Providing some water storage to even out the flow and head (mill-pond)
- Using a machine to convert the energy in the water into rotational mechanical energy (waterwheel)
- Controlling the system inflow (sluice gate) and mechanical output (gear engagement)

These essential features exist in some of the earliest historical descriptions of hydropower systems and are still recognisable in the most sophisticated of modern-day installations.

1.2 Waterwheels and Turbines

There is literary evidence that the virtues of hydropower were known to the ancient Greeks, and historians believe that waterwheels probably appeared independently in China, roughly during the same period [1]. Technology advances were already apparent during Roman times, with hydropower subsequently expanding to become one of the most important prime movers available to the medieval world. Hydropower technology continued to improve, even surviving the coming of steam during the industrial revolution as the waterwheel gradually mutated into the hydraulic turbine.

G.A. Munoz-Hernandez et al., *Modelling and Controlling Hydropower Plants*, Advances in Industrial Control, DOI 10.1007/978-1-4471-2291-3_1,

Fig. 1.1 The 'Lady Isabella' waterwheel at Laxey in the Isle of Man (Courtesy GWEFR Cyf)

The zenith of waterwheel technology is well represented by the 'Lady Isabella' at Laxey in the Isle of Man, which is now the world's largest working installation (Fig. 1.1).

Built in 1854, it was used to pump water from the nearby mine that produced lead, copper, silver and zinc. The wheel itself is of pitch back-shot type and has a diameter of 22 m and width of 1.8 m. Revolving at 2–3 rpm and rated at about 150 kW, the wheel's power was transmitted to the top of the mineshaft by a wooden crank atop a 400 m long viaduct. An inverted T-shaped rocker changed the horizontal movement of the crank to a vertical movement of the pump rods. Somewhat ironically, the forthcoming demise of the traditional waterwheel is also represented at Laxey where, in 1856, the machine house was equipped with a water turbine of the Fourneyron type to provide power for raising ore up the mineshafts. This is one of the earliest recorded uses of this type of turbine in the British Isles.

A simple mathematical model of an idealised waterwheel reveals its limitations. Consider an overshot waterwheel whose buckets are free to pivot around a horizontal axis and fed at the top so that the momentum of the input flow does not contribute torque to the wheel. Assume that each bucket is precisely filled at the top and held horizontal as the wheel rotates until it reaches the bottom, at which point the water is emptied. Denny [2] shows that the total input power (P_{in}) to the wheel is then

$$P_{in} = \rho g Q H, \tag{1.1}$$

where ρ is the density of water, g the acceleration due to gravity, Q the volume flow rate of water and H the diameter of the wheel.

The first two parameters in Eq. 1.1 are fixed. Increasing the diameter of the wheel makes it heavier and is ultimately limited by structural considerations. The flow rate can be increased by making the buckets bigger but this means a wider wheel, leading once more to structural limitations. So there are fundamental limitations on how much input power a basic waterwheel can accept, even in its idealised form.

An alternative method of increasing the flow rate in (1.1) is to make the wheel spin faster. However, the faster the wheel spins, the more water is lost from the buckets due to centrifugal force, causing its efficiency to suffer as Denny shows by means of a more realistic model [2].

The efficiency of waterwheels was studied by John Smeaton. In 1759, he presented the results of experiments that showed the maximum efficiency then achieved by an overshot wheel to be 63%. This inspired research into methods for improving their efficiency, such as the introduction by Sir William Fairbairn in 1828 of ventilated buckets that allowed more rapid filling and emptying of the compartments. However, the first step to circumventing the limitations of the waterwheel had already been taken by Bélidor, who in 1737 described vertical axes waterwheels with curved blades that reduced hydraulic losses – a crucial move towards the turbine. In 1827, a revolutionary design was introduced by Fourneyron – a 4.5 kW reaction-type turbine that had two sets of blades, curved in opposite directions, which impart a swirl component of force to the spinning runner. This machine:

- Worked with a wide range of head, from a few meters to hundreds of meters
- Could accommodate a large flow of water
- Was physically small
- Could spin at 2,000–3,000 rpm
- Achieved 80–85% efficiency

Improvements to the Fourneyron turbine (Latin *turbo, turbinis*, meaning 'whirling' or 'vortex') followed rapidly, now aided by a growing understanding of the theoretical foundations of fluid mechanics. The work of Lazare Carnot (1753–1823) and Jean Charles de Borda (1733–1799) had led to the Borda–Carnot equation of fluid dynamics which relates the energy loss (ΔE) in a fluid due to a sudden change in the velocity of the flow, such as at an expansion or contraction in a pipe:

$$\Delta E = \frac{1}{2} \xi \rho (v_1 - v_2)^2, \tag{1.2}$$

where ρ is the density of the fluid, v_1, v_2 the flow velocities before and after the expansion ($v_1 > v_2$) and ξ the empirically determined loss factor ($0 \leq \xi \leq 1$).

Fig. 1.2 Photograph of a modern Francis turbine runner showing the intricate geometry of the inflow and outflow vanes (Courtesy First Hydro Company)

The implication of Eq. 1.2 is that hydraulic losses can be reduced by avoiding sudden changes in flow velocity. This insight led to the influential contributions made by Uriah Boyden in the USA between about 1840 and 1850. By considering the detailed path or 'streamline' of water through the turbine, Boyden was able to modify its geometry such that hydraulic losses caused by friction and nonlaminar flow were significantly reduced. Boyden collaborated extensively with Francis and their work led to the realisation that the Fourneyron and Boyden turbines have a fundamental problem. Both of these are of the outward-flow type, meaning that the water is fed in axially at the centre of the turbine and exhausted in a radial direction. As a consequence, the channels through which the water flows get wider as the water moves towards the circumference of the runner. This causes the flow to separate, which results in eddies and loss of energy. The answer was an inward flow machine, patented by Howd in 1838 and much improved by James Francis in 1849, to the extent that it is now known universally as the *Francis turbine*. It is the most common turbine in use today – a photograph of a typical runner is shown in Fig. 1.2. Francis's contribution to hydraulics and science in general extends well beyond turbine design as described in an essay by Layton [3]. The work of these pioneers still resonates in today's studies on the effects of cavitation, vortices and wakes, although the tools of 3D computational fluid dynamics and high-power computers to solve the Navier–Stokes equations are very different.

The final major type is the *Kaplan turbine*, an axial flow reaction turbine developed around 1913. Like the Francis turbine, water enters through a spiral casing but it is diverted axially onto a propeller whose blade pitch can be controlled. They tend to be used at sites where the head is relatively low but high water flow is available.

In the earlier discussion of the idealised waterwheel, it may be recalled that the momentum of the input flow was ignored, so that the torque was produced only by the gravitational potential energy of the water. Dropping the water onto the wheel from a height (effectively an increase in head) would impart an additional torque, provided the point of impact is not directly above the axle. However, the work of de Borda and Carnot had shown that the most efficient waterwheels were those that operated smoothly, the water entering the buckets smoothly at the top from a flume placed at the same level and streaming smoothly from the buckets at the bottom into the tailrace. Equation 1.2 explains this observation because dropping the water onto the buckets causes splashing, i.e. turbulent changes of velocity, and a consequent hydraulic loss. Although the increased head produces a net increase in the input power, the overall efficiency is less.

The preceding discussion seems to discourage any machine that involves water impact so it is perhaps surprising that the impulse turbine was developed at all. The modern impulse turbine originates in the California gold rush where in 1866 Samuel Knight steered a high-velocity jet of water tangentially onto a wheel fitted with small buckets on its rim. The change in momentum (impulse) which occurs as the jet strikes the bucket imparts a force and therefore a turning moment on the wheel. The high-velocity jet is produced by a nozzle fitted to a high-head water supply, thus converting virtually all its potential energy to kinetic energy in the jet. Just as in the water wheel and reaction turbine, the hydraulic losses are minimised by ensuring a uniform flow of water. A vital improvement to Knight's design was made by Pelton who introduced a double bucket with a semi-circular profile that smoothly reverses the water jet and exhausts the water to either side with a minimum of splashing. This was further refined by Doble who used an elliptical bucket profile capable of extracting almost all of the jet's kinetic energy and achieving efficiencies over 90%. Figure 1.3 shows the runner for a 1.0 MW Pelton turbine used at the Cwm Dyli power station in North Wales from 1906 to 1988 [4].

By the mid-nineteenth century, the energy conversion efficiency of the hydraulic turbine was better than that of the steam engine – and the fuel was more or less free too. Its big drawback was that it had to be near a water supply. A steam engine could be placed next to the factory which would use its power. Coal – a concentrated source of energy – could be brought to the same site. This allowed the factory to be placed conveniently for the transport of both raw materials and manufactured goods, next to a railway or port for instance. Using water power often required the factory to be sited in some relatively remote upland location, where transportation costs undermined the benefits of cheap power. Excellent technology notwithstanding, the growth of hydropower was constrained by economic factors – until electricity came along.

Fig. 1.3 The runner for a Pelton turbine (Courtesy GWEFR Cyf)

1.3 Hydroelectricity

Electricity separated the point of power generation from the point of consumption. Hydropower could now be produced at the most advantageous sites – mountains and lakes offering an ample supply of water and a high head – and transmitted to the towns and cities where it would be used. It is no coincidence that 98.5% of Norway's electricity supply in 2008 came from hydroelectricity!

It is difficult to pin down exactly where and when the first hydraulic turbine and generator were connected together to create hydroelectricity. However, a fair claim to this distinction can be made by Sir William Armstrong, an eminent industrialist from the north east of England whose country residence, Cragside House, became in 1878 the first to be lit by hydroelectricity – a single arc lamp in the picture gallery. The system consisted of a Siemens series-wound dynamo driven by a Vortex turbine, supplied by Williamson Brothers of Kendal.

The inventor of the *Vortex* turbine was James Thomson (1822–1892) who had trained with Sir William Fairbairn [5]. Patented in 1850, the *Vortex* was an inward-flow reaction turbine with characteristics similar to those of a Francis turbine. It worked on any head of water from 1 to 100 m, had an efficiency of 70–75% and was compact with a 0.9 m diameter. In 1881, Williamson Brothers was bought by Gilbert Gilkes & Gordon Ltd. who went on to manufacture 440 Vortex turbines for export all over the world. They remain a prominent supplier of turbines to this day.

In a parallel development in the USA, a hydraulic turbine driving a brush dynamo (DC) was used to provide theatre and storefront lighting using an arc lamp in Grand Rapids, MI, in 1880. A similar arrangement was lighting the streets of Niagara Falls, NY, by 1881. In 1882, the world's first industrial hydroelectric plant began operation on the Fox River in Appleton, WI. The plant used two dynamos supplied by Thomas Edison's company and had an output of about 12.5 kW at 110 V (DC). They were used to light the Appleton and Vulcan paper mills as well as the mill owner's home.

Another landmark development was the construction by the Niagara Falls Power Company in 1895 of the world's first commercial hydroelectric power plant to deliver networked electricity. It used 5,000 kW Westinghouse alternators to produce two-phase AC, an innovation that had been recently patented by Nikola Tesla, and was capable of supplying power to the city of Buffalo, 32 km away. From the beginning, hydropower has been intimately associated with the transmission of electricity over long distances and the foundation of a large-scale electricity grid.

One clear trend over the course of the following century was the construction of hydroelectric stations of increasing capacity in all parts of the world where favourable conditions existed. This culminates in the giant stations at Itaipu on the border between Brazil and Paraguay, rated in 2003 at 14 GW peak, and the Three Gorges complex in Hubei, China, which is rated at 22.5 GW peak. In 2011, feasibility studies were being performed for an even larger project which would lead to construction of a fourth dam across the Congo River at Inga Falls in the Democratic Republic of Congo. Known as the Grand Inga Dam, this scheme would have a capacity of 39 GW peak and be the largest energy-producing source ever built. It is estimated [6] that in 1980 the hydroelectric power installed capacity for the world was 480 GW, rising to 647 GW in 1990 and 921 GW by 2007, representing 21% of all installed electrical capacity.

It is generally recognised that hydroelectricity is a clean, efficient and reliable form of energy whose contribution to greenhouse gas emissions is negligible compared to fossil fuels. It is also an effective technology for flood protection and stabilisation of water supply. Hydroelectric stations tend to be reliable, long-lived and have relatively low operating costs. However, large-scale hydroelectric schemes do have some well-known drawbacks:

- Establishing the reservoirs often requires large-scale displacement of the human population, which leads to criticism and local opposition and particularly so if inadequate compensation is offered.
- The environment is affected – animal and plant life is disrupted and delicate ecosystems can sometimes be put in peril.
- The course and flow of river systems can be altered, which can be contentious if the watercourse crosses national boundaries.

Once the technical feasibility of a site has been established and environmental factors resolved, there remains the question of economic viability. Conventional

Table 1.1 Examples of major hydroelectric projects under construction in 2011

Country	Project	Planned capacity (MW)	Planned end date
Brazil	Belo Monte	11,200	2011
Russia	Boguchan	3,000	2011
India	Siang Upper	11,000	2024 (phased)
China	Xiluodu	12,600	2015

hydroelectric schemes have a high initial capital cost and long construction period. Naturally, most of the best sites have already been used, so the continued expansion of hydroelectricity depends on a combination of factors:

- Acceptable construction costs and availability of investment finance
- Cost of competing energy sources
- Development of sizeable power markets reasonably near to the proposed hydroelectric site
- Prices that support the cost of transmission to distant markets

In 2011, most of the major projects under construction were in the developing world, especially the so-called BRIC countries (Brazil, Russia, India and China) as shown in the examples in Table 1.1.

1.4 Pumped-Storage Hydroelectricity and Grid Control

The special characteristics of hydroelectric power became apparent almost as soon as different types of power plant were connected to a common power network. It was recognised that that they were especially well suited to regulating the network as well as fulfilling the normal role of a power source. As a result, a specialised form of hydroelectric plant known as *pumped storage* was devised.

It is generally thought that the principle of pumped hydro was first demonstrated in 1882. It consisted of a hydro-mechanical storage system that operated a pump and turbine for nearly a decade at a small reservoir near Zurich, Switzerland [7]. In 1908, Voith installed the first pumped hydroelectric scheme in Germany at their R&D establishment in Heidenheim. The first commercial plant, rated at 1.5 MW, was opened in 1909 near Schaffhausen, Switzerland. The Rocky River plant, constructed in 1929 on the Housatonic River in Connecticut, was the first such installation in North America.

The essential idea is to store some of the electricity produced at night, when the load is light, and to use it during the day, when the load is heavy. Water is pumped to a high reservoir using relatively cheap base-load electricity and then released back to a low reservoir for generation, when demand on the power network is high. Losses in the conversion process mean that a pumped storage scheme is a net consumer of electricity but the difference between day and night prices allows the operator to make a profit. This strategy reduces the need to change the output power of thermal plant (coal and nuclear) from their optimum settings, which leads to inefficient production.

Hydroelectric plants are more responsive than thermal plant – their output can be varied between zero and full load in a matter of minutes, or even seconds in some cases. They are therefore used to compensate for sudden load transients on the power network, sometimes known as 'peak-lopping'. Modern pumped storage plant are constructed for fast response duties and are equipped with advanced control systems for continuous automatic control of network frequency (sometimes known as 'part-load' operation). These control methods will be one of the main topics of this book. More recently, an extended role for pumped storage as a means of compensating for the intermittent nature of renewable power sources is foreseen [8]. Today, the global capacity of pumped hydro storage plants totals more than 95 GW and at the moment it remains the only large-scale form of energy storage.

Significant advances in pumped storage technology have taken place over the last century. Early installations were equipped with a motor and pump on one shaft and a generator and turbine on a separate shaft – an expensive and clumsy arrangement. This was replaced with a tandem system, composed of a single vertical shaft supporting a motor-generator, pump and turbine. This is a more compact arrangement which saves on electrical plant and the civil works needed for housing.

In his pioneering theoretical work, Carnot had explained that an imaginary perfect waterwheel would neither waste nor dissipate any of the water's energy and all the motion would be completely reversible. In other words, if the perfect waterwheel were run backward, it would become the perfect pump. As noted previously, the efficiencies of turbines in the mid-nineteenth century already approached 90% and inventors were inspired to imitate Carnot's theoretical machine by using the turbine in reverse to pump water. By the 1930s, the reversible pump-turbine was a reality and the first known installation was at Baldeney, Germany, in 1933, involving a 1,330 kW machine operating under 8.5 m head [9]. Further progress in applying the reversible pump-turbine occurred in 1953 with the installation of a 15 MW unit at Pedreira, Brazil. In 1956 the Tennessee Valley Authority commissioned a 59.5 MW unit on the Hiwassee River in North Carolina, USA [10] which used wicket gates (guide vanes) to control the turbine output power. Improved hydraulic design, mechanical engineering methods and materials have led to increased size and better efficiency of the pump-turbine. Of particular importance is the increased capability for pumping against a high head which was crucial to its use at many sites.

The most efficient operating point of a pump-turbine varies with head, flow and speed. As a consequence, the efficiency of a machine rotating at synchronous speed and subject to head variation (as the upper reservoir drains) will vary. This has stimulated the development of adjustable-speed pump-turbines whose speed can be varied to track the most efficient operating point. Of course, this means that the mechanical rotor speed is no longer an integer multiple of the power network frequency. This problem is effectively overcome by the doubly fed asynchronous generator. High-power electronics in the form of a voltage source inverter is used to inject a slip frequency current into the rotor winding so that the electrical frequency of the stator is made synchronous with the network. Another advantage of the technique is that it allows pumped storage plants to provide a frequency control service while in 'pumping' mode.

Much of the research on adjustable-speed pump-turbines was done in Japan, where the first installation became operational in 1990.

1.5 Small-Scale and Hydrokinetic Systems

Over the last couple of decades, increasing concern about the environment has stimulated a revival in small hydro projects. For instance, it is estimated that in 1820 there were about 60,000 waterwheels in France alone [2]. However, as in most countries, the vast majority of these fell into disuse as relatively cheap, reliable, high-quality electricity was offered by centrally controlled power networks. Nevertheless, their potential as a source of energy remains – the question is whether they can be developed (or in many cases refurbished from older installations) in a way that is both economically and environmentally sound. To this end, the European Small Hydropower Association has produced a handbook for prospective developers [11].

The IEEE Standard 1020-1988 [12] admits that the term 'small hydroelectric power plant' does not lend itself to rigorous definition. It lists a number of attributes that help to classify them (though the guidelines do vary from one country to another):

- Have unit rating between 100 and 5,000 kW and generated voltages between 480 V and 13.8 kV. (The term 'micro-hydro' refers to unit rating below 100 kW)
- Have either synchronous or induction AC generators
- Are commonly applied where associated civil construction work and costs are minimal
- Are equipped for unattended operation
- Are generally connected to a distribution circuit for supplying a local load, with surplus generation sold to a power company

Most small hydro plants are 'run-of-river' schemes, meaning that they have little or no water storage and the turbine only produces power when the river level allows. They are also relatively low head schemes, the ultimate case being *hydrokinetic power*, referring to the extraction of energy from moving water which occurs in:

- Tidal flow
- Ocean currents
- Ocean waves
- Rivers

In contrast to conventional hydro schemes, it eschews impoundment or any substantial diversion of water from its natural path. Because hydrokinetic turbines are small and operate in-stream, there is no need for the construction of dams, barrages or lagoons or any large civil works. Their main advantages are claimed to be [13]:

- Minimal environmental impact
- Relatively short time from construction to operation
- Competitive energy price
- Predictable and reliable energy production

Like windmills, hydrokinetic turbines extract only the kinetic energy of the flow and they are sometimes known as 'ultra-low-head' or 'free-flow' turbines because there is no pressure drop across them. Many different types have been proposed, including:

- Submerged, part-submerged, floating
- Vertical or horizontal axis
- With or without diffuser (to augment inflow)
- Uni-directional or bi-directional (for tidal flow)
- Turbine/nonturbine technology

Three recent surveys [14–16] review the state of the state of the art in 2009/2010. An illustrative example is shown in Fig. 1.4. The SeaGen prototype marine current turbine takes advantage of the 2.4 m/s tidal race in Strangford Lough, Northern Ireland. The twin 16 m turbines generate 1.2 MW and feed 10 MWh into the UK national grid at every tide.

One of the problems of hydrokinetic turbines is their relatively poor efficiency, stated by Lago et al. as having a typical value of around 35%. This falls well below that of their conventional counterparts. The maximum power available in the flow (P_{in}) is

$$P_{\text{in}} = \frac{1}{2}\rho A v^3, \tag{1.3}$$

where ρ is the density of water, A the area of the inlet (rotor blades) and v the velocity of the flow.

The output power, according to momentum theory, is

$$P_{\text{out}} = \frac{1}{2}\rho A v^3 C_{\text{p}}. \tag{1.4}$$

The parameter C_{p} is known as the power coefficient, the fraction of power extracted from the flow. This is maximised when $C_{\text{p}} = 16/27 = 0.593$, commonly known as the Betz maximum; so the theoretical maximum efficiency for a hydrokinetic turbine is about 60%.

Fig. 1.4 The SeaGen tidal turbine (Copyright SeaGen Ltd.)

In fact, an undershot waterwheel is a basic hydrokinetic turbine and Smeaton's measurements in 1759 had shown that their best measured efficiency was about 22%. In a simplified analysis of an undershot waterwheel with radial vanes, Denny [2] applies momentum theory to derive its power coefficient as

$$C_{\mathrm{p}} = 2c(1-c)^2 \qquad \text{where } c = \frac{v'}{v} \quad 0 < c < 1, \tag{1.5}$$

where v' is the tangential velocity of the vanes. Differentiation shows that setting the velocity ratio to $c = 1/3$ achieves the maximum value of $C_{\mathrm{p}} = 8/27 = 0.296$, which is only half the Betz maximum. The challenge for designers is to at least exceed the performance of the undershot waterwheel and hopefully approach the Betz maximum.

As with other renewable sources, the intermittent nature of small hydro makes it difficult to integrate large numbers of producers onto a power network while maintaining the stability and control which balances production and consumption and ensures quality of supply. By and large, this form of distributed generation only becomes possible when the correct framework is put in place, including:

- Government feed-in tariffs
- Improved electronics and change of legislation to allow small generators to feed power into the grid
- Upgraded transmission and distribution networks
- Communications, information and control technology to allow implementation of a 'Smart-Grid'

1.6 Conclusions

This chapter has given a brief overview of the development of hydropower, from its ancient origins to today's technology, which takes every advantage of progress in science and engineering to sustain and advance a huge and vital global industry. It does not pretend to be a balanced view; the authors are not professional historians (or even amateur ones) and there is just so much history that objective balance would likely be impossible in a whole book, let alone a single chapter. For instance, we are acutely aware that hydropower depends critically on civil engineering – the turbine/generator may be at the 'sharp-end' of a hydroelectric system but it can only succeed with the 'blunt instrument' support of reservoirs, dams and tunnels. But we have made little or no mention of this or indeed many other important aspects of hydropower.

In a busy world, it may be tempting to dismiss historical discourse as irrelevant to the solution of immediate problems. This would be a mistake. First, the scientific principles do not change – de Borda and Carnot's insight into turbulent losses are fundamental, even though the numerical tools now available to analyse water flow are vastly superior. Secondly, history has shaped the technology – we are where we are, technically speaking, because of the economic, social and environmental constraints and opportunities that existed over the last two millennia. Who knows what we may have missed? New conditions, such as may be presented by climate change, will need new and innovative solutions – and the seeds of these are often already present in the historical record, as an 'idea ahead of its time'.

Chapter 2
The Form and Function of Hydroelectric Plant

2.1 Introduction

The foremost principle of modern architecture in the twentieth century was 'form follows function', the idea being that the shape of a building should be largely determined by its intended purpose or function. This principle can be extended to computer architecture – the internal structure and components of an integrated circuit intended for digital signal processing differ significantly from those of a general purpose computer. It may not be the case that this principle applies quite as firmly to hydroelectric plant but they undoubtedly possess both 'form' and 'function'. The purpose of this chapter is to provide an overview of different types of hydroelectric plant (their 'form') and what they are used for (their 'function').

Building a hydroelectric power station requires numerous complex and often conflicting aspects to be taken into account, ranging from global issues to local concerns. For instance, at any prospective location for a hydroelectric plant, the type of installation and its purpose will be strongly influenced by both economic and environmental factors. Every hydroelectric station will therefore have some unique features in its design and construction, so it is convenient here to confine our discussion to broad categories of plant.

2.2 Types of Hydroelectric Plant

A broad distinction can be made between hydroelectric plant that have a large reservoir and substantial energy storage capability and those that do not. The latter are often known as 'run-of-river' schemes and either have no storage at all or only minimal storage in a pond for the purpose of short-term smoothing of the water flow. Storing energy in large reservoirs is very attractive because it provides a dependable and regular source of power. When the reservoir is a natural lake that has an extensive catchment area, building a hydroelectric plant can have minimal

G.A. Munoz-Hernandez et al., *Modelling and Controlling Hydropower Plants*,
Advances in Industrial Control, DOI 10.1007/978-1-4471-2291-3_2,

environmental impact. However, building a dam across a river and flooding a valley to create an artificial reservoir is often unacceptable. The same can be said for partial diversion of water from rivers and streams where the reduced flow may adversely affect plant and wildlife habitat in and around a stretch of the natural waterway.

2.2.1 *Run-of-River Hydroelectric Plant*

These can be sub-divided into:

- ‘Pure’ run-of-river with no pond
- ‘Hybrid’ run-of-river with a small pond

The principal characteristic of the first type is that the running water of the river is used to power the turbine directly – this is sometimes termed ‘hydrokinetic’ power. This type of hydropower plant is totally dependent on the instantaneous state of the river. It is usually the case that a national environment agency will stipulate that the river must have some minimum level of flow before water can be withdrawn, so power generation is restricted to these periods. Some small stand-alone installations will ameliorate uneven production with battery storage. Typically, intrusion into the river or stream is minimal, often just a weir at the point of abstraction. This type of hydroelectric station produces most of its energy during the rainy seasons.

A key design decision is selecting the correct size of power unit. A small turbine will not be able to extract all the power available when the river is high and will capture less energy over the year than a larger one. However, a large and expensive turbine will only be working at capacity for a small fraction of the year and may not justify its capital cost. The optimum size of power unit will be a balance between total energy production and capacity factor. A typical capacity factor of 50–70% will give a satisfactory rate of return on investment for mini-hydro schemes [17].

A ‘hybrid’ run-of-river scheme that includes a pond has several advantages over a ‘pure’ scheme. Although the pond will be limited in capacity, it will smooth short-term flow variation at the turbine. There may even be sufficient capacity to allow increased generation at peak demand during the day with overnight replenishment of the pond when consumption ceases at night.

If there is a significant vertical fall in the river as depicted in Fig. 2.1, then the pond may be placed upstream of the powerhouse and water delivered through a closed pipe in order to increase the head at the turbine.

In practice, there are many variations on the basic layout of a run-of-river scheme.

2.2.2 *Reservoir Hydroelectric Plant*

The principal characteristic of this type of hydroelectric power plant is a reservoir located in an upland or mountainous region. Normally the reservoir has the capacity to store very large quantities of water and its potential energy is available for use

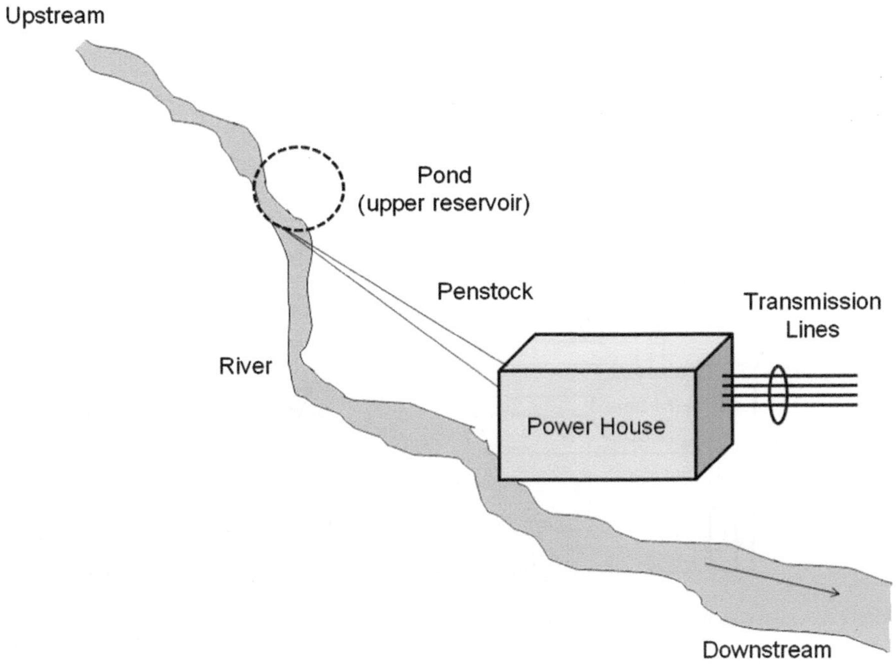

Fig. 2.1 Run-of-river hydroelectric plant with upstream pond

throughout the year. The reservoir accumulates water during the rainy season and drains during the dry season. The water in the reservoir is delivered to the turbine through a combination of tunnel and penstocks, the water flow being controlled by a gate or valve. Large hydroelectric power stations around the world are almost exclusively of this type. When the powerhouse is distant from the reservoir, a single tunnel will often be used to deliver water to a manifold that divides the flow into multiple penstocks, each feeding an individual turbine/generator unit, as shown in Fig. 2.2.

Pumped storage stations are a special case of this form of hydroelectric scheme where there are two reservoirs located at significantly different vertical levels. At times of peak consumption, water is released from the upper reservoir so that the turbine/generators can contribute power to the national grid, the water being collected in the lower reservoir. At times of low demand, water is drawn back from the lower reservoir by pump/motors using electricity drawn from the national grid and deposited into the upper reservoir. The price difference between peak and off-peak electricity allows the scheme to make a financial return despite the inefficiencies involved. Pumped storage schemes act on a *closed cycle*. Natural inflow to the upper reservoir offsets water losses (mostly due to evaporation and seepage) and at times of high rainfall may decrease the pumping requirement slightly but, by and large, it is necessary to maintain the downstream flow at its natural levels and the total water volume within the twin reservoir system remains constant.

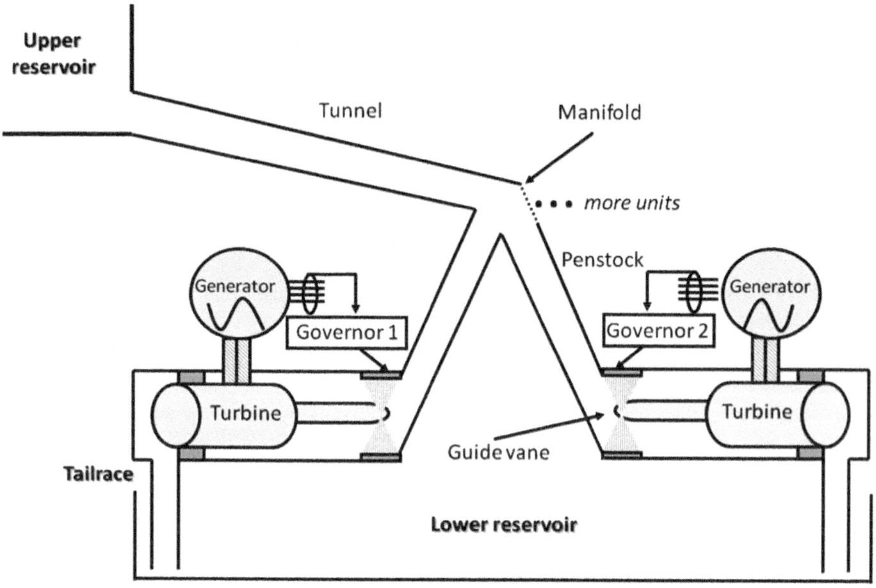

Fig. 2.2 Reservoir hydroelectric power plant

An example of this kind of pumped storage scheme is the Dinorwig power station, which is owned by the First Hydro Company (part of International Power/GDF Suez). Located in north-west Wales and commissioned in the early 1980s, this 1,800 MW rated station is an important supplier of reserve margin, fast response and ancillary services and is integral to the operation of the UK national grid. It will be used as a rolling case study throughout this book.

Three kinds of hydropower stations can be distinguished based on the height of the reservoir above the powerhouse: low, medium and high head plant.

- *Low head*: To be considered a low-head station, the head of water at the turbine will be less than 30 m (although this figure is not consistent in the literature). Typically, the dam in this type of scheme is small and in some installations can be only a few metres in height. The available head largely determines the type of turbine to use. A combination of low head and high flow will usually favour an axial flow propeller type of reaction turbine, such as the well-known Kaplan turbine which has pitch-controlled blades.
- *Medium head*: A medium head station will typically have between 30 and 300 m of head at the turbine. This type of hydroelectric station will generally be situated in a mountainous region and be equipped with a Francis (reaction) turbine or an impulse turbine of the cross-flow type, whose rather flat efficiency curve is less sensitive than other types to variation in flow.

- *High head*: When the available head exceeds 300 m, the station is considered to be high head. Impulse turbines such as the Pelton or Turgo wheels are often found in these installations, although Francis turbines are more usual at the larger power levels.

2.3 The Purpose of Hydroelectric Plant

2.3.1 Grid Requirements

Despite their common underlying principle, it is clear from the previous section that hydroelectric plant come in an immense variety of forms. We now consider their various functions, with a particular focus on the role played by pumped-storage stations. Hydroelectric schemes can be separated into two broad classes according to the service they supply:

- Base load plant
- Peak load plant

Base load plant are intended simply to supply power consistently and reliably to the national grid, ideally throughout the day and throughout the year, with breaks only for maintenance. In countries with an abundance of natural lakes at high altitudes (such as Norway), it is possible for hydropower to produce a very high proportion of the total consumption. Because the fuel is naturally sourced and free, it is a relatively cheap and environmentally acceptable method of producing power. A more common situation is where the base load is supplied mostly by a mixture of fossil-fired and nuclear plant, supplemented by relatively small amounts of renewable sources (mostly wind and solar). The load is not constant as shown in Fig. 2.3, which is a plot of the power consumed on the UK Grid during March 2001. This averages at about 40 GW and has a peak-to-peak variation of about 10 GW [18]. Note the downward trend in the first third of the plot, probably correlated to weather conditions at the time.

Two cyclical components are clearly evident in Fig. 2.3:

- A diurnal variation, usually exhibiting a maximum during the day and a minimum at night
- A weekly variation, with a smaller load at weekends than during the working week

There also exists a seasonal variation, typically with increased consumption during the winter, although this pattern tends to be reversed in tropical countries due to the use of air conditioning in the summer.

Computing the spectrum of this plot reveals other cycles as shown in Fig. 2.4. The relationship of these cycles to physical events is less clear though it can be speculated that the 7.7 h cycle is related to the length of the working day.

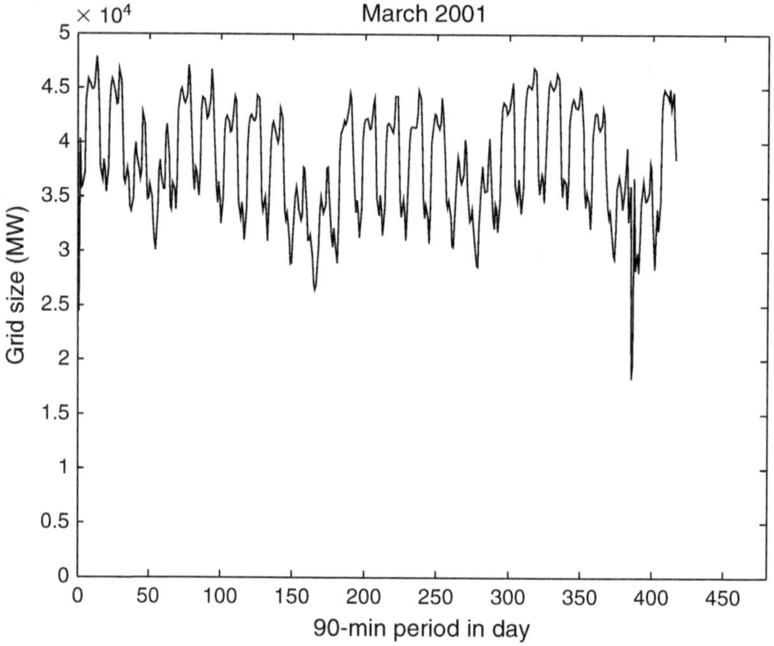

Fig. 2.3 Variation of total system load on the UK Grid during March 2001

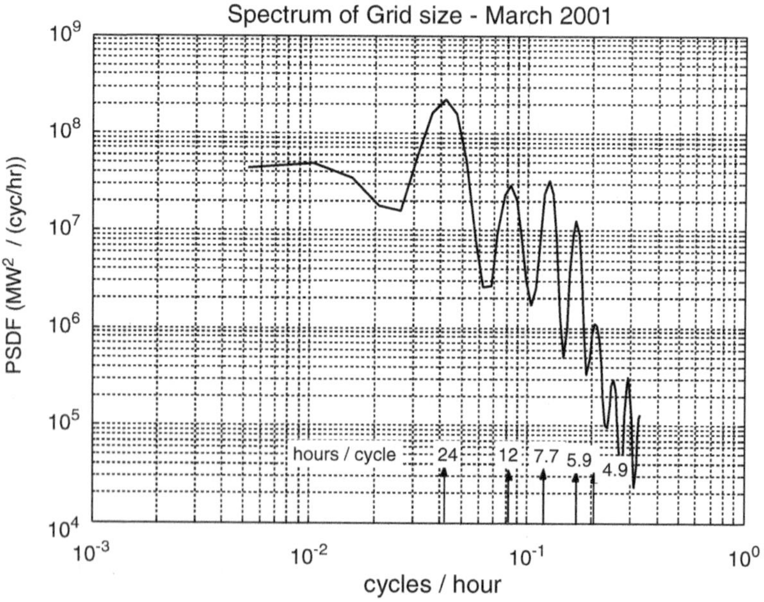

Fig. 2.4 Spectral density function for the grid load

It is essential that the generated power be matched to the load demand in a timely fashion so that the grid frequency remains always within a designated band – this is known as *load frequency control* [19]. In the UK, this band is set by the Electricity Supply Regulations (1998) which require '... the system frequency to be maintained at 50 Hz $\pm 1\%$ save in exceptional circumstances'.

In the UK, the National Grid company is responsible for managing the flow of power on the transmission network. Each day, a forecast of the load for the following day is prepared using a statistical model. The predictions are based partly on historical trends and cycles and partly on the estimated effect of various factors such as the weather, expected usage of different classes of customer and the timing of prominent public events. The forecast is used to plan the dispatch schedules for generating plant, in order to meet the anticipated load and to ensure that there is sufficient reserve to cover deviation between predicted and actual demand. The planned despatch of extra generation during times of heavy demand, such as the early evening surge which occurs as workers go home and night falls, is known as 'peak-lopping'.

2.3.2 Controlling Grid Frequency

Hydroelectric plant are well suited to peak-lopping because they can be switched in and out of production on a scale of minutes compared to the hours required for thermal-based plant. Pumped storage stations on the transmission network are also used for *load shifting* and *frequency control*. The term 'load-shifting' refers to the use of pumped storage plant to compensate for the inflexibility of nuclear and large coal-fired power stations. These base load plant operate most economically if their output is fixed at its optimum value, so attempting to vary their output to match the load leads to inefficient operation. It is preferable to use the excess nighttime generation to pump water in a pumped storage scheme so that it may be released for generation during the day. In effect, this shifts the load from daytime to nighttime.

A risk of unpredictable disturbances exists on any network, such as unexpected outage of a power station or interconnector. The risk level is often set at the capacity of the largest generator on the grid – about 1.3 GW in the UK. Frequency control measures are used to keep the network frequency within the limits set out in the Grid Code[1] if a rapid and large loss of generation or (more rarely) an increase in load occurs. This requires sufficient operating reserves to be present on standby, and

[1]The Grid Code is under continual review and the current version is available on the website of the National Grid company (http://www.nationalgrid.com).

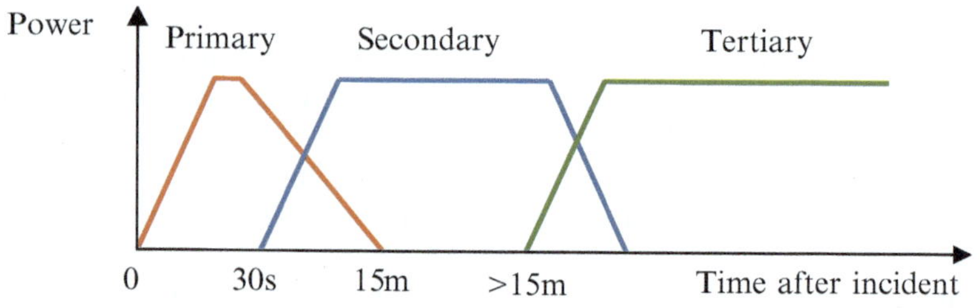

Fig. 2.5 Response of network frequency control providers (not to scale)

coal, oil-fired and combined-cycle gas turbine (CCGT) stations may be required to declare 10–15% of their registered capacity as available for this purpose [20]. There are two types of reserve:

- *Spinning reserve*, consisting of generators that are rotating synchronously with the grid but not producing active power
- *Fast-start reserve*, consisting of generators that are at standstill but can be rapidly brought on line

The response of the network to a large and rapid frequency deviation is dealt with in three stages, known as the primary, secondary and tertiary responses, which act on different time scales to return the network to its nominal operating point [20, 21]. This is illustrated in Fig. 2.5.

The primary response is an initial automatic increase in power output intended to contain the frequency deviation over the first 30 s or so after the event and allow time for the secondary measures to be initiated. As these progressively take over the supply, the primary measures are freed back into the reserve. Once the secondary generators have returned the frequency to an acceptable value, they too can be taken offline and replaced by slow-acting base load generation. The network is now established at its new operating point. Figure 2.6 shows the profile that this strategy is intended to achieve, where Δf_p is the peak frequency deviation.

Some power networks also employ another fine scale control of frequency during normal operation to continuously regulate the frequency towards its nominal value. This is known as the 'regulating reserve' or 'dynamic frequency response' and is provided by placing the governors of several generating plant into closed loop frequency control mode. Typically, this could be implemented by setting a thermal power station to generate at 80–90% capacity with the steam boiler full but the steam valve only partially open. The governor automatically adjusts the valve in response to small frequency deviations and thus reduces the power imbalance. Because several generators are acting in concert, an individual generator supplies only a fraction of the imbalance power and their governors use nonunity feedback (known as 'droop' gain) to achieve stable load sharing [19]. Regulating reserve is prevalent on large island grids, such as UK and Ireland. However, the normal frequency

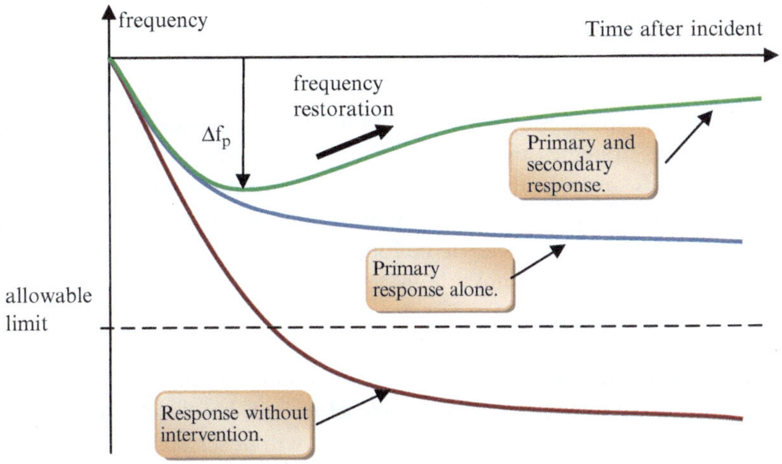

Fig. 2.6 Planned frequency profile in response to large and sudden loss of generation

deviations on large interconnected systems with multiple balancing regions, such as North America and Continental Europe, are sufficiently small that this function is not required.

The primary response to a frequency event is initiated when the frequency deviation exceeds a preset threshold. This is sometimes known as an ‘under-frequency’ or ‘low-frequency (LF)’ trip and the governor responds by quickly ramping the generator to full power. For a steam plant operating with three or four generators active, this may yield around 500 MW in 25–30 s. A pumped storage plant such as Cruachan in Scotland would supply about 200 MW in a similar timeframe.

An important difference between these cases is that the hydroelectric plant does not need to be operated in a partly loaded state to achieve this response time. Instead, with the main inlet valve closed, high-pressure air is used to push water out of the turbine blades. The dramatic drop in viscous friction allows the generator to be run as a synchronous motor, drawing only a modest amount of electricity from the grid – a mode known as ‘spinning-in-air’. When a LF trip occurs, the main inlet valve is opened and as the governor opens the guide vanes the air bubble is quickly flushed via the draft tube. A rapid transition then occurs to generation at full power – at Dinorwig it takes 12–15 s to ramp a unit on spinning reserve from drawing about 10 MW to delivering 330 MW to the grid. This is one of the main reasons for pumped storage plant being attractive as providers of primary response.

An extended discussion of the management of spinning reserve, which adds insight into the methodology of frequency control and the type of plant needed to implement it, is given by Koessler et al. [22]. They point out that tuning the governor on a hydroelectric plant for fast response can compromise stability, particularly when connected to a small, lightly loaded grid. This problem is addressed in the later chapters of this book.

2.4 Conclusions

This chapter has focused on the form and function of hydroelectric plant, explaining how a diversity of types has evolved in order to fulfil various tasks. This chapter has also outlined the strategies that managers use to keep network frequency within acceptable bounds, even in the event of quite severe disturbances. Hydroelectric stations that supply base load electricity are important in many countries but the more difficult control problems are to be found in the provision of primary response. Pumped storage plant play a vital role in delivering primary response and it is clear that fast, accurate and robust governors are required in order to realise the full potential of these expensive capital assets. How this can be achieved on a plant which is nonlinear, multivariable and possesses nonminimum phase characteristics is the subject of the later parts of the book. However, before commencing on the more advanced control systems, the next chapter will describe how the basic speed governors found in early hydroelectric installations evolved into today's electronic governors.

Chapter 3
Overview of Hydropower Control Systems

3.1 Introduction

As we saw in Chap. 1, hydropower was known in ancient Greece and Rome where the waterwheel was common machinery; similarly the use of hydropower in China goes back at least 2,000 years. In the Middle Ages (roughly between the ninth and thirteenth centuries), the variety and applications of waterwheels increased greatly, though it was not until the eighteenth century that the first comprehensive theoretical studies of waterwheels were undertaken. As the first turbines appeared in the nineteenth century, the need to control their relatively high speed quickly became evident. It was natural to extend the use of the centrifugal regulators found in windmills and steam engines to hydraulic turbines.

The function of the governor is to manipulate the actuator which opens and closes the guide vanes, in order to modulate the flow of water in the penstock. Early governors used a direct mechanical linkage for this purpose. Electronic controllers measure the difference between the demanded and actual speed of the turbine and calculate the signal to be sent to an electro-hydraulic actuator that drives the guide vanes. Whatever the implementation, the basic theory of feedback applies. It is also possible to extend the governor to control the generated power and thereby contribute to the regulating reserve, as discussed in Chap. 2.

It is impossible to separate any discussion of hydropower control systems from the technological and theoretical developments that took place in the general field of control engineering over the same period. This will become clear in the brief historical review given in Sect. 3.2.

G.A. Munoz-Hernandez et al., *Modelling and Controlling Hydropower Plants*,
Advances in Industrial Control, DOI 10.1007/978-1-4471-2291-3_3,

3.2 Historical Review

3.2.1 Early Development

For many decades, water turbines were controlled almost exclusively by 'flyball' governors. Figure 3.1 shows a simple schematic of the flyball governor. The principle of this most famous of control devices is straightforward, although, as the early users quickly realised, this concealed profound truths about closed loop feedback which were only exposed by the insight of a succession of brilliant mathematicians, scientists and engineers. The rotation of the flyballs is proportional to turbine speed. In Fig. 3.1, the shaft that carries the pantograph levers and flyballs is attached to the turbine. As the shaft rotates, centrifugal force pushes the flyballs outwards and the central piston is drawn up, thus inhibiting the flow of water to the turbine. The spring compression force opposes the piston movement and when this is equal to the flyball force the valve maintains an equilibrium position. This is now recognised as negative feedback. For an absorbing account of the flyball governor and its use for hydro turbine speed control, the reader is referred to the article by Karl Heinz Fasol [23], who shows how the early devices patented by Mead in 1787 and Hooper in 1789 developed into intricate, precision instruments which remained in common use well into the twentieth century.

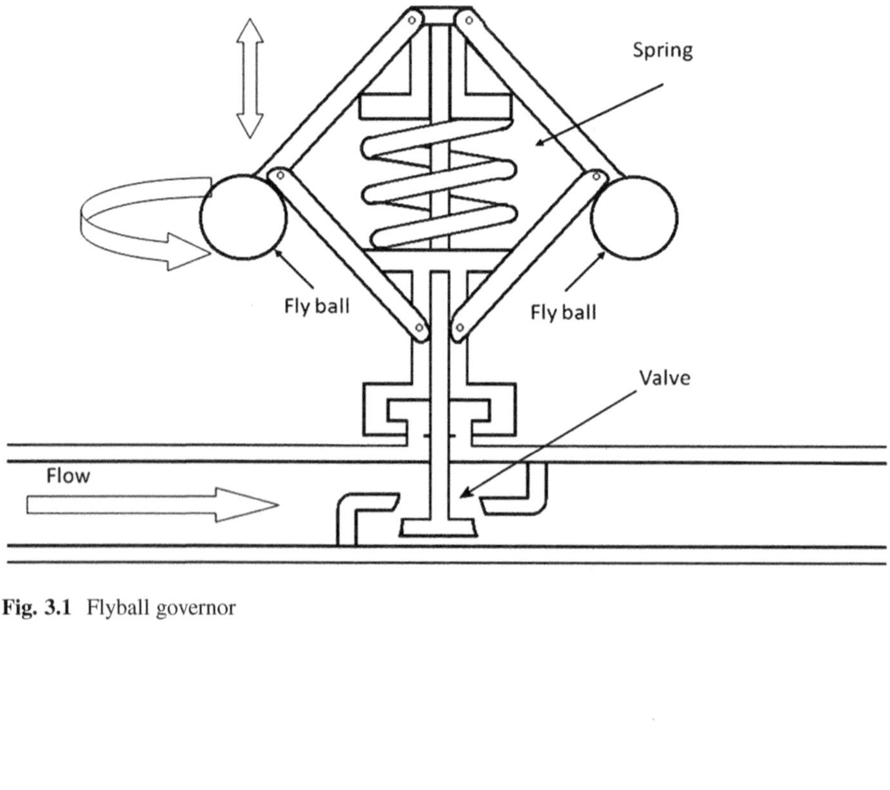

Fig. 3.1 Flyball governor

As the technology of flyball governors evolved, it became apparent that their performance had a number of deficiencies [24]:

- Basic flyball governors are capable only of proportional control and, having no integral action, cannot eliminate steady state error when the turbine speed changed.
- Recovery from load disturbance was slow because of the limited response rate of the governor.
- A large number of governors exhibited limit cycling (known as 'hunting').

Little was known about the effect of the mechanical design of the governor on its performance, even in steady state let alone during transient conditions. For instance, better manufacturing tolerances led to mechanisms with less friction, which improved the speed of response but seemed to make them more prone to hunting. It became imperative to address these problems by a better understanding of the governor's dynamics and how this related to the design parameters.

One of the first scientists to formally address the question of feedback dynamics was the English Astronomer Royal, George B. Airy, who was motivated by the problem of rotating a telescope to counteract the earth's motion while observing a star. His analysis, published in 1840, showed mathematically that a closed loop system could become unstable. Similarly, it was the practical motivation of maintaining an electromagnetic coil at a constant speed that prompted James Clerk Maxwell to undertake the first methodical study of the closed loop stability problem. It culminated in his pioneering paper 'On governors', published in 1868 [24]. Here, he derived the stability conditions for a third-order differential equation that described his control loop. The conditions set boundaries on the equation's coefficients which in turn were related to the physical parameters of his system. He also attempted another analysis based on a fifth-order differential equation but was unable to find a general solution. However, his frustration gave the required stimulus for other mathematicians to study the problem, and by 1877 E. J. Routh had established the necessary criterion.

Independently, similar work was being undertaken by I. A. Vyshnegradskii in Russia who analysed a system consisting of a steam engine controlled by a flyball governor. He obtained a third-order differential equation, which included the influences of mass and friction, to describe the behaviour of the system. He expressed this model in terms of two generalised parameters, which eventually came to bear his name. By drawing loci in the plane of these two parameters, he was able to isolate regions associated with particular types of transient behaviour of the model, including whether or not its response was stable [24].

The implication of these results for hydraulic turbine control was recognised by A. B. Stodola in 1893 at the Federal Polytechnic, Zurich, where he was investigating the stability of a high-head hydropower plant. He successfully applied Vyshnegradskii's method to a third-order plant model but, like Maxwell, found a more realistic seventh-order model to be mathematically intractable. In a manner reminiscent of Maxwell and Routh, Stodola too sought the help of colleagues and this resulted in the development of a general stability criterion by Adolf Hurwitz. Although derived independently and in different forms, the work of Routh and Hurwitz was eventually shown to be equivalent; hence the well-known Routh–Hurwitz stability criterion. Stodola was then able to use the Hurwitz criterion to establish guidelines that guaranteed stability and even a well-damped transient response, when designing a hydropower plant [23].

Much of the improved understanding and expanding body of knowledge was encapsulated in a German textbook written by Max Tolle. This was one of the first books to consider speed governors in the specific context of hydraulic turbines and included a description of Tolle's graphical method for analysing stability. Engineers could use his published design curves to predict whether a governor of known dimensions would be stable, even taking into account the effects of inertia and friction. As Fasol points out [23], the ability to do this was becoming increasingly important in order to predict the performance of more difficult systems, typically high head hydroelectric stations during isolated operation.

3.2.2 Mechanical Governors

Mechanical governors are the direct descendants of flyball governors with the crucial addition of a 'power-assisted' actuator for moving the flow control valve or the guide vanes on the turbine. The earliest versions employed levers and gears to implement the power assistance, such that the turbine's power not only rotated the flyballs but was also used to open and close the guide vanes. In effect, this increased the gain available in the control loop making possible a faster response. In a further development, the mechanical linkages were replaced with an oil-hydraulic actuated piston which was supplemented with springs, throttles, and oil-hydraulic dashpots to produce the required dynamic performance. This developed into the staple mechanical oil-hydraulic governor which was prevalent throughout the first half of the twentieth century. The increasing number of hydropower installations made it economic for manufacturers to mass-produce governors as separate units that integrated all the required components and could be installed on many different types of plant. The quality control that could be imposed on production items allowed them to be developed into sensitive and precise devices [23]. Nevertheless, the control structure of these mechanical governors remained fixed, usually proportional-integral (PI) or proportional-integral-derivative (PID), and this was one of the factors that stimulated the adoption of electronic governors.

3.2.3 Modern Governors

To all intents and purposes, the flyball had by now become a speed measurement device, having lost its original actuating function. It was only a matter of time before it would be replaced by an electronic or magneto-electrical sensor. The signal processing in governors was, for a rather brief period, accomplished by analogue electronics. This reflected the general situation for control system implementation that persisted between the advent of high-performance semiconductor operational amplifiers and their displacement by digital controllers. The main advantages of these analogue electronic governors were the improvement in transient response, flexibility of physical layout and the possibility of combining single controls into one overall power plant regulator [25]. Although they allowed the control parameters to be changed with relative ease, electronic analogue governors had rigid PI or PID structures, like their mechanical predecessors. The arrival of digital electronic governors solved this limitation [23, 26].

Digital governors have the capability to be adapted for different operational conditions. For example, changing from standstill to generation mode requires the speed to be matched to the frequency of the power system and then an electrical phase match to achieve synchronisation. Once this has been achieved, it is straightforward to switch a digital governor into an entirely different mode for automatic frequency regulation. Digital governors are implemented on programmable logic controllers (PLCs) and were generally programmed to mimic the action of the electrical relay circuits that they replaced. The computing power of the PLC module responsible for direct digital control was modest and offered little more scope than a PI or PID controller. However, the increased computing power of later PLCs offers the possibility of implementing more advanced control methods.

Digital computers have been used to study unsteady fluid flow since the 1960s and the design of a hydraulic installation of any size would require an analysis of pressure surges ('water-hammer') before being approved. The basic nonlinear partial differential equations for transient flow and the numerical methods for solving them are described in the classic book by Wylie and Streeter [27]. The technology for complete nonlinear plant simulation has also been available for many years and was summarised in 1992 by a IEEE Working Group in an influential paper [28]. However, whereas computational fluid dynamics is well known and commonly utilised in the hydroelectric power industry, systems level simulation has yet to realise its full potential. This is despite the fact that it is a method advocated by IEEE Standard 1207-2004 (see Sect. 3.4) and being the central technique of model-based design, whose benefits have been proven in the aerospace, automotive and other industries. In the case of 'small hydro', simulation is rarely used during planning, even though improved software now makes it cheap enough for its advantages to be available to smaller installers.

The vast majority (if not all) of the control systems now running on hydroelectric plant will have been designed on the basis of single-input, single-output (SISO) linearised models. In fact, many will not truly have been 'designed' but will instead have been adjusted in situ using a standard PI or PID tuning procedure provided by the governor's vendor. Yet, the operating point of the turbine is known to vary with the water flow and hydraulic head, so a fixed parameter controller will only be optimal at the chosen operating point; there should be ample scope for better controllers. As well as the previous observation about the limited real-time computational capability of PLCs, other factors which may have inhibited experimentation with new control methods are:

- (In the past) a lack of a compelling market incentive to improve performance
- Conservative attitudes based (with justification) on an over-riding necessity for safety and continued plant operation
- Lack of applicable research

Nevertheless, advances have taken place as discussed in the following section.

3.2.4 Control System Development

In this section, we attempt to give the reader a sense of the new approaches that have been proposed for hydropower control over the last couple of decades. Space only allows a sample of the literature to be mentioned here. For an extended review, the reader is referred to the paper published recently by Kishor et al. [29].

The behaviour of power stations controlled by a PID type governor has been studied by several authors. Echoing the pioneering work of Stodola, the stability regions of a hydraulic turbine generating have been studied by Thorne and Hill [30]. Dhaliwal and Wichert conducted a SISO study of the effect of derivative gain on the stability of a system supplying an isolated load, their work being extended to cover the case of twin machines [31]. Similarly, Hagihara applied the root locus method to investigate the effect of the PID parameters on the stability boundaries of a hydraulic turbine generating unit supplying an isolated load [32]. In the same vein, Mansoor investigated the behaviour and operation of Dinorwig power station [33], using the Routh–Hurwitz stability criteria and root locus method to determine the optimal set of PID parameters.

Methods for improving performance across the operating envelope have been developed which adapt the controller according to the operating condition. For instance, Orelind et al. [34] demonstrated the use of a gain-scheduled controller which selects the parameters of a PID compensator as a function of the guide vane angle. Ye et al. [35] also described a controller whose parameters vary over the plant's operating envelope as a function of the static head, guide vane angle and turbine speed. The parameter values (e.g. the PI gains) are pre-determined from linearised analyses at various operating points and stored for online access. They also observed that the dynamics are affected by the guide vane rate limit and proposed a nonlinear gain term to compensate this effect. Finally, they proposed that the structure of the controller should change in order to accommodate various operational modes, e.g. frequency control or speed regulation.

Adaptive controllers have also been considered. Mansoor et al. [36] showed that open loop gain scheduling according to the number of units online is a simple, but quite effective, action. Lansberry and Wozniak [37] have suggested the use of a genetic algorithm to perform the adaptive function, so that the gains of a PI governor are made to continuously track changes in either the water starting time or the grid stiffness (i.e. the sensitivity of grid frequency to load). In a recent paper, Eker and Tumay [38] used the method of H_∞ optimisation to design a robust controller which is insensitive to uncertainties in some plant parameters, including the water starting time and the wave travel time for an inelastic water column. Their simulation results indicate that the H_∞ controller, although linear and fixed, gives better rejection of both electrical load disturbances and 'water' disturbances (such as may occur if the guide vane on another unit is opened or closed rapidly), than does a PID controller. Designing on the basis of a SISO model, with every governor tuned for the worst-case interaction (all units online), leads to conservative tuning [39]. The need for caution is understandable because exceeding the stability boundary can cause highly undesirable frequency oscillation on the grid [40]. It is common, however, for hydroelectric plant to spend considerable periods with only one or two units active, when the conservative tuning leads to sub-optimal performance.

One way of addressing the effect of coupling is to use multivariable control design methods. For instance, Jones [41] has shown that a two-input, two-output linearised model gives a more accurate representation of the dynamics than a SISO model and demonstrates how hydraulic cross-coupling leads to the loss of stability margin. It is also shown how a decoupling controller can be designed, using the direct Nyquist array (DNA) technique, to counter this effect.

Other methods for advanced control are the subject of the final part of this book but for the moment we will introduce the basic feedback loops typically used for controlling power and frequency.

3.3 The Basic Control Loops

As we saw in Chap. 2, generators connected to a synchronous network must manage their individual power contribution and may also be required to act as a regulating mechanism on frequency. Power stations normally accomplish this by means of two control loops, one for power and another for frequency [19]. A 'droop gain' strategy is normally used to adjust the speed reference of individual governors so that stable power sharing between inter-connected generators is established. This will be discussed further in Chap. 6. Figure 3.2 shows the classical configuration for load/frequency control.

3.3.1 Power Control Loop

As discussed in Chap. 2, one of the primary aims of the network manager is to maintain a balance of active power so that the frequency remains within set limits. Another aim is to operate the network at a flat voltage profile, which means that the voltages at all nodes in the system will be almost equal [42], thus ensuring that consumers' equipment remains within safe voltage limits. It also minimises

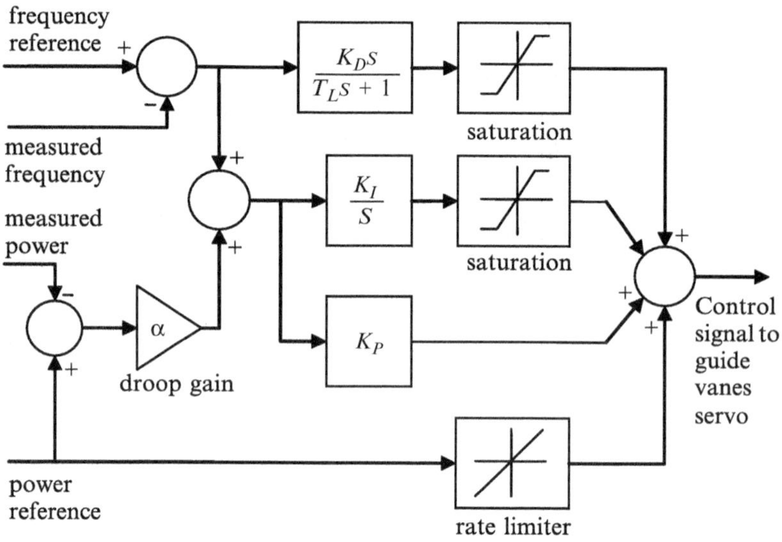

Fig. 3.2 Block diagram of a generic PID governor for controlling power and frequency

voltage gradients on the transmission lines, which minimises the transmission of reactive power and associated losses. As a consequence, reactive power can only be controlled locally [19] and a range of devices are distributed around the network to accomplish this, such as:

- Shunt capacitors
- Regulating transformers
- Generator excitation (field current)
- Power electronics-based equipment (FACTS devices)

This monograph deals exclusively with the control of active power. The flow of active power on a transmission line depends mainly on the angle by which the sending end voltage leads the receiving end voltage [19]. Bearing in mind that the magnitudes of these two voltages are deliberately made to be nearly identical, the active power flow then becomes almost exclusively a function of the generator's load angle. A measurement made at a generator's point of connection alone is therefore sufficient to control its contribution of active power to the network. The measurement is usually done by means of current and voltage transformers located on the busbars.

The inner power control loop in Fig. 3.2 forms an error signal as the difference between the measured and reference values (the latter being held constant in steady state operation). The error is multiplied by the speed regulation droop (α) and input to a PI controller whose output is used to adjust the turbine's guide vanes via an electro-hydraulic servo system. Note that there is also feed-forward of the power reference to provide a direct demand signal to the turbine guide vanes. This allows the operator to set the power operating point. The rate limiter prevents the guide vanes from closing or opening too rapidly with the consequent risk of hydraulic over-pressure in the penstock. The negative feedback also serves to linearise the relationship between guide vane opening and generated power.

3.3.2 Frequency Control Loop

Small load variations that occur in day-to-day operation can be automatically countered by the regulation reserve provided by a generating unit run in part load. This is the function of the outer frequency loop shown in Fig. 3.2. Deviation of the frequency from nominal produces an error signal which adds to (or subtracts from) the set-point and therefore produces incremental contributions of power into the network. The outer frequency control loop effectively provides a reference to the power control loop. A derivative feed-forward loop is included as an aid to rapid response when the network frequency is changing quickly.

Large frequency deviations, occurring possibly as a result of power plant outages or line tripping, will cause a primary response. Under these circumstances, the frequency feedback is effectively disconnected and the generator set-point is ramped to maximum output.

3.4 Applicable Industrial Standards

Among other activities, the Hydroelectric Power IEEE Subcommittee has elaborated standards for the hydroelectric community. These are intended to help the user with the processes involved in the operation, purchase and maintenance of hydroelectric power plants. Some of these standards are briefly reviewed in the following sections; they can be purchased directly from the IEEE.

3.4.1 IEEE Std. 125-2007

IEEE Std. 125-2007 is named *IEEE Recommended Practice for Preparation of Equipment Specifications for Speed-Governing of Hydraulic Turbines Intended to Drive Electric Generators* [43]. It was sponsored by the IEEE Energy Development and Power Generation Committee of the IEEE Power Engineering Society and approved by the IEEE-SA Standards Board (June 7, 2007).

This standard is a useful guide intended for prospective purchasers of hydraulic turbine governors. The guide includes performance criteria, for instance temperature range or fluid velocity; although they are representative of North American practice, these can be applied worldwide. Terms and functions are defined. Also, the specific components of a governor system are described, including guide specifications on power supply, transient immunity, electronic components, test facilities and accessories such as generation control circuits. The guide also describes specifications of equipment, for example, components for pump turbines, cabinet construction, rotor creep detector, fire protection system, emergency stop controls and hydraulic pressure supply system accessories. The performance characteristics of a 'good' governor system, as well as the adjustments and tests to obtain and confirm the desired performance, are examined. Those specifications include damping requirements and the tests comprise stability studies. Also, the process to get information necessary for maintenance purposes is examined.

3.4.2 IEEE Std. 1010-2006

IEEE Std. 1010-2006 is named *IEEE Guide for Control of Hydroelectric Power Plants* [44]. It was sponsored by the IEEE Energy Development and Power Generation Committee of the IEEE Power Engineering Society and approved by the IEEE-SA Standards Board (March 30, 2006).

This guide provides information on existing industry practices for the control of hydroelectric power plants. The standard examines basic requirements and characteristics of hydroelectric power plant control systems, such as architecture, reliability, redundancy, control level, location and control modes. This guide also reviews the centralised and off-site control and their specific requirements for hydroelectric plants. Logical diagrams to show the flow and sequence of the control are included; these tools facilitate the description of requirements for control and monitoring the plant equipment. Furthermore, the control of hydroelectric generating units is described, both conventional generating and pumped-storage units.

3.4.3 IEEE Std. 1020-1988

IEEE Std. 1020-1998 is named *IEEE Guide for Control of Small Hydroelectric Power Plants* [12]. It was sponsored by the IEEE Energy Development and Power Generation Committee of the IEEE Power Engineering Society and approved by the IEEE Standards Board (March 10, 1988).

This standard is a guide with the objective of assisting in the planning, design, development and operation of control systems for small hydroelectric power plants. Although it does not cover every event that may be encountered, it helps to teach the terminology and principles involved.

This document is basically a tutorial that examines the control requirements from an electrical point of view. For this reason it does not have a detailed discussion of the constructional, hydraulic and mechanical considerations that must also be taken into account when planning a hydroelectric project. Also, the document excludes treatise topics such as economics, environmental factors, financing and licensing.

3.4.4 IEEE Std. 1147-2005

IEEE Std. 1147-2005 is named *IEEE Guide for the Rehabilitation of Hydroelectric Power Plants* [45]. It was sponsored by the IEEE Energy Development and Power Generation Committee of the IEEE Power Engineering Society and approved by the IEEE Standards Board (September 22, 2005).

The standard describes some alternatives that should be considered before carrying out a rehabilitation of hydroelectric power plants. These alternatives are intended to assure that the improvements will not overlook the current process. Also, this standard is a guide for practicing engineers, which provides directions to the decision-making processes for the rehabilitation of hydropower stations.

3.4.5 IEEE Std. 1207-2004

IEEE Std. 1207-2004 is named *IEEE Guide for the Application of Turbine Governing Systems for Hydroelectric Generating Units* [46]. It was sponsored by the IEEE Energy Development and Power Generation Committee of the IEEE Power Engineering Society and approved by the IEEE-SA Standards Board (June 24, 2004).

This standard is a guide for practitioner engineers that need to apply turbine governing systems for hydroelectric generating units. This document contains four major clauses, which are directly related to the subject matter addressed in IEEE Std. 125-2007, discussed in Sect. 3.4.1. The characteristics and functions of the turbine governing system and the equipment related to the design of this system are examined in Clause 4. A tutorial that discusses the major elements of the turbine governing system, from a control theory perspective, is presented in Clause 5. Clause 6 presents some applications close to specify a turbine governing system. Stability aspects of the turbine governing system are discussed in Clause 7. Also, various bibliographic citations related to the subjects discussed in this guide are provided. Examples are included to clarify many of the systems and concepts discussed in the document. The guide also includes annexes where more specialised information, relating to the impact of turbine characteristics, system modelling and tuning, and performance auditing, is provided.

3.4.6 IEEE Std. 1248-1998

IEEE Std. 1248-1998 is named *IEEE Guide for the Commissioning of Electrical Systems in Hydroelectric Power Plants* [47]. It was sponsored by the IEEE Energy Development and Power Generation Committee of the IEEE Power Engineering Society and approved by the IEEE-SA Standards Board (June 25, 1998).

This standard is a guide directed towards people who are involved in the commissioning of electrical systems of hydroelectric plant, such as owners, designers and contractors. This guide recommends inspection and tests to be used following the completion of the installation of components and systems through to commercial operation. Although this standard was developed following the general North American terminology and practice, with minor adaptations it can be used in other areas. This guide was developed to assist engineers involved in the commissioning of electrical equipment to observe specific electrical equipment tests and the testing programme for placing the equipment in operation. The assignment of electrical equipment could be for:

- A new hydroelectric plant installation
- Rehabilitation of an existing hydroelectric plant
- Replacement and upgrade of existing electrical equipment

The standard explains the development of a start-up organisation, and then the commissioning phases of a hydropower plant are explained. A matrix format is used to illustrate the major information for each major type of electrical equipment, including the various tests associated with the equipment. For each specific test the following is included:

(a) A brief description
(b) Supporting documents
(c) Equipment required
(d) Duration or time required

Taking the above information as a basis, guidance is provided for planning, developing and documenting an assignment programme. This guide addresses conventional hydropower stations but portions of the guide are relevant also to pumped storage plants, although the special features of pumped storage plants are not particularly covered. A bibliography of industry standards, recommended practices and guides, which may be used as resources by the engineer engaged in the commissioning of electrical equipment, is included.

3.4.7 IEEE Std. 1249-1996

IEEE Std. 1249-1996 is named *IEEE Guide for Computer-Based Control for Hydroelectric Power Plant Automation* [48]. It was sponsored by the IEEE Energy

Development and Power Generation Committee of the IEEE Power Engineering Society and approved by the IEEE-SA Standards Board (December 10, 1996).

The standard is a guide to the application of digital control systems to hydroelectric plants. The guide is planned to be useful for practicing engineers who have some knowledge of computer-based control systems. The guide includes a review of the functional capabilities of computer-based control systems and covers data acquisition, alarm processing, report generation and operator training. Furthermore, various system architecture and network bus configurations are reviewed and also system performance and testing are discussed. Finally, four case studies of actual computer-based hydroelectric control applications are presented.

3.5 Conclusions

In the first part of this chapter, the historical development of governors for hydraulic turbines was discussed. It is clear that considerable advances have been made from the early days, in both the technology and theory of turbine control. However, it is also clear that this process is not at an end. Opportunities exist for more advanced control systems that will help realise the full potential of the plant, achieving optimal speed and accuracy of response across the whole operating envelope. Part III of this book (Chaps. 9, 10, 11, 12 and 13) is an account of some of the methods that have been proposed by the authors. This chapter also described the basic power and frequency control loops, as a prelude to the detailed treatment of Chap. 6. Finally, a list and summaries have been provided of various IEEE standards, of which many readers may not be aware. These are relevant to both practicing engineers and academic researchers seeking information about best practice in the hydropower industry.

Part II
Modelling the Power Plant

Chapter 4
Hydraulic Models

4.1 Introduction

This chapter examines the characteristics of the hydraulic prime mover based on the fundamental relationships between the elements of the system. Models of the hydraulic system are developed systematically with increasing complexity suitable for their representation in system dynamic studies. Section 4.2 describes the effects of the characteristics of the water column on the performance of the hydraulic turbine. These include the effects of water inertia, water compressibility and pipe wall elasticity in the penstock or pressure tunnel feeding the turbine. Water column inertia causes changes in the turbine flow to lag behind changes in the turbine guide vane opening. This introduces phase lag into the speed-governing loop and hence has a destabilising effect on the generating unit. Pipe wall elasticity causes travelling waves of pressure and flow in the pipe. These are of little consequence when the penstock is short in relation to the wave velocity but can build up to destructive levels in cases where resonance between the penstock and the control system causes standing waves and local magnification of pressure oscillation.

The work then proceeds to analyse the hydraulic coupling due to a common tunnel supplying multiple penstocks. The pressure variation of the system due to guide vane movement is introduced into the model. In Sect. 4.4, linearised models of the hydraulic system are obtained for use in the control design study in subsequent chapters. The Dinorwig pumped storage scheme is constructed as shown in Fig. 4.1. It has a simple layout consisting of the upper reservoir, a low-pressure tunnel and a high-pressure tunnel. The high-pressure tunnel is divided into six individual penstocks, each supplying a turbine unit. Hence, the total flow in the common tunnel will depend upon the number of units in operation. Figure 4.1 shows the data used for evaluation of the system parameters. These parameters are utilised in simulation studies.

G.A. Munoz-Hernandez et al., *Modelling and Controlling Hydropower Plants*, Advances in Industrial Control, DOI 10.1007/978-1-4471-2291-3_4,

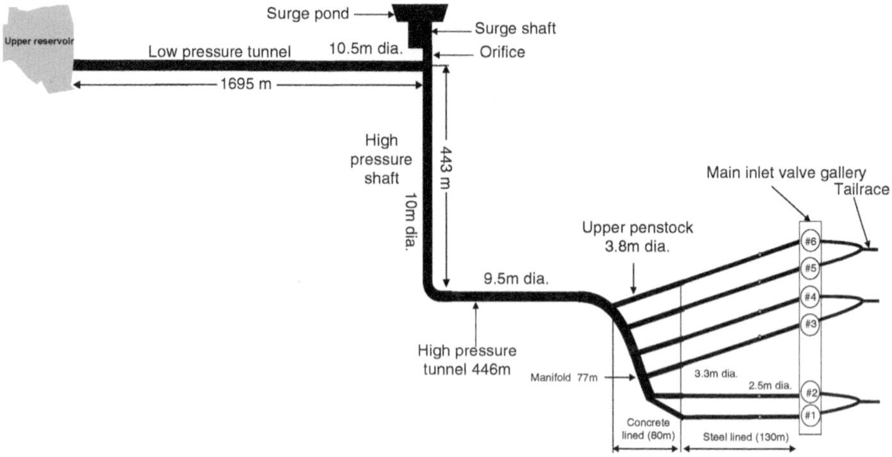

Fig. 4.1 Dinorwig hydraulic system

4.2 Turbine Model

The oldest form of energy conversion is by the use of waterpower; the turbine converts the potential energy of the water into the rotational kinetic energy of the turbine. In the traditional hydroelectric scheme, the energy is obtained free of cost as the water comes from a high-level reservoir into the turbine in which the water energy is converted directly to mechanical energy. In the turbine, the tangential momentum of the water passing through a runner's blade will be changed in direction and a tangential force on the runner is produced. The runner therefore rotates and the energy is transferred from the water to the runner and hence to the output shaft. The water is discharged with reduced energy. The hydraulic turbine may be classified into one of two general categories: impulse and reaction [49].

4.2.1 Impulse Turbine

The impulse turbine has one or more fixed nozzles, in each of which the pressure is converted to the kinetic energy of a water jet. The jets of water then impinge on the moving plates of the runner where they lose practically all their kinetic energy. The velocity of the water at discharge is only sufficient to enable it to move clear of the runner. The important feature of the impulse turbine is that there is no change of static pressure across the runner. The only hydraulic turbine of the impulse type is the Pelton wheel, as shown in Fig. 4.2. The rotor consists of a circular disc with several 'buckets' evenly spaced around its periphery. The splitter ridge in the centre of each bucket divides the oncoming water jet into two equal portions,

Fig. 4.2 Impulse turbine

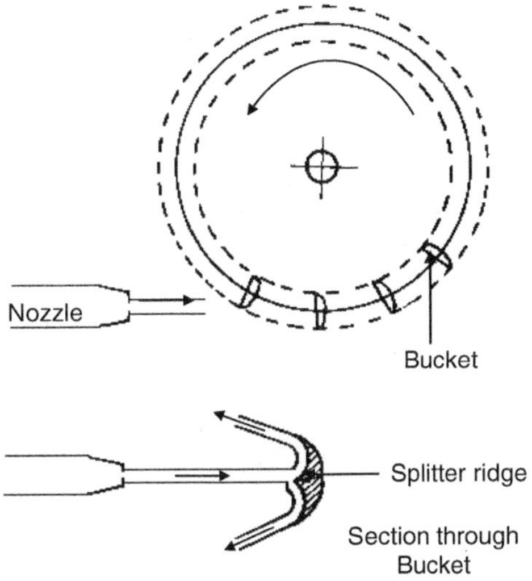

and after flowing around the inner surface of the bucket the water leaves with a velocity opposite in direction to the original jet. The flow partly fills the buckets, and the water remains in contact with the atmosphere. Thus once the jet has been produced, the static pressure of the water is atmospheric throughout the turbine. Impulse turbines are best suited to applications with high hydraulic head, and they are characterised by high reliability, low maintenance cost and over 90% efficiency.

4.2.2 *Reaction Turbines*

In the reaction turbine, the water is fed to the runner all around the circumference from a volute casing through a ring of guide vanes. To keep the pressure losses in the runner at a minimum, it is necessary for the runner to be always full with water (in contrast to the impulse turbine where only a few of the runner blades are in use at any moment). Therefore, reaction turbines are able to deal with a larger quantity of water for a given runner size.

The classifications of turbines are based on the predominant direction of water flow through the runner. In a radial-flow turbine, the path is wholly or mainly in the plane of the rotation, the water enters the rotator at one radius and leaves at a different radius; the Francis turbine is an example of this type. If however the main flow direction is parallel to the axis of rotation, then the turbine is said to be an axial-flow turbine; the *Kaplan* or *propeller* turbines are examples of this type. The number of runner blades varies with the hydraulic head; the higher the head the more blades there are in the runner. The *propeller* turbine has fixed blades while the blade

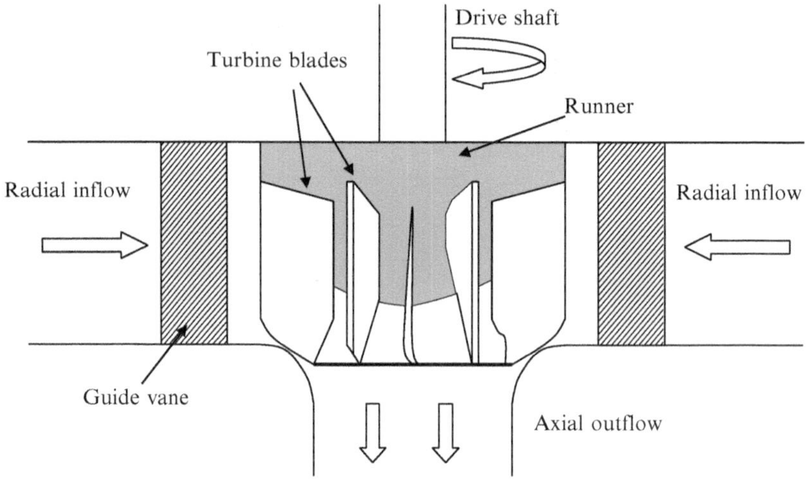

Fig. 4.3 Simplified diagram of a Francis turbine, cross-sectional view

pitch in *Kaplan* turbines is automatically adjusted by the means of an oil-driven piston located in the main shaft. The control is designed such that the blade angle varies automatically with the guide vane opening to produce the maximum possible efficiency for a given operating condition. The fixed *propeller* turbine has a good efficiency at the optimum design point but it decreases sharply with the reduction of flow. Because of the adjustable blades feature, the *Kaplan* turbine has a relatively flat efficiency curve over a wide range of flow. If the flow is partly radial and partly axial, the term mixed flow turbine is used.

Dinorwig power station is equipped with reversible pump-turbines of the Francis type, as illustrated in Fig. 4.3. The water enters a spiral casing (volute) which surrounds the runner, whose cross-sectional area decreases along the water path in such a way as to keep the water velocity constant in magnitude. Departing the volute the water is directed onto the runner by the guide vanes mounted all around the periphery of the runner. Each vane is pivoted and all will be turned in synchronism to alter the flow rate throughout the turbine, and hence the power output as required by governor action. The runner blades deflect the water so that its angular momentum is changed. From the centre of the runner, the water is turned into the axial direction and flows to the tailrace via the draft tube. To ensure the hydraulic turbine is full of water, the lower end of the draft tube is always submerged below the water level in the tailrace.

The mechanical power (P_m) available from an ideal hydraulic turbine is the product of hydraulic head available (h) and mass flow rate (q) but in practice this is reduced by an efficiency factor, η, to account for power losses. The turbine torque at rated speed and head is almost linearly related to guide vane position for most turbines in the range from no-load to rated load but only approximately in the range from fully closed guide vane to no-load guide vane, as shown in Fig. 4.4. The turbine

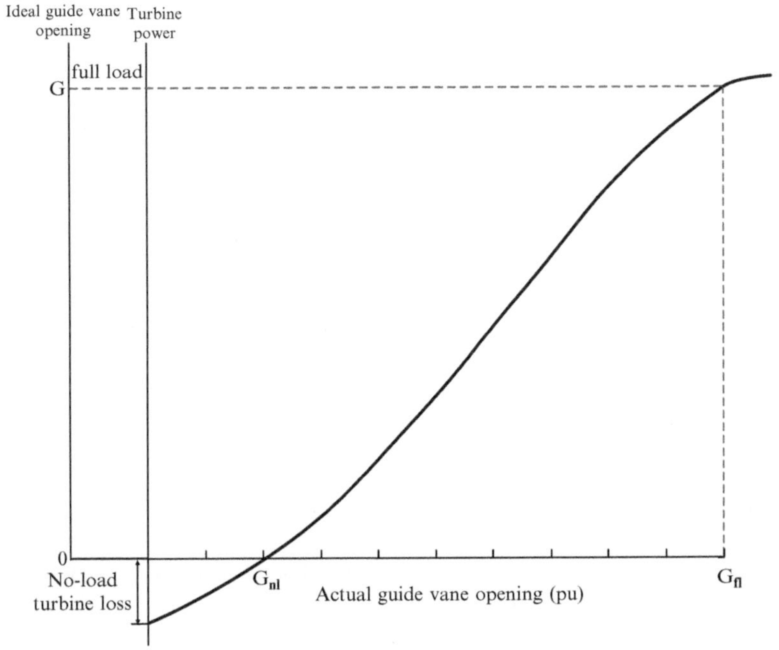

Fig. 4.4 Typical torque/guide vane characteristic

model is based on the equation for steady state operation relating the output power to water flow and head [50]:

$$P_{\mathrm{m}} = \eta q \rho g_{\mathrm{a}} h, \tag{4.1}$$

where P_{m} is the turbine output power, η the turbine efficiency, ρ the water density, g_{a} the acceleration due to gravity (m^2/s), h the head at the turbine admission (m) and q the actual turbine flow (m^3/s).

The fact that the turbine is not 100% efficient is taken into account by subtracting the no-load flow q_{nl} from the net flow to give the effective flow which, when multiplied by the head, produces mechanical power. There is also a turbine damping effect, which is a function of guide vane opening, to be included. Therefore, the per unit turbine power, P_{m}, can be expressed as

$$\overline{P_{\mathrm{m}}} = A_{\mathrm{t}} \overline{h} (\overline{q} - \overline{q_{\mathrm{nl}}}) - D_n \overline{G} \Delta \overline{n}. \tag{4.2}$$

The turbine MW rating is used as power base, q_{base} is chosen as the turbine flow rate, with guide vanes fully open (guide vane position $=1$) and h_{base} is equal to the static head of water column h_0. The base values used throughout this study are shown in Table 4.3. The parameter D_n accounts for the effect of the speed variation Δn on the turbine efficiency; typical values of D_n fall in the range $0.5 \leq D_n \leq 2.0$.

The turbine gain A_t is obtained from the ratio of effective gate position to the actual gate position and can be calculated using Eq. 4.3:

$$A_t = \frac{1}{G_{fl} - G_{nl}} \times \frac{\text{Turbine MW rating}}{\text{Generator MW rating}}, \quad (4.3)$$

where G_{fl} is the guide vane position at full load and G_{nl} is the guide vane position at no-load; both are calculated at rated speed and head. The relationship between idealised and real guide vane position is shown in Fig. 4.4.

The turbine characteristics define base flow through the relationship between the flow (q), guide vane position (G), and head (h). The per unit flow rate through the turbine using q_{base} as the base flow rate and assuming h_{base} is equal to the static head h_0, is given by its valve characteristic,

$$\overline{q} = \overline{G}\sqrt{\overline{h}}. \quad (4.4)$$

A linearised representation of the turbine can be established using data obtained by testing the system. The power developed by the turbine is a function of flow through it, the guide vane position and the utilised head. Therefore, the behaviour of the turbine may be characterised by the variation in flow (q) and output torque (m) relative to speed (n), guide vane opening (g) and the head (h). Figure 4.5 shows the four-quadrant operating characteristics for one of Dinorwig's pump turbines [51]. These curves are obtained by testing the turbine at a particular guide vane opening; the speed is varied and the turbine flow and torque are measured. The tests are repeated for various guide vane openings; from these curves it can be established what the speed of the turbine should be at any guide vane opening in order to give the best efficiency for that guide vane position.

For small variations around an operating point, the turbine can be represented by the following linearised Taylor series approximations relating the flow and torque to head, speed and guide vane position [52]:

$$\Delta q = a_{11}\Delta h + a_{12}\Delta n + a_{13}\Delta g \quad (4.4)$$

$$\Delta m = a_{21}\Delta h + a_{22}\Delta n + a_{23}\Delta g. \quad (4.5)$$

The parameters a_{ij} are the partial derivatives of flow and torque with respect to head speed and guide vane position, respectively. They remain constant for variation near the operating point (q_0, m_0). Their values depend upon the initial steady state point of the turbine and they can be measured accurately by experiment.

Changes in turbine power output are essentially determined by two components, one related to turbine guide vane position and the other to changes in speed. Since the extent of the anticipated frequency variations (and hence the associated speed changes) is generally very small, especially when the plant is connected to the power system, it may be assumed to be insignificant. Variation of flow with the rated speed a_{12} is very small, as shown in the turbine characteristic of Fig. 4.5; therefore, it is possible to neglect it. The deviation of the mechanical torque with speed a_{22} is known as turbine self-regulation, which is negative with an absolute value near

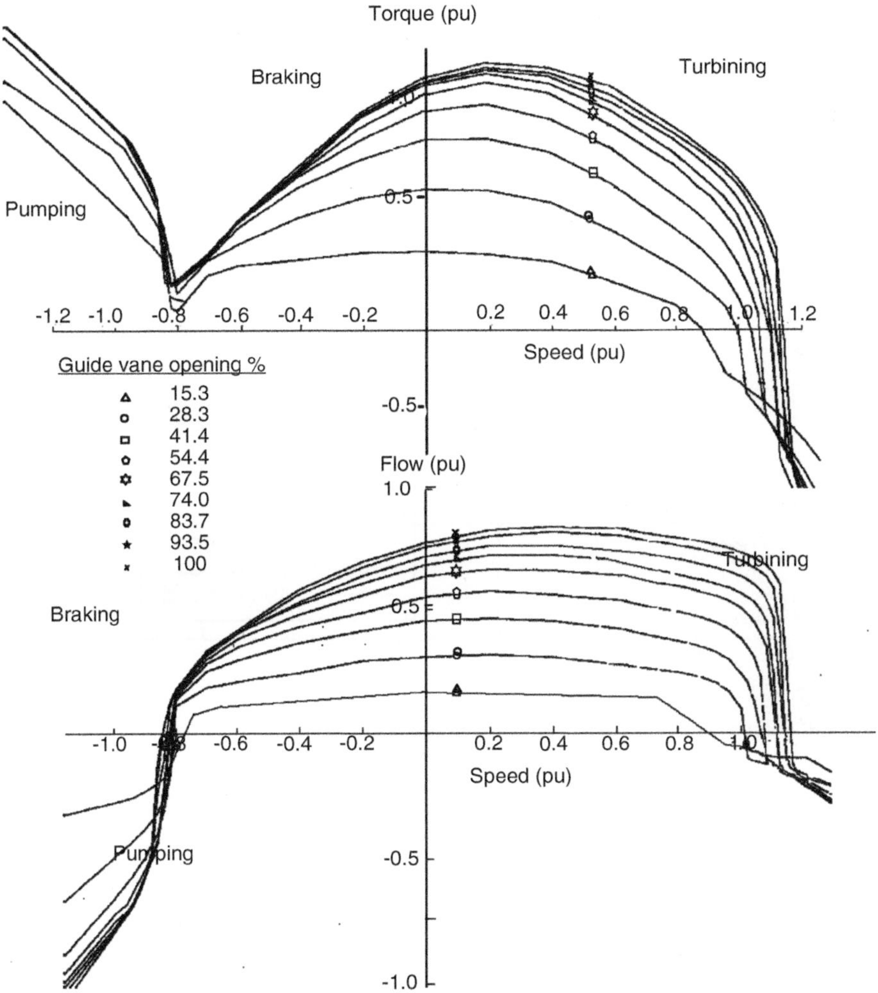

Fig. 4.5 Dinorwig's pump-turbine characteristic

unity. For an ideal lossless turbine operating at rated speed and head, the partial derivatives versus load are deduced as [53]

$$\begin{aligned} a_{11} &= 0.5 \quad a_{12} = 0 \quad a_{13} = 1 \\ a_{21} &= 1.5 \quad a_{22} = -1 \quad a_{23} = \frac{\partial m}{\partial g}. \end{aligned} \tag{4.6}$$

The turbine representation depends mainly on the coefficient a_{23}. This coefficient is a critical parameter for an accurate approximation of the unit dynamics; a_{23} varies widely from the ideal value of unity. It can be measured precisely from the torque/guide vane characteristic, a curve that is readily found by test [54]. At

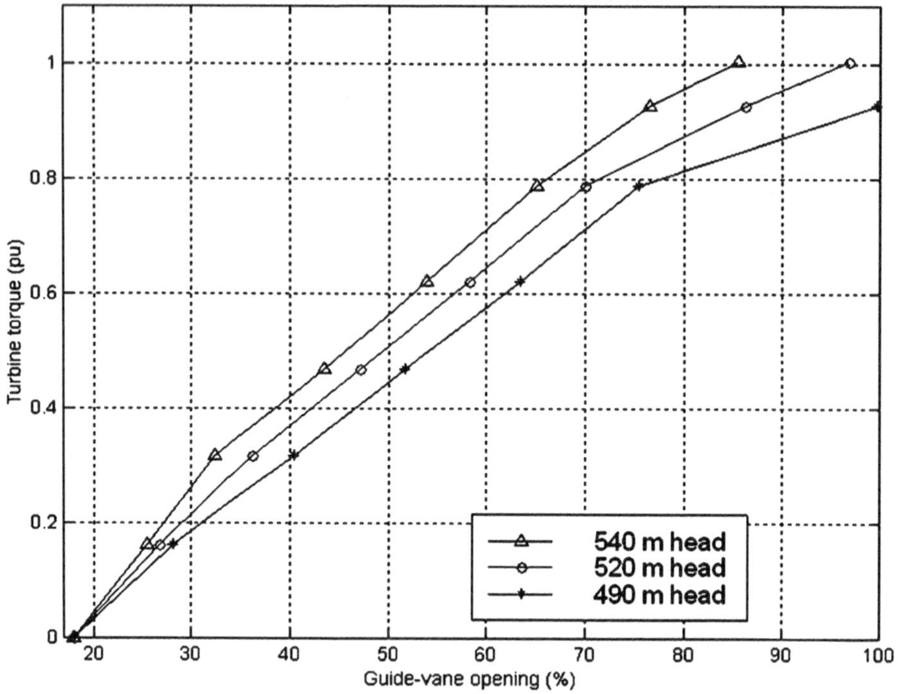

Fig. 4.6 Torque/guide vane characteristic for Dinorwig

Dinorwig, a test was carried out to determine the power output versus guide vane opening for different head conditions as shown in Fig. 4.6 [55]. This relationship was used to evaluate the Dinorwig turbine gain when operating at rated speed and head. Dinorwig turbine rating is 310 MW and the generator rating is 330 MW; thus using Eq. 4.3 results in $A_t = 1.12$, which is used in simulation studies.

4.3 Modelling the Water Column

The performance of a hydraulic turbine is greatly influenced by the characteristics of the water column which feeds it; these include the effects of water inertia, water compressibility and pipe wall elasticity in the penstock. The effect of water inertia is to cause changes in turbine flow to lag behind changes in the guide vane opening. In fact, the power has a transient response which is initially in the opposite sense to that intended by changing the guide vane position. Although the turbine guide vane opening may change rapidly, the water column inertia prevents the flow from changing as rapidly. Consequently, after a rapid increase in guide vane opening, and before the flow has had time to change appreciably, the velocity of water into the wheel drops because of the increased area of the guide vane opening. The power transfer to the wheel actually drops before it increases to its required steady state value. This is the most prominent factor that makes a hydraulic turbine such an uncooperative component in a speed control system [56].

Pipe wall elasticity causes travelling waves of pressure and flow in the water – a phenomenon commonly referred to as water hammer. Water hammer occurs when a change in pressure, above or below normal pressure, is caused by a sudden change in the rate of water flow. Because sudden changes in the demand for water occur during load fluctuation, water hammer occurs at all points in the penstock between the forebay and the turbine. Under severe conditions, the effect can damage or destroy valves, turbine guide vanes and the penstocks.

4.3.1 Single Penstock Modelling

A simple hydropower plant consists of a single conduit supplying a turbine-generating unit, and the initial model development is restricted to this case with an inelastic water column. The turbine and penstock characteristics are determined by three basic equations relating to the velocity of water in the penstock, acceleration of the water column under the influence of gravity and the production of mechanical power in the turbine. First, a nonlinear representation is developed which is appropriate when large changes in speed and power are to be considered, such as in islanding, load rejection and system restoration studies.

The basic water column model represents a single penstock with a very large or no surge tank. The penstock is modelled on the assumption that the water acts as an incompressible fluid so that the water hammer effect may be neglected. Consider here a rigid conduit of length l and cross-section area A, where the penstock head losses h_f due to the friction of water against the penstock wall are proportional to flow (q) squared:

$$h_f = f_p q^2, \tag{4.7}$$

where f_p is the head loss coefficient in the penstock due to friction [28]. The loss coefficients in different parts of the Dinorwig hydraulic system are calculated in Sect. 4.6.3.

Assuming that the water in the penstock can be treated approximately as a solid mass, the rate of change of flow can be related to the head of water using Newton's second law of motion. This law states that "An object's acceleration is proportional to the net force, and the object's mass is the proportionality factor between the force and the acceleration". The force on the water mass is

$$(h_0 - h - h_f)\rho g_a A = \rho A l \frac{dv}{dt}, \tag{4.8}$$

where h_0 is the static head of water column (m), h the head at the turbine admission (m), h_f the head loss due to friction (m), f_p the head loss coefficient $[\mathrm{m}/(\mathrm{m}^3/\mathrm{s})^2]$ and v the water velocity (m/s).

The rate of change of the flow in the penstock can be determined as

$$\frac{\mathrm{d}q}{\mathrm{d}t} = (h_0 - h - h_{\mathrm{f}})\frac{g_{\mathrm{a}}A}{l}. \tag{4.9}$$

Equation 4.9 can be written in per unit form in order to normalise system representation. Compared to the use of physical units, the per unit format offers computational simplicity by eliminating units and expressing the system quantities as dimensionless ratios. The base values are chosen so that the principal variables will be equal to one per unit under rated conditions. Here, the base head h_{base} is chosen to be the available static head h_0 which is equal to the reservoir head minus the tailrace head, and the base flow q_{base} is equal to the turbine flow with guide vane fully open. Expressing Eq. 4.9 in per unit yields

$$\frac{\mathrm{d}\bar{q}}{\mathrm{d}t} = (\bar{1} - \bar{h} - \bar{h}_{\mathrm{f}})\frac{h_{\mathrm{base}}g_{\mathrm{a}}A}{l q_{\mathrm{base}}} \tag{4.10}$$

$$\frac{\mathrm{d}\bar{q}}{\mathrm{d}t} = \frac{(\bar{1} - \bar{h} - \bar{h}_{\mathrm{f}})}{T_{\mathrm{w}}}, \tag{4.11}$$

where $T_{\mathrm{w}} = l q_{\mathrm{base}}/h_{\mathrm{base}}g_{\mathrm{a}}A = l v_{\mathrm{base}}/h_{\mathrm{base}}g_{\mathrm{a}}$ is the water inertia time constant, sometimes known as the water starting time.

The water starting time represents the time required for a head h_{base} to accelerate the water in the penstock from standstill to the velocity v_{base}. This is calculated between turbine inlet and the forebay or the surge tank if a large one exists [57]. Consider a simple penstock supplied from an open reservoir discharging into the atmosphere as shown in Fig. 4.7. Opening the guide vane in a time Δt causes the velocity of the water in the penstock to increase by Δv and the head at the turbine inlet to drop by Δh.

The acceleration of water due to change in head at the turbine, characterised by Newton's second law of motion, may be expressed as

$$\rho A l \frac{\mathrm{d}\Delta v}{\mathrm{d}t} = -\rho g_{\mathrm{a}} A \Delta h. \tag{4.12}$$

The acceleration equation can be converted to per unit form by dividing by v_{base} and h_{base} to give

$$\left(\frac{l v_{\mathrm{base}}}{g_{\mathrm{a}} h_{\mathrm{base}}}\right)\frac{\mathrm{d}\Delta\bar{v}}{\mathrm{d}t} = -\Delta\bar{h}. \tag{4.13}$$

Writing in terms of per unit variables:

$$T_{\mathrm{w}}\frac{\mathrm{d}\Delta\bar{v}}{\mathrm{d}t} = -\Delta\bar{h}. \tag{4.14}$$

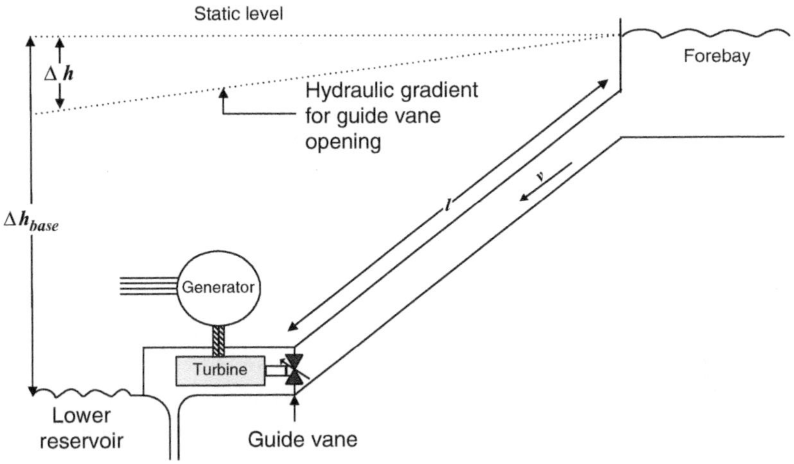

Fig. 4.7 A schematic diagram of pump storage plant

This equation represents an important characteristic of the hydraulic plant. Inspection of Eq. 4.14 shows that, if the guide vane is closed, a back pressure will arise causing the water to decelerate. That is, if there is a positive pressure change, there will be a negative acceleration change. Similarly, a negative pressure change will cause a positive acceleration change. The maximum acceleration occurs immediately after the guide vane opening because the entire difference in pressure is available for accelerating the water. For a nonuniform penstock with different cross-sectional areas, the water inertia time constant is calculated as [58]

$$T_{\mathrm{w}} = \frac{\sum lv}{g_{\mathrm{a}}h}, \tag{4.15}$$

where $\sum lv$ is the summation of length and velocity of sections in the water passage.

4.3.2 Elastic Water Column Model

The majority of pumped storage stations operate with a long penstock, and a high head and an accurate description of penstock dynamics must account for the water hammer phenomenon caused by the characteristic of the hydraulic line [59]. Water hammer is the result of a pressure change in the penstock caused when flowing water is decelerated or accelerated by closing or opening the guide vane or changing the velocity of the water rapidly in some other manner. The phenomenon is characterised by a series of positive and negative pressure waves, which travel back and forth in the penstock until they are damped out by friction.

To investigate the effect of the water elasticity on the hydraulic system, consider the case of water flowing with a certain velocity in a penstock being brought to rest by closing the turbine guide vane. If the water was entirely incompressible and the penstock wall perfectly rigid, then all the particles in the entire water column would have to decelerate together. From Newton's second law of motion, the more rapid the deceleration the greater would be the corresponding force, and with an instantaneous closure of the guide vane all the water would be stopped instantaneously and the force would be infinite, which is clearly impossible. In fact, even the water is to some extent compressible and so its particles do not decelerate uniformly. A rapid closure of the guide vane would not bring the entire column of water to a halt immediately. Only those particles of water in contact with the guide vane would be stopped at once and the others would come to rest later.

Consider a rigid penstock, when a guide vane is suddenly closed the pressure head immediately upstream of the guide vane is increased, causing a high-pressure wave to propagate to the upper reservoir (or the surge tank if one exists). The water particles nearest to the guide vane are compressed by the water above it. The water will continue to move at the original velocity and successive elements of water are compressed. The action of compression moves upstream as a wave until it reaches the open water surface. The pressure wave moves at velocity, a, which is the velocity of sound in water. The time taken for the pressure wave to travel the length of the penstock to the open surface is called the wave travel time T_e, which is given as

$$T_e = \frac{l}{a}. \tag{4.16}$$

The kinetic energy of the moving water is converted to elastic energy in compressing the water and stretching the penstock. The last water element will expand at the open water surface to its original state followed by other elements, causing a negative pressure wave. As the wave travels downstream, conditions change from an increased water pressure head back to the normal pressure when the wave reaches the guide vane at time, $t = 2T_e$. The water moving away from the guide vane will cause a reduction in pressure and a negative pressure wave moves upstream to the open water surface [60]. The periodic fluctuation following a sudden guide vane closure is shown schematically in Fig. 4.8. In practice, water friction will act within the water and at the boundaries so that the pressure wave will be attenuated.

Pressure waves in the penstock can be modelled by treating it as a hydraulic transmission line terminated by an open circuit at the turbine end and a short circuit at the reservoir [61]. Assuming the penstock to be a uniform conduit supplied from a large reservoir, the incremental head and flow at the turbine inlet are related as shown in the transfer function (4.17):

$$\frac{H(s)}{Q(s)} = -\frac{T_w}{T_e}\tanh(T_e s + F), \tag{4.17}$$

where F is the friction losses in the penstock and s the Laplace complex frequency variable.

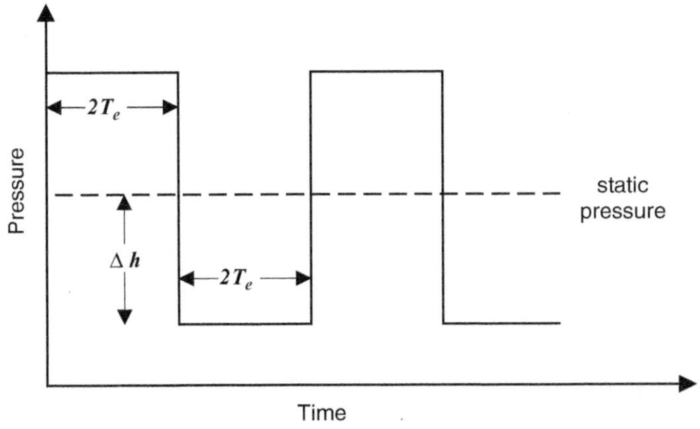

Fig. 4.8 Pressure wave at the guide vane due to sudden closure

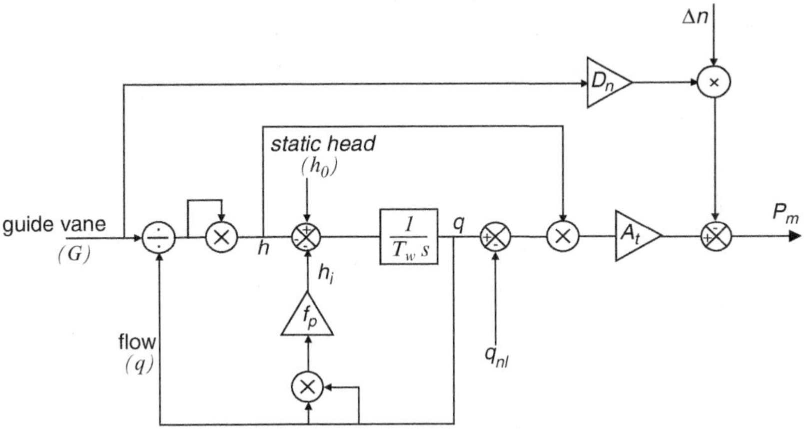

Fig. 4.9 Inelastic water column model of penstock and turbine

4.3.3 Combined Turbine/Penstock

4.3.3.1 Inelastic Water Column

The hydraulic system can be modelled by combining Eq. 4.2 for the turbine and Eq. 4.11 for the inelastic water column. The block diagram of Fig. 4.9 is a nonlinear representation showing how the generated power depends on the guide vane position. Note that the power also depends on additional inputs n, h_0 and q_{nl} but that these change slowly compared to the primary control input. The value for water starting time of the penstock (T_{w}) is obtained at rated conditions using rated head and rated flow as the base values.

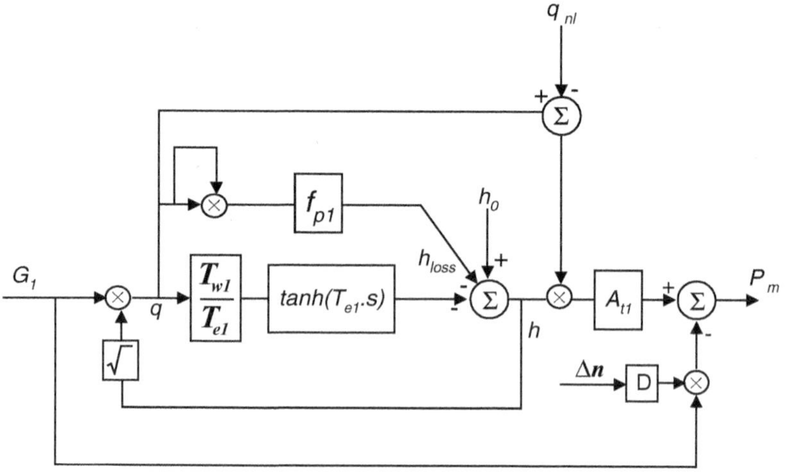

Fig. 4.10 Nonlinear model of turbine with elastic water column

4.3.3.2 Elastic Water Column

Whilst modelling the hydraulic system using the assumption of inelastic water columns is adequate for a short penstock, there is sometimes need to consider the effects which cause travelling waves of pressure in the penstock. The nonlinear hydraulic system including water hammer effects can be modelled by combining the turbine equation (4.11) and the equation for the elastic water column (4.17). Figure 4.10 shows the block diagram of the system, where the dynamics of the turbine power are functions of the head across the turbine and the guide vane opening.

4.3.4 Multiple Penstock Model

The hydraulic turbine models presented hitherto are based on a single conduit analysis. There are cases, however, where all the turbines of a hydropower plant share a common tunnel, which is subsequently bifurcated to form the penstocks leading to individual turbines. This is the case at Dinorwig. Due to the common tunnel, a hydraulic coupling between the individual units is introduced and the dynamics of each turbine depends on the control action taken on every unit of the plant. This is of great importance because the generators of the same plant are also electrically coupled. Therefore, a closer examination of the dynamics of the coupled hydraulic turbines is required.

To probe into the nature of the coupling effects, consider the case of multiple units operating at full load connected to an infinite bus and one unit being taken off load. This is accomplished at Dinorwig by setting the power generating reference to zero, causing the guide vane to be ramped down. As the guide vane of the unit coming offline is being closed, the turbine head for that unit increases. With the guide vanes of the remaining units almost stationary, the initial rise in the turbine head for the unit coming offline will result in a decreased flow to this unit. At the same time, it will produce an increase in the flows for the remaining units, because the total flow in the common tunnel cannot be changed instantaneously. The increase of head at the turbines will simultaneously cause a decrease in the rate of flow through the tunnel until the final flow conditions reach a new steady state [62]. Qualitatively, then, it is clear that hydraulic coupling leads to interaction between the individual machines. Again, two modelling approaches can be taken depending on the characteristic of the water columns. The initial approach neglects water compressibility while the second approach assumes an elastic water column.

4.3.4.1 Inelastic Water Column

For short-medium penstocks, it is well known [62] that the dynamics of the hydraulic turbine can be analysed with little loss of accuracy by neglecting the compressibility of the water. The flow at the common tunnel can be calculated using the continuity equation:

$$q = \sum_{i=1}^{j} q_i,$$

where j is the number of units and q_i the flow in the penstock i.

The total flow in the common tunnel must be equal to the sum of the flows in the individual penstocks. The momentum equation for the water at the common tunnel can be written using the representation of Eq. 4.9:

$$h_0 - h = \frac{l}{g_a A}\left(\frac{\mathrm{d}q_1}{\mathrm{d}t} + \frac{\mathrm{d}q_2}{\mathrm{d}t} + \cdots + \frac{\mathrm{d}q_n}{\mathrm{d}t}\right). \tag{4.18}$$

Writing (4.18) in per unit notation based on the rated flow q_{i0} yields

$$h_0 - h = \frac{l}{g_a A}\left(q_{10}\frac{\mathrm{d}\bar{q}_1}{\mathrm{d}t} + q_{20}\frac{\mathrm{d}\bar{q}_2}{\mathrm{d}t} + \cdots + q_{n0}\frac{\mathrm{d}\bar{q}_n}{\mathrm{d}t}\right), \tag{4.19}$$

where $\bar{q}_n = q_n / q_{n0}$.

The momentum of water in an individual penstock is

$$h - h_\mathrm{f} = \frac{l_\mathrm{i}}{A_\mathrm{i} g}\left(q_{\mathrm{i}0}\frac{\mathrm{d}\bar{q}_\mathrm{i}}{\mathrm{d}t}\right). \tag{4.20}$$

Using Eq. 4.20 to eliminate h in Eq. 4.19 and expressing the head in per unit form by dividing by the rated static head h_0, a matrix expression representing the special case of Dinorwig where $q_{10} = q_{20} \cdots = q_{n0}$, can be written as shown in Eq. 4.21:

$$\begin{bmatrix} 1-\overline{h_1} \\ 1-\overline{h_2} \\ \vdots \\ 1-\overline{h_n} \end{bmatrix} = \begin{bmatrix} T_{\mathrm{wT}}+T_{\mathrm{w1}} & T_{\mathrm{wT}} & \cdots & T_{\mathrm{wT}} \\ T_{\mathrm{wT}} & T_{\mathrm{wT}}+T_{\mathrm{w2}} & \cdots & T_{\mathrm{wT}} \\ \vdots & \vdots & \vdots & \vdots \\ T_{\mathrm{wT}} & T_{\mathrm{wT}} & \cdots & T_{\mathrm{wT}}+T_{\mathrm{w}n} \end{bmatrix} \begin{bmatrix} \dot{\overline{q_1}} \\ \dot{\overline{q_2}} \\ \vdots \\ \dot{\overline{q_n}} \end{bmatrix}, \tag{4.21}$$

where $T_{\mathrm{w}n}$ is the water starting time for the penstock of the nth unit, the rated flow for the nth unit is used to calculate this value; T_{wT} the water starting time for the common tunnel only, the rated flow for the nth unit is used to calculate this value.

4.3.4.2 Elastic Water Column

The effects of water compressibility can be introduced into a multiple penstock model in a similar manner to the single penstock representation. The model now incorporates the nonlinear single penstock model shown in Fig. 4.10. The coupling effect of the tunnel is included by using the same form of transfer function between the head and the flow that, for the tunnel, is the sum of the flows in the individual penstocks. The nonlinear model of the hydro-turbine including the hydraulic interaction model is illustrated in Fig. 4.11. The head loss in the upper tunnel is proportional to the coefficient f_{t} times flow rate times absolute value of flow rate to maintain direction of head loss where the flow can reverse.

4.4 Linearised Models

Much insight into the salient dynamic characteristics of this model can be obtained by considering a linearised representation. This is used for small signal analysis and control design studies.

4.4.1 Inelastic Water Column

The inelastic water column transfer function is obtained by linearising the basic penstock-turbine equations (4.4) and (4.11) [33, 63]. This results in the first-order transfer function of Eq. 4.11 relating small changes in the mechanical power to changes in the guide vane opening. Note that the water time constant T_{w} here corresponds to the operating condition rather than the rated condition. Thus to model the unit correctly in stability simulations, it is necessary to adjust the value of T_{w} each time the initial operating conditions are changed [64]:

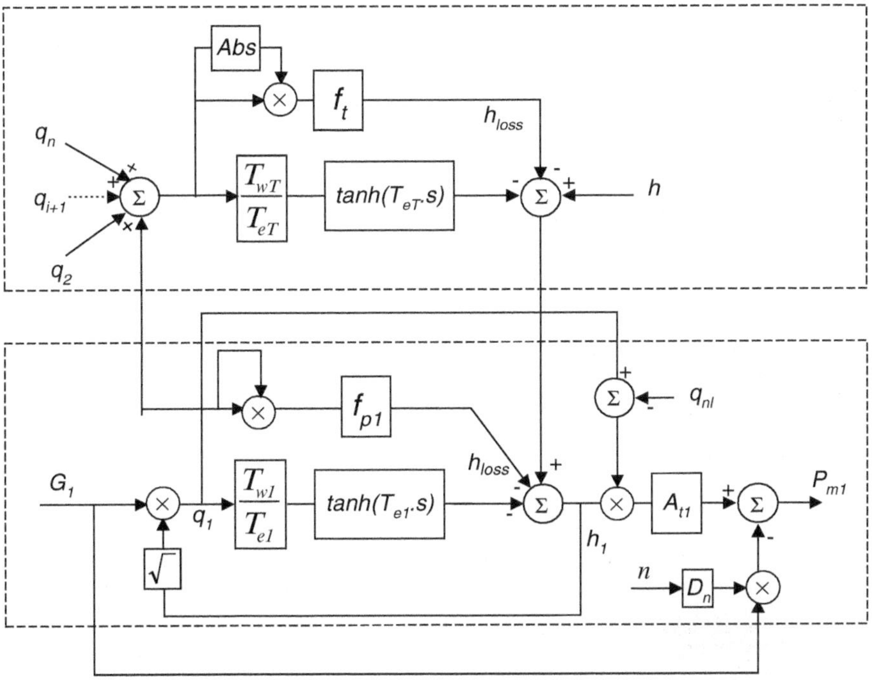

Fig. 4.11 Nonlinear model of multiple penstocks supplied from a common tunnel

$$\frac{\Delta P_{\mathrm{m}}(s)}{\Delta G(s)} = \frac{1 - T_{\mathrm{w}}s}{1 + 0.5T_{\mathrm{w}}s}. \tag{4.22}$$

The transfer function (4.22) has a zero is in the right half of the s-plane located at $s = 1/T_{\mathrm{w}}$, which represents a nonminimum phase system [65]. The system's outstanding dynamic characteristic is illustrated in Fig. 4.12, which shows the change in the turbine mechanical power for a step change in the guide vane position applied at $t = 0$ for a system with a water time constant $T_{\mathrm{w}} = 3$ s. A transient change in power occurs, which is opposite to the direction of change in guide vane position, and the change in the turbine power is twice as large and in the opposite direction to the final change. The subsequent power increase depends on the value of T_{w}, as the water accelerates until the flow reaches the new steady state value that establishes the new power output.

The initial and final power values for a unit step change in guide vane position can be determined as follows. The initial value theorem gives [66]

$$P_{\mathrm{m}}(0) = \lim_{s\to\infty} s\frac{1}{s}\left[\frac{1 - T_{\mathrm{w}}s}{1 + 0.5T_{\mathrm{w}}s}\right] = -2,$$

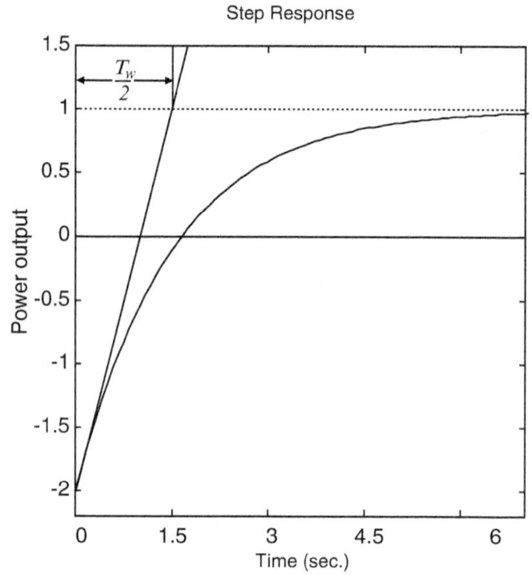

Fig. 4.12 Turbine power change due to step guide vane opening

while the final value theorem gives

$$P_{\mathrm{m}}(\infty) = \lim_{s \to 0} s \frac{1}{s} \left[\frac{1 - T_{\mathrm{w}} s}{1 + 0.5 T_{\mathrm{w}} s} \right] = 1.$$

The step time response can be calculated from the inverse Laplace transform of Eq. 4.22:

$$P_{\mathrm{m}}(s) = \mathscr{L}^{-1} \left[\frac{(1 - T_{\mathrm{w}} s)}{s(1 + 0.5 T_{\mathrm{w}} s)} \right]$$

from which the power generated for unity guide vane opening is given by

$$\Delta P_{\mathrm{m}}(t) = 1 - 3e^{-\frac{2t}{T_{\mathrm{w}}}}.$$

4.4.2 *Elastic Water Column*

A linearised model for the turbine with elastic water column can be assembled by combining Eq. 4.17 with Eqs. 4.5 and 4.6 to yield Eq. 4.23, the transfer function relating the incremental torque or power output to changes in guide vane position:

$$\frac{\Delta P_{\mathrm{m}}(s)}{\Delta G(s)} = \frac{a_{23} + (a_{11} a_{23} - a_{21} a_{13}) \frac{T_{\mathrm{m}}}{T_{\mathrm{e}}} \tanh(T_{\mathrm{e}} s + F)}{1 + a_{11} \frac{T_{\mathrm{m}}}{T_{\mathrm{e}}} \tanh(T_{\mathrm{e}} s + F)}. \tag{4.23}$$

A close examination of the transfer function reveals that its gain varies between two limiting values:

$$\text{Low limit} \left| \frac{\Delta P_{\text{m}}(0)}{\Delta G(0)} \right| = a_{23}$$

$$\text{High limit} \left| \frac{\Delta P_{\text{m}}(\infty)}{\Delta G(\infty)} \right| = \frac{|a_{11}a_{23} - a_{21}a_{13}|}{a_{11}}$$

Using the ideal turbine parameter values summarised in Sect. 4.2 and ignoring the friction losses by setting the friction coefficient $F=0$, then Eq. 4.23 can rewritten to yield the transfer function:

$$\frac{\Delta P_{\text{m}}(S)}{\Delta G(S)} = \frac{1 - Z_0 \tanh(T_{\text{e}}S)}{1 + 0.5Z_0 \tanh(T_{\text{e}}S)}, \tag{4.24}$$

where $Z_0 = T_{\text{w}}/T_{\text{e}}$ is the normalised impedance of the penstock.

Equation 4.24 is a distributed-parameter model and it is difficult to use it for system stability studies; therefore, the general approach is to approximate it with a reduced-order model. The first approach is to use a Maclaurin series to obtain an nth-order rational transfer function, which approximates the irrational transfer function of Eq. 4.23. Series approximation methods must be applied with caution as the order of the approximation is increased in pursuit of model accuracy. A fourth-order series is found to be the highest useful approximation because including further terms makes the transfer function unstable, although the original transfer function of Eq. 4.20 is stable [67]. The fourth-order transfer function can be written as

$$\frac{\Delta P_{\text{m}}(s)}{\Delta G(s)} = \frac{1 - T_{\text{w}}s + 0.5T_{\text{e}}^2 s^2 - 0.167T_{\text{e}}^2 T_{\text{w}} s^3 + 0.04167T_{\text{e}}^4 s^4}{1 + 0.5T_{\text{w}}s + 0.5T_{\text{e}}^2 s^2 + 0.0834T_{\text{e}}^2 T_{\text{w}} s^3 + 0.04167T_{\text{e}}^4 s^4}. \tag{4.25}$$

The representation of Eq. 4.23 could alternatively be modified by forming the lumped parameter approximation. This is performed by expanding the transfer function into the general nth-order model, using the relationship:

$$\tanh(x) = \frac{1 - e^{-2x}}{1 + e^{-2x}}. \tag{4.26}$$

This leads to the finite approximation [67]:

$$\tanh(T_{\text{e}}s) = \frac{sT_{\text{e}} \prod_{n=1}^{n=\infty} \left[1 + \left(\frac{sT_{\text{e}}}{n\pi}\right)^2\right]}{\prod_{n=1}^{n=\infty} \left[1 + \left(\frac{2sT_{\text{e}}}{(2n-1)\pi}\right)^2\right]}. \tag{4.27}$$

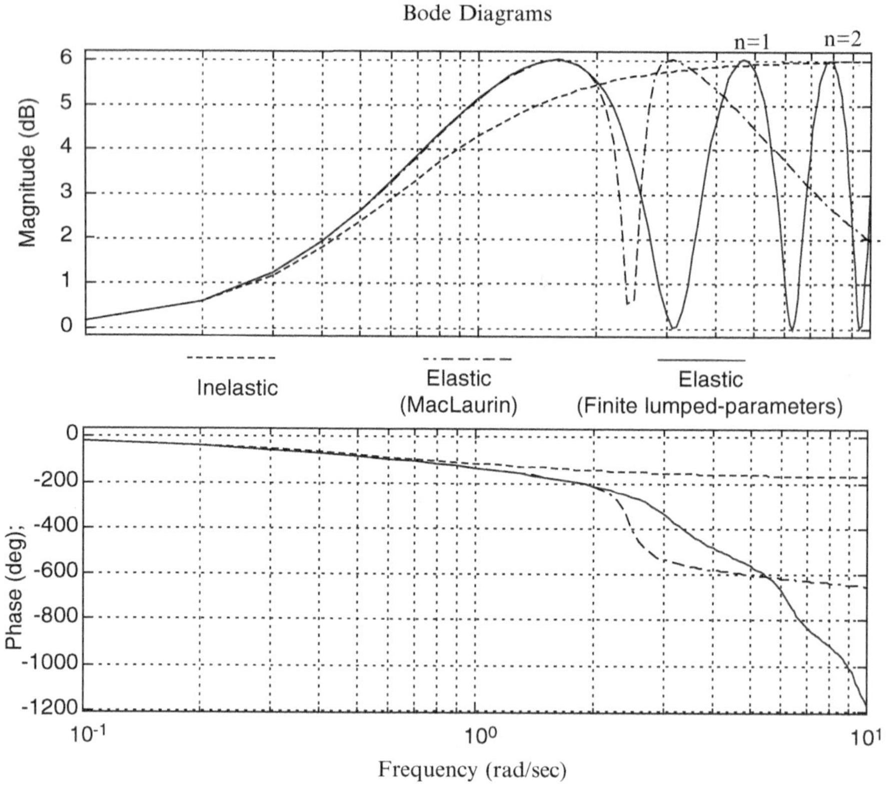

Fig. 4.13 Open loop frequency responses for the penstock-turbine

For $n = 1$, Eq. 4.27 can be written in the form of Eq. 4.28, which is a more accurate system representation:

$$\frac{\Delta P_{\mathrm{m}}(s)}{\Delta G(s)} = \frac{1 - T_{\mathrm{w}}s + \frac{4}{\pi^2}T_{\mathrm{e}}^2 s^2 - \frac{T_{\mathrm{w}}T_{\mathrm{e}}^2}{\pi^2}s^3}{1 + 0.5T_{\mathrm{w}}s + \frac{4}{\pi^2}T_{\mathrm{e}}^2 s^2 + 0.5\frac{T_{\mathrm{w}}T_{\mathrm{e}}^2}{\pi^2}s^3}. \tag{4.28}$$

Figure 4.13 shows the Bode plots for the transfer functions of Eqs. 4.22, 4.25 and 4.28 assuming a water starting time $T_{\mathrm{w}} = 2$ s and wave travel time $T_{\mathrm{e}} = 1$ s. It is clear that the linearised first-order model is accurate at very low frequencies with significant errors at high frequencies compared to the representation of the lumped parameter model. The Maclaurin approximation matches the lumped parameter response for frequencies below 2 rad/s. The gain limits of the transfer function lie between 0 and 6.02 dB (gain range 1–2), which agrees with the earlier prediction.

The classical representation of the system of Eq. 4.22 that represents a short-medium penstock can be derived by assuming T_{e} is very small. Thus, Eq. 4.25 can be rewritten as Eq. 4.28 and the first-order reduced model can be obtained:

$$\frac{\Delta P_{\mathrm{m}}(s)}{\Delta G(s)} = \frac{a_{23} - (1.5 - 0.5a_{23})T_{\mathrm{w}}s}{1 + 0.5T_{\mathrm{w}}s}. \tag{4.29}$$

4.5 Pressure Control Systems

4.5.1 Surge Tanks

Hydraulic transients and pressure changes can be controlled in several ways. Two simple solutions are to ensure that the guide vane can only change position slowly and to provide a pressure relief valve.

- The simplest solution to reducing the pressure surge is to slow the closing time of the guide vane. Guide vane controls and governor regulation can limit the guide vane closure time so that there is no damaging pressure head rise. The pressure surge is a function of the inverse of the closing time. If a pipeline full of water is about 1,000 m long and it takes 10 s to close the valve, both the pipe and valve experience a relatively mild rise in pressure. If the valve is closed over a period of 5 s, the rise in pressure is higher [68].
- Another way to control the pressure is to install a pressure regulator valve near the turbine, which is used to dump water out of the hydraulic system into the lower reservoir. The relief valve can be connected to the turbine spiral case and controlled by the turbine guide vane mechanism to prevent excessive pressure by maintaining a nearly constant water velocity in the penstock. The relief valves are designed to open at a rate that limits pressure rise to an acceptable value [60].
- A common method of controlling pressure is to construct a surge tank. The surge tank prevents excessive pressure occurring by providing a storage volume into which the flow can pass. Surge tanks act as a forebay and shorten the distance for relief from the pressure wave. It provides flow stabilisation to the turbine, pressure regulation and improvement of speed control. If the load on the turbine is suddenly reduced, the governor will act to close the guide vane aiming to decrease the rate of water flow in the tunnel. The water flow in the tunnel cannot drop immediately to the new flow rate and the temporary surpluses of water go into the surge tank. Consequently, the water level in the surge tank rises and the back pressure increases causing the flow in the tunnel to be decelerated gradually. The surge tank also serves as an auxiliary supply to ensure that sufficient water is immediately available close to the turbine when there is a sudden increase in the electrical load. Three slightly different types of surge tanks, which are illustrated in Fig. 4.14, are used in hydro plant [69].
- Figure 4.14a is a simple surge tank consisting of a vertical standpipe connected to the penstock with an opening large enough so that there is no appreciable loss in head as the water enters the surge tank. This is the most efficient surge tank to provide a ready water supply to the turbine, but it is the most hydraulically unstable.
- Figure 4.14b is a restricted orifice surge tank. This is connected in such a way that there is a restricted passage which resists the flow of water back and forth through the tank and the penstock. Appreciable head losses will develop in the water that

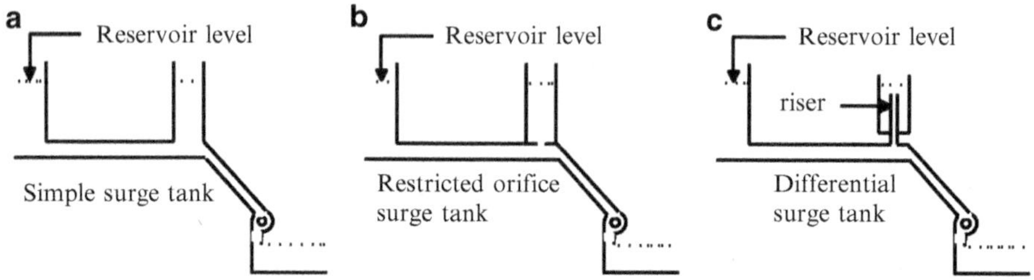

Fig. 4.14 Types of surge tanks

flows into or out of the tank. Thus, the orifice tank has the disadvantage that it does not supply or accept excess water flow from the penstock, but it is more hydraulically stable.

- Figure 4.14c is a differential surge tank and is a combination of the two previous types. It consists of an internal riser of smaller diameter than the penstock, freely connected to it and a storage tank of larger diameter surrounding the riser pipe and connected to it by a restricted passage. The level of the stored water in the larger tank is independent of the accelerating head and the head acting on the turbine, both of these heads being determined by the water level in the smaller inner riser.

4.5.2 *Modelling of the Surge Tank*

At Dinorwig, the hydraulic transients and pressure changes are controlled by the use of a surge tank. This is in the form of an open reservoir located immediately above the high-pressure shaft, as shown in Fig. 4.1, having a capacity of 27,000 m^2. An integral parapet in the pond floor leads into a 30-m-diameter surge shaft. This connects the high-pressure shaft and pond via a 10-m-diameter orifice shaft. Thus, the surge tank can be considered as a restricted orifice (throttled) type. Although large, the tank is still treated as a conduit for modelling purposes. Consequently, the system is composed of three-way compound conduits: the low-pressure tunnel, the high-pressure tunnel and the surge tank including the throttling orifice [70].

The surge tank model is derived from the continuity relation of flow at the tank junction, which can be written as [69]:

Flow down the upper tunnel q_t = flow into the surge tank q_s + flow to turbines q_p

The flow into the surge tank depends upon the area of the tank (A_s) and the rate of change of the tank level (h_s):

$$q_s = A_s \frac{dh_s}{dt}. \tag{4.30}$$

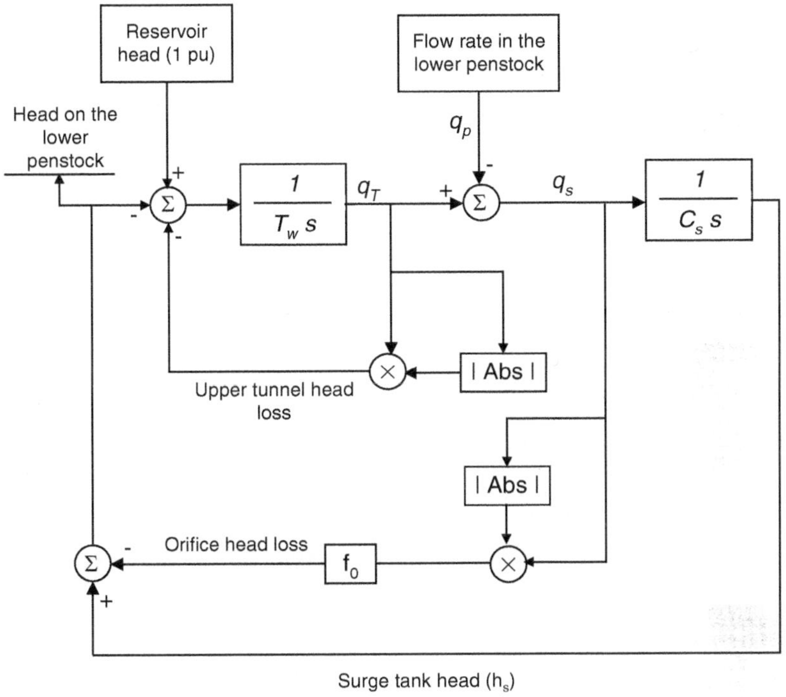

Fig. 4.15 Surge tank model

The surge tank water level in per unit is calculated as

$$h_s = \frac{q_s}{C_s s}, \tag{4.31}$$

where C_s is the storage constant of the surge tank, which is defined as

$$C_s = \frac{A_s h_{\text{base}}}{q_{\text{base}}} \text{ seconds.} \tag{4.32}$$

The head losses in the orifice to the surge tank are proportional to the coefficient f_0 times flow rate times the absolute value of the flow rate (to maintain the direction of the head loss). Hence, the head on the lower penstock is the surge tank level (head) minus the head losses due to the orifice. The level of the surge tank defines the head across the lower penstock. The inclusion of the surge tank effect is warranted in cases where dynamic performance is being simulated over a few minutes of "real" time. Figure 4.15 shows the surge tank model as a block diagram.

The addition of the surge tank will give rise to a poorly damped oscillation between the tank and the reservoir. These oscillations are generally quite slow, of the order of a few minutes per cycle, and can be neglected in studies of governing and load frequency control. In the lumped system (rigid column) theory, if the head loss in both the tunnel and the orifice is neglected, the surge oscillation in the surge tank has a period (T_{st}), which is the time interval between the turbine load change and the occurrence of the maximum surge. It can be computed from the following expression [71]:

$$T_{st} = 2\pi\sqrt{\frac{l A_s}{gA}}, \tag{4.33}$$

where l is the length of the tunnel from the reservoir to the surge tank, A_s the cross-sectional area of the surge tank and A the cross-sectional area of the tunnel.

The surge follows a damped sine wave and the maximum change in the water level occurs at $1/4T_{st}$ after the start. At Dinorwig, the surge tank period is 240 s and the storage constant of the surge tank $C_s = 960$ s.

4.6 Evaluation of Hydraulic Parameters for Dinorwig

4.6.1 *Water Starting Time*

The water starting time for rated conditions is calculated from the plant drawings and measured flows between the surge tank and the turbine inlet. Consequently, the water time constant becomes a function of the number of units that are online [64]. Starting at the surge tank, the water is brought to the power station by a single tunnel of about 994 m in length. At the end of this tunnel are individual penstocks of 210 m in length. The static head used is 513 m. The maximum generating flow (six units operating at full load) in the common tunnel is measured to be 390 (m^3/s). Table 4.1 summarises the calculation of water starting time for each section of the hydraulic system based on Eq. 4.15. The water starting time for a unit varies from 0.69 s when one unit is online to 2.72 s when all the six units are online.

Table 4.1 Dinorwig water starting time

Section	One unit	Two units	Three units	Four units	Five units	Six units
L.P. shaft	0.24	0.48	0.73	0.97	1.22	1.46
H.P. shaft	0.07	0.14	0.21	0.28	0.36	0.42
H.P. tunnel	0.078	0.156	0.234	0.312	0.390	0.468
Tunnel T_{wT}	0.388	0.776	1.174	1.562	1.97	2.348
Manifold concrete	0.054	0.054	0.053	0.063	0.083	0.11
Penstock concrete	0.034	0.034	0.04	0.048	0.051	0.057
Penstock steel	0.166	0.166	0.166	0.166	0.166	0.166
Main inlet valve	0.0526	0.0526	0.0526	0.0526	0.0526	0.0526
Penstocks T_{wn}	0.3066	0.3066	0.3116	0.3296	0.3526	0.3856

4.6.2 Wave Travel Time

The wave travel (propagation) time T_e is dependent on the length of the tunnel and the velocity of sound in the water. Its value is determined by Eq. 4.34:

$$T_e = \frac{l}{a}. \tag{4.34}$$

According to Parmakian [72], the velocity of pressure wave propagation in a pipe is given by the following formula:

$$a = \sqrt{\frac{1}{\rho\left(\frac{1}{k} + \frac{d}{eE}\right)}}, \tag{4.35}$$

where ρ is the water density, k the bulk modulus of the water, d the penstock diameter, e the thickness of the penstock wall and E the elastic modulus of the penstock material.

For a perfectly rigid pipe, the velocity with which a wave is propagated relative to the liquid is the same as the velocity of sound in an infinite expanse of the liquid. Thus, Eq. 4.35 can be rewritten to represent a rigid pipe condition:

$$a = \sqrt{\frac{k}{\rho}}. \tag{4.36}$$

The bulk modulus of water is 2.05×10^9 N/m^2 and so the wave velocity for a rigid pipe is

$$a = \sqrt{\frac{2.05 \times 10^9\,\mathrm{N/m^2}}{10^3\,\mathrm{kg/m^3}}} = 1,432\,\mathrm{m/s}.$$

The tunnels at Dinorwig are concrete lined in the rock, and the thickness of the tunnel wall is much greater than the diameter of tunnel. Thus, it is possible to apply the rigid pipe approximation of the wave velocity to calculate the wave's travel time for each section of the hydraulic system. This was calculated and found to be:

Wave time from the manifold to the surge tank $T_{et} = 0.642$ s
Wave time from the turbine end to the manifold $T_{ep} = 0.146$ s

4.6.3 Head Loss Coefficients

The flow in the penstock is usually turbulent and hence highly complex. Water flowing through a penstock will cause a pressure decrease in the direction of flow.

Table 4.2 Head loss for Dinorwig hydraulic system

Section	Diameter, m	Length, m	Velocity, m/s^3	f_r (Moody)	h_f, m	f_p, m/(m^3/s)^2
L.P. tunnel	10.5	1,695	0.75	0.0151	0.05	1.65e-5
H.P. shaft	10	412	0.82	0.0147	0.0277	4.9e-6
H.P. tunnel	9.5	446	0.917	0.0154	0.031	7.33e-6
Manifold	9.5	77	0.917	0.0154	0.0053	1.25e-6
Penstock concrete	3.8	50	5.73	0.0189	0.416	9.85e-5
Penstock steel	3.3	110	7.6	0.0108	1.059	2.5e-4
M.I.V.	2.5	20	13.24	0.0117	0.75	1.78e-4

Table 4.3 Dinorwig data used in the simulation studies

Total head loss (1 unit)	2.36 m
Total head loss (2 unit)	2.7 m
Total head loss (3 unit)	3.34 m
Total head loss (4 unit)	4.2 m
Total head loss (5 unit)	5.3 m
Total head loss (6 unit)	6.7 m
Head loss coefficient f_t common tunnel	2.873e-5 [m/(m^3/s)^2]
Head loss coefficient f_p individual penstock	5.2e-4 [m/(m^3/s)^2]

The pressure drop over a length of the penstock can be expressed in terms of the head loss h_f as [73]

$$h_f = f_r \left(\frac{l}{d}\right)\left(\frac{v^2}{2g_a}\right), \tag{4.37}$$

where f_r is the friction factor and d is the penstock diameter.

This expression is called the D'Arcy–Weisbach form and the factor $(v^2/2g)$ is called the velocity head. The head loss depends upon the water velocity in the penstock and consequently on the number of units in operation if they are supplied from a common tunnel. The head loss coefficient f_p, used earlier in the modelling of the water column is calculated using expression (4.7).

The head losses calculated for Dinorwig are given in Table 4.2, which represent the hydraulic pressure losses for the plant when a single unit operating at full load draws a water flow of 65 (m^3/s). The friction factor f is a function of the Reynolds number and of the relative roughness of the pipe and is calculated graphically. The data for the friction factor were supplied by the mechanical department at the First Hydro Company (FHC) [74].

The flow in the common tunnel depends upon the number of units dispatched, and the total hydraulic loss therefore increases with the number of units running. The total hydraulic loss and the head loss coefficients corresponding to the number of units operating are summarised in Table 4.3.

4.7 Distributed Parameter Models

4.7.1 The Water Hammer Equations

The transfer function (4.17) which represents the travelling pressure wave in the penstock is derived from the basic hydraulic equations [19] that relate the unsteady pressure head (H) and velocity (V) of liquid flowing in a pipe at some point x and time t:

$$\frac{\partial H}{\partial x} = -\frac{1}{g}\frac{\partial V}{\partial t}$$
$$\frac{\partial V}{\partial x} = -\frac{g}{a^2}\frac{\partial H}{\partial t}. \tag{4.38}$$

Equations (4.38) are similar to the 'transmission line' equations, well known to the electrical engineer. As Kundur explains [19], it is possible to obtain (4.17) by applying Laplace transforms to (4.38), on the assumption that the variables are small deviations from steady state values and that the pressure deviations at the reservoir are zero. Taking the partial derivative of (4.38) with respect to both t and x:

$$\frac{\partial^2 V}{\partial t^2} + g\frac{\partial^2 H}{\partial t\,\partial x} = 0$$
$$\frac{\partial^2 V}{\partial t\,\partial x} + \frac{g}{a^2}\frac{\partial^2 H}{\partial t^2} = 0 \tag{4.39}$$

and

$$\frac{\partial^2 V}{\partial x\,\partial t} + g\frac{\partial^2 H}{\partial x^2} = 0$$
$$a^2\frac{\partial^2 V}{\partial x^2} + g\frac{\partial^2 H}{\partial x\,\partial t} = 0. \tag{4.40}$$

Subtracting (4.40) from (4.39) yields

$$\frac{\partial^2 V}{\partial t^2} - a^2\frac{\partial^2 V}{\partial x^2} = 0$$
$$\frac{\partial^2 H}{\partial t^2} - a^2\frac{\partial^2 H}{\partial x^2} = 0, \tag{4.41}$$

which are the classic one-dimensional wave equations. These linear hyperbolic partial differential equations can be solved individually to obtain the flow velocity and head. An analytical solution to (4.41) is due to d'Alembert:

$$H(x,t) = \frac{f(x-at)+f(x+at)}{2} + \frac{1}{2a}\int_{x-at}^{x+at} g(s)\mathrm{d}s, \tag{4.42}$$

where $H(x,0)$ is the initial pressure profile across the pipe and $g(x) = \partial H(x,0)/\partial t$ the initial rate of change of pressure.

The d'Alembert solution is the sum of two waves travelling in the opposite direction at the constant wave velocity a. A complete solution is obtained by specifying initial conditions (at $t=0$) and boundary conditions (at both ends of the pipe, $x=0$ and $x=L$), where the boundary conditions are either 'free' or 'fixed'. This yields a mathematical expression for the travelling wave which is consistent with the physical description of Sect. 4.3.2. Although the d'Alembert solution is valuable as a concept for visualising and understanding water hammer, its use is limited to cases that have a single dominant conduit and very low levels of friction.

4.7.2 Numerical Solution Methods

Practical solutions to the hydraulic equations are usually obtained numerically. Recalling that they are similar to the transmission line equations, one approach is to adopt the idea of hydraulic impedance (see Chapter 12 in [27]). A particular conduit can then be represented as an equivalent transmission line with specified inductance, capacitance and resistance per metre of length. In turn, this can be modelled by a lumped parameter approximation of the transmission line composed of many cascaded T-filter sections. This idea has been developed extensively by Nicolet et al. at EPFL in Switzerland [75], who have embodied it within a numerical simulation tool called SIMSEN. This software has a modular structure that can represent many of the common components used in power system studies and allows simultaneous representation of electric, hydraulic and control equations [76].

Probably the best known method for computing hydraulic transients is the method of characteristics (MoC). This has been suggested by an IEEE Working Group [28] as an alternative to the elastic water column methods described in Sect. 4.3. It is particularly applicable when investigating the conduit pressure pulsations that occur during full load rejection and the poorly damped travelling waves that are sometimes seen when the guide vanes are almost closed. The method has been described by many authors [27, 77] and only an overview is included here.

Essentially, MoC transforms the partial differential form of the water hammer equations into ordinary differential equations (ODEs) that are much easier to solve. However, the validity of these ODEs is constrained to a set of *characteristic curves* that are related to the velocity of the travelling wave. The treatment below is limited to describing this distinctive feature of MoC. Once the transformation has been done, the method of finite differences can then be applied to discretise the differential equations (both temporally and spatially) so that numerical integration can be used to compute the solutions.

A version of the water hammer equations that includes the nonlinear loss of head due to friction is given in Eq. 4.43, where the coefficient K_f depends on the pipe diameter and the D'Arcy–Weisbach friction coefficient:

$$\frac{\partial V}{\partial t} + g\frac{\partial H}{\partial x} + K_f V\,|V| = 0$$
$$a^2\frac{\partial V}{\partial x} + g\frac{\partial H}{\partial t} = 0. \tag{4.43}$$

Introduce the Lagrange multiplier λ:

$$\lambda\left\{\frac{\partial V}{\partial t} + g\frac{\partial H}{\partial x} + K_f V\,|V|\right\} + \left\{a^2\frac{\partial V}{\partial x} + g\frac{\partial H}{\partial t}\right\} = 0. \tag{4.44}$$

Gather terms:

$$\left(\lambda\frac{\partial V}{\partial t} + a^2\frac{\partial V}{\partial x}\right) + \left(\lambda g\frac{\partial H}{\partial x} + g\frac{\partial H}{\partial t}\right) + \lambda K_f V\,|V| = 0. \tag{4.45}$$

Now $\lambda\frac{\mathrm{d}V(x,t)}{\mathrm{d}t} = \lambda\frac{\partial V}{\partial t} + \lambda\frac{\partial V}{\partial x}\frac{\mathrm{d}x}{\mathrm{d}t}$ so, by comparison of coefficients, the first bracketed term in (4.45) can be made equal to $\lambda\frac{\mathrm{d}V}{\mathrm{d}t}$ by letting:

$$\lambda\frac{\mathrm{d}x}{\mathrm{d}t} = a^2. \tag{4.46}$$

Also $\frac{\mathrm{d}H(x,t)}{\mathrm{d}t} = \frac{\partial H}{\partial x}\frac{\mathrm{d}x}{\mathrm{d}t} + \frac{\partial H}{\partial t}$ and the second term in (4.45) can be made equal to $\mathrm{d}H/\mathrm{d}t$ by letting:

$$\lambda = \frac{\mathrm{d}x}{\mathrm{d}t}. \tag{4.47}$$

So to satisfy both (4.46) and (4.47) we need $\lambda^2 = a^2$ with the solution:

$$\lambda = \pm a, \tag{4.48}$$

where λ is always positive but the wave velocity can be positive or negative.

Applying the constraint in Eq. 4.48 makes the first two terms in (4.45) become full differentials so that

$$\lambda\frac{\mathrm{d}V}{\mathrm{d}t} + g\frac{\mathrm{d}H}{\mathrm{d}t} + \lambda K_f V\,|V| = 0. \tag{4.49}$$

The PDEs (4.42) have now been transformed to the single ODE (4.49) which applies at points where $\lambda = \frac{\mathrm{d}x}{\mathrm{d}t} = \pm a$. Choosing $\lambda = +a$:

$$\frac{\mathrm{d}V}{\mathrm{d}t} + \frac{g}{a}\frac{\mathrm{d}H}{\mathrm{d}t} + K_f V\,|V| = 0, \tag{4.50}$$

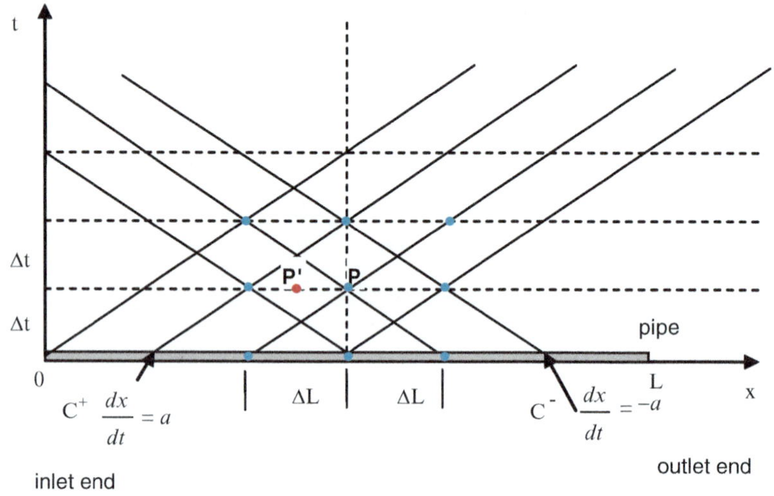

Fig. 4.16 The characteristic curves C^+ and C^- on which the solution to the ODE is valid

which is known as the C^+ equation. Similarly, choosing $\lambda = -a$:

$$\frac{\mathrm{d}V}{\mathrm{d}t} - \frac{g}{a}\frac{\mathrm{d}H}{\mathrm{d}t} + K_f V\,|V| = 0, \tag{4.51}$$

which is known as the C^- equation.

Equations 4.50 and 4.51 are the basis for the finite-difference solution of the water hammer problem:

$$\text{For } C^+,\ \frac{\mathrm{d}x}{\mathrm{d}t} = a \qquad \therefore x = at + c' \qquad \therefore t = \frac{x}{a} - \frac{c'}{a} = \frac{1}{a}x - k'$$

$$\text{For } C^-,\ \frac{\mathrm{d}x}{\mathrm{d}t} = -a \qquad \therefore x = -at + c'' \qquad \therefore t = -\frac{x}{a} - \frac{c''}{a} = -\frac{1}{a}x - k''$$

These represent a set of straight-line relationships between t and x on which C^+ and C^- are valid, as shown in Fig. 4.16.

This means that, if we wish to know the values of V and H at some selected point P along the length of the pipe at some time t, then it must be done on the basis of the values of V and H at $(x-\Delta L)$ and $(x+\Delta L)$ at time zero. On the other hand, if we wanted to work out the values of V and H at P' at (say) time $\Delta t/2$ then we would need to do this on the basis of their values at $(x-\Delta L/2)$ and $(x+\Delta L/2)$ at time zero.

So the smaller the time step is made, the finer the spatial grid must be and *vice versa*: the relationship $\Delta L = a\Delta t$ must hold. This constraint means that the computation time will vary with the square of either the number of time-steps or spatial segments chosen. If ΔL is adjusted so that it is an integer divisor of the length of the pipe, then the time increment Δt is unlikely to be a convenient value. On the other hand, if Δt is fixed, then the length of the pipe must be adjusted to be an integer multiple of ΔL; this is an acceptable approximation if Δt is made sufficiently small.

Two more-or-less conventional stages then remain before MoC can be programmed. The first consists of applying the method of finite differences to digitise the C^+ and C^- ODEs. The second stage is to specify the boundary conditions that constrain the solutions, examples of which are:

- Constant pressure at the upstream end of the pipe (reservoir)
- Downstream valve
- Change in cross-sectional area of the pipe
- Nonlinear functional relationship for a pump/turbine
- Surge tank
- Branching pipe

A full account of these stages can be found in [27, 77] and similar textbooks.

4.7.3 Comparison with the Inelastic Model

Figure 4.17 shows the results when MoC was used to simulate the head, flow and power of two units that share a common tunnel and the guide vanes on one unit are ramped open. Initially, Units #1and #2 are producing 280 and 288 MW, respectively. The guide vanes on Unit #1 are opened from 86% to 95% over a period of 2.6 s while those on Unit #2 remain fixed, as shown in Fig. 4.17a. The variation of head at the units' respective inlets is shown in Fig. 4.17b where the cross-coupling to Unit #2 is clear. There are some quite sharp changes of slope to be seen as the travelling pressure wave impinges and is then reflected at the turbine inlets. In contrast, Fig. 4.17c shows that the flow changes in a fairly smooth manner and essentially follows the profile of guide vane opening. The effect of pressure oscillation on the power produced is shown in Fig. 4.17d where the initial fall over the first 0.2 s should be noted.

It is instructive to compare the MoC results with those of an inelastic hydraulic model of the type described in Sect. 4.3.4. This is shown in Fig. 4.18.

Figure 4.18 shows that the steady state values and the major features of the transient agree well. However, the inelastic model cannot reproduce the power oscillations of the MoC simulation, although the initial dip in power is present. More discrepancy between the two models occurs as the rate of guide vane opening is increased, thus causing a larger travelling pressure wave.

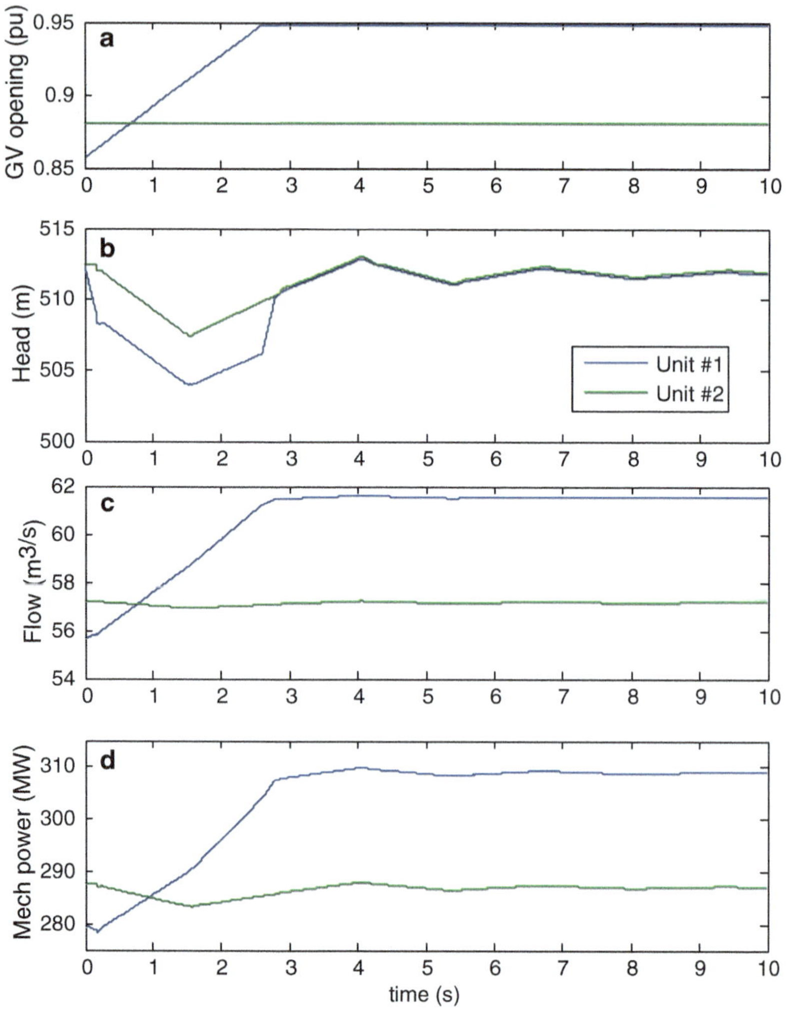

Fig. 4.17 (**a**) Guide vane openings, (**b**) Head at the turbine inles, (**c**) Pestock flows and (**d**) Mechanical power (product of head and flow)

4.8 Conclusions

Mathematical models are of fundamental importance in understanding any physical system. Any mathematical description of a system implies some degree of approximation or some qualifying assumptions so knowledge of these assumptions and the range of conditions over which the models are valid is equally important. In this chapter, the hydraulic system was analysed where the specific area of discussion included the water hammer phenomenon, its effect on the water column dynamics,

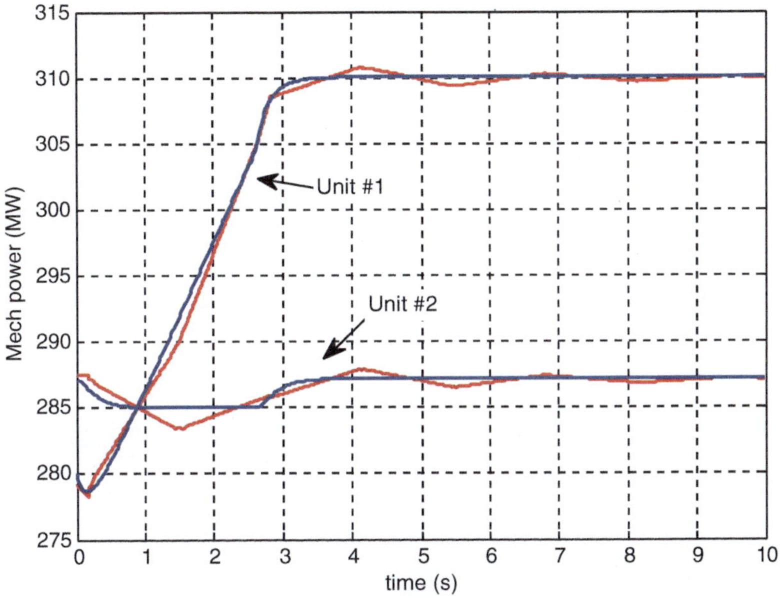

Fig. 4.18 Comparison of power production for the MoC and inelastic models

the hydraulic coupling between units connected to a common tunnel and the method used to calculate the hydraulic parameters for Dinorwig.

The nonlinear models are required for simulation studies where the speed and the power changes are large, such as governor performance evaluation, islanding, load rejection and system restoration studies. Linearised models are useful for control system tuning using linear analysis techniques (frequency response, root locus, etc.). They are rarely adequate for plant simulation because their validity is confined to small perturbations of the unit's head and flow around the chosen operating point; even if this condition is satisfied, the value of the water starting time parameter must be changed as other units are dispatched or taken off-line. Which type of model to use depends on its intended purpose and the selection requires the exercise of engineering judgement.

Chapter 5
Power System Dynamics

5.1 Introduction

This chapter focuses on the characteristics and modelling of the individual elements of the power system. For reliable service, a power system must remain stable and capable of withstanding a wide range of disturbances. Power systems are identified by the physical layout of the generators and loads in addition to commercial boundaries. The flows of active and reactive power in the transmission lines are independent of each other as the active power depends on the angle by which the sending end leads the receiving end, while the reactive power depends on the voltage magnitudes. Since we are interested in modelling the effect of Dinorwig on the stability of the power system, this chapter concentrates on active power control. Therefore, the power system models constructed here are used for frequency control.

Section 5.2 deals with a single unit supplying an isolated load where a generator model based on the equation of motion of a synchronous machine is introduced. This is followed by an explanation of the different types of load and their representation for stability studies, while Sect. 5.3 considers an unit operating in parallel with a large power system. Operating within a large system induces an electrical coupling between the unit and the power system; this effect is also investigated and included in the model. The remainder of the chapter examines the control of active power for a large power system, which is commonly referred to as load frequency control.

5.2 Isolated Operation

The simplest power system representation is the case of a single generator supplying an isolated load (islanding). This can occur in emergencies such as power system restoration, or dividing the power system into small islanding systems. In the case

G.A. Munoz-Hernandez et al., *Modelling and Controlling Hydropower Plants*,
Advances in Industrial Control, DOI 10.1007/978-1-4471-2291-3_5,

of large interconnected power systems, all generating units operating as frequency regulators will contribute to the overall change in generation irrespective of the location of the load change. In the case of small, isolated load operation, the unit must hold the system frequency constant. This is an important distinction because the criteria for system stability differ from those for a unit connected to a large power system. This will allow an understanding of the dynamics of the generator for different loading conditions.

5.2.1 Mechanical Model of the Generator

The dynamic behaviour of the generators within a power system is of fundamental importance to the overall quality of the power supply. The synchronous generator converts mechanical power to electrical power at a specific voltage and frequency. The source of the mechanical power, the prime mover, may be a diesel engine, a steam turbine or a water turbine as in the case of Dinorwig. Whatever the source, it must have the basic property that its speed is almost constant regardless of the power demand. The analysis of any power system to determine its transient stability involves the mechanical properties of the machines because, after any disturbance, they must adjust the angle of their rotors to meet the conditions of power transfer imposed. The electric dynamics have very short time constant compared to hydrodynamics and can be ignored.

Dinorwig power station is mainly used for load/frequency control and since the system frequency is dependent on the active power balance, the generator model is based on its response to frequency changes. However, if voltage stability is to be investigated, this involves the addition of AVRs, excitation system and PSS (power system stabilizer) into the generator model.

The mechanical equations of a rotating machine are very well established and they are based on the swing equation of the rotating inertia. For the purpose of control analysis, the generating unit is modelled by linear differential equations, which describe their response to small perturbations. The swing equation relates the machine's rotor torque angle to the acceleration torque, which is the difference between the shaft torque and electromagnetic torque [78]. Constant shaft speed for a given machine is maintained when there is equilibrium between the mechanical shaft and braking electrical torques. Any imbalance between the torques will cause the acceleration or deceleration of the machine according to the laws of motion of a rotating body:

$$T_{\mathrm{acc}} = J\frac{\mathrm{d}^2\delta_{\mathrm{m}}}{\mathrm{d}t^2} = T_{\mathrm{mech}} - T_{\mathrm{elec}}, \tag{5.1}$$

where T_{acc} is the accelerating torque (N·m), J the combined moment of inertia of the generator and turbine (kg m^2), δ_m the mechanical torque angle of the rotor (rad), t the time (s), T_{mech} the mechanical torque (N m) and T_{elec} the electromagnetic torque (N m).

The mechanical angular velocity ω_m is the time derivative of the torque angle. Rewriting Eq. 5.1 yields

$$J\frac{d\omega_m}{dt} = T_{mech} - T_{elec}. \tag{5.2}$$

The kinetic energy of a rotating body is equal to $1/2J\omega_m^2$, thus Eq. 5.2 can be normalised in terms of the per unit inertia constant H, which is defined as the kinetic energy of the machine at rated speed per machine volt-ampere rating. Using ω_{m0} to denote rated angular velocity in mechanical radians per second gives

$$\frac{2HVA_{base}}{\omega_{m0}^2}\frac{d\omega_m}{dt} = T_{mech} - T_{elec}. \tag{5.3}$$

The angular velocity of the rotor in electric rad/s ω is related to the rotor mechanical angular velocity by $\omega = \omega_m/p_n$, where p_n is the number of generator poles. The equation of motion in per unit form can be written using the angular velocity of the rotor in electric rad/s:

$$2H\frac{d\bar{\omega}}{dt} = \bar{T}_{mech} - \bar{T}_{elec}. \tag{5.4}$$

Noting that $T_{base} = VA_{base}/\omega_{m0}$.

It is preferable to express the relationship of Eq. 5.4 in terms of mechanical and electrical power rather than torque. Since the power is equal to torque times angular velocity, $P = T\omega$, expanding for small oscillations around the operating point and neglecting the second order terms gives

$$\Delta\bar{P} = \bar{\omega}_0\Delta\bar{T} + \bar{T}_0\Delta\bar{\omega}. \tag{5.5}$$

Therefore,

$$\Delta\bar{P}_m - \Delta\bar{P}_e = \bar{\omega}_0(\Delta\bar{T}_{mec} - \Delta\bar{T}_{elec}) + (\bar{T}_{mec0} - \bar{T}_{elec0})\Delta\bar{\omega}. \tag{5.6}$$

At steady state, the mechanical torque is equal to the electrical torque ($T_{mec0} \cong T_{elec0}$). Combining Eqs. 5.4 and 5.5, the deviation in per unit speed $\Delta\omega$ of the rotor as a function of deviations in the mechanical and electrical powers can be represented as

$$\Delta\bar{P}_m - \Delta\bar{P}_e = 2Hs\Delta\bar{\omega}. \tag{5.7}$$

5.2.1.1 Mechanical Starting Time

The mechanical starting time of the machine, T_m, can be calculated using Eq. 5.4, where

$$\frac{d\bar{\omega}}{dt} = \frac{\bar{T}_{acc}}{2H}.$$

Let T_m be the time required by the rated torque to accelerate the rotor from stand still to rated speed. Integrating with respect to time with $\omega = 1.0$ p.u., and $T_{acc} = 1.0$ p.u. results in

$$1.0 = \frac{1}{2H}\int_0^{T_m} 1.0\,dt.$$

Therefore, the mechanical starting time $T_m = 2H$.

As defined earlier the machine inertia constant H is given by

$$H = \frac{\text{Stored energy at rated speed}}{\text{MVA}_{\text{rating}}},$$

where the stored energy at rated speed $= 1/2J\omega_{m0}^2 \times 10^{-6}$MW s, and the rated speed $\omega_{m0} = 2\pi n_m/60$ rad/s; where n_m is the rotor speed in revolution per minute.

Therefore,

$$H = 5.4775 \times 10^{-9}\frac{Jn_m^2}{\text{MVA}}.$$

At Dinorwig, the total rotating inertia of the turbine generator is $H = 3.995$; hence, the machine starting time is $T_m = 7.99$ s [55].

5.2.2 *Load Modelling*

The term load can be defined as a device connected to a power system (bus) that consumes reactive or active power. Load modelling is qualitatively different from generator modelling. It is relatively simple to construct models of any of the typical load component such as lamps, heaters and refrigerators. However, this is only a small part of the problem because the exact composition of the load is often very difficult to estimate. Load composition changes continuously reflecting customer patterns of using various appliances and devices. It depends on weather, consumer life style and many other factors. Even if the load composition were known exactly, it would be impractical to represent each individual load component.

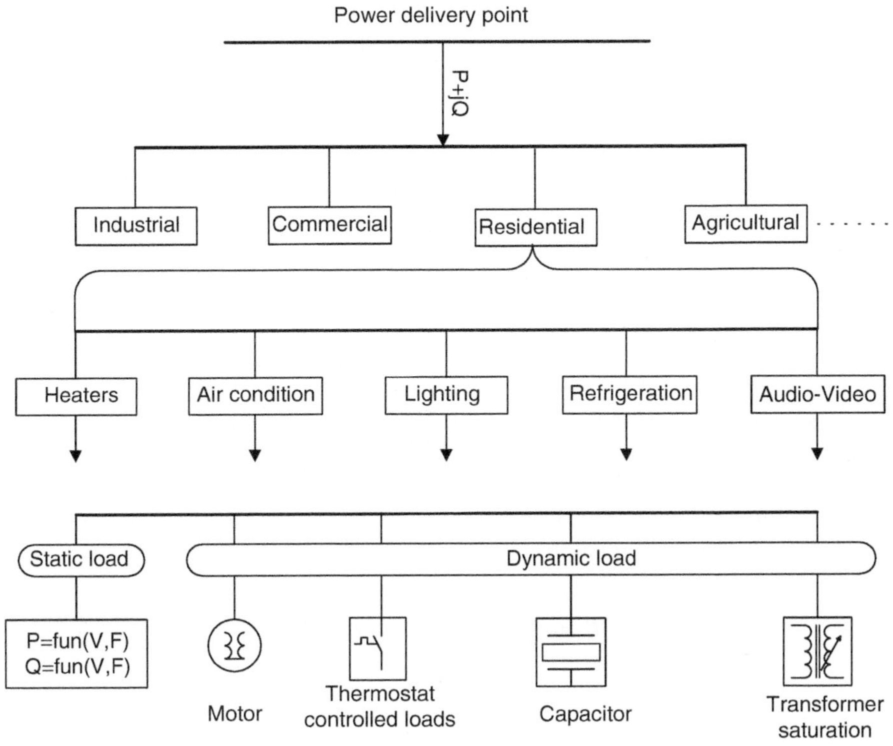

Fig. 5.1 Component-based load modelling

In power system stability, the common practice is to represent the composite load characteristic as seen from bulk power delivery points. As illustrated in Fig. 5.1, the aggregated load is categorised into load classes and each category is represented in terms of load component. Historically, load characteristics are divided into two categories: static and dynamic [79]. Their effects on system stability are discussed separately in this section.

5.2.2.1 Static Model

A static model expresses the active and reactive powers as functions of the bus voltage and frequency at any instant of time. It is common to represent the load by separately considering the active power (P) and reactive power (Q); both can be represented by a combination of constant impedance, current and power elements. Polynomials or other algebraic functions can reasonably represent static loads. The representation is based on the frequency and voltage dependence of the load observed over a rather limited range of variation and often is based only on the measured slopes ($\partial P/\partial f$) and ($\partial Q/\partial f$). This is because these quantities are still the best-known and most generally available data. Traditionally, the model has represented the relationship between power and voltage as an exponential equation, usually in the following form:

$$P = P_0\left(\frac{V}{V_0}\right)^a$$
$$Q = Q_0\left(\frac{V}{V_0}\right)^b. \tag{5.8}$$

The exponents are $a = \partial P/\partial V$ and $b = \partial Q/\partial V$, at $V = V_0$, where the subscript '0' identifies the values of the respective variables at the initial operating conditions. Berg [80] has identified the parameters for different loads and established that a lies between 0 and 2 and b lies between 0 and 3.

The static load model that includes frequency dependence is usually represented by multiplying the exponential load model by a frequency sensitivity factor, which yields the general load equations:

$$P = P_0\left(\frac{V}{V_0}\right)^a (1 + D_{pf}\Delta f)$$
$$Q = Q_0\left(\frac{V}{V_0}\right)^b (1 + D_{qf}\Delta f), \tag{5.9}$$

where P_0, Q_0, are the connected loads at normal frequency, D_{pf} and D_{qf} are the frequency sensitivity parameters of the model and D_{pf} ranges from 0 to 3 and D_{qf} ranges from −2 to 0, depending on the type of load.

The static load models can be classified into three different representations [81], as demonstrated graphically in Fig. 5.2.

- *Constant power model* – the power does not vary with changes in voltage magnitude, which is also called the constant MVA model; this is obtained by setting the exponents to '0'.
- *Constant impedance model* – the power varies directly with the square of the voltage magnitude; the model is constructed by setting the exponents to '2'.
- *Constant current model* – the power varies directly with the voltage magnitude; the model is constructed by setting the exponents to '1'.

The range of frequency and voltage excursions depends on the nature of the disturbances. Accidental loss of generation will lead to a collapse in the frequency and voltage, whereas loss of the load will cause an increase in the frequency and the voltage of the power system.

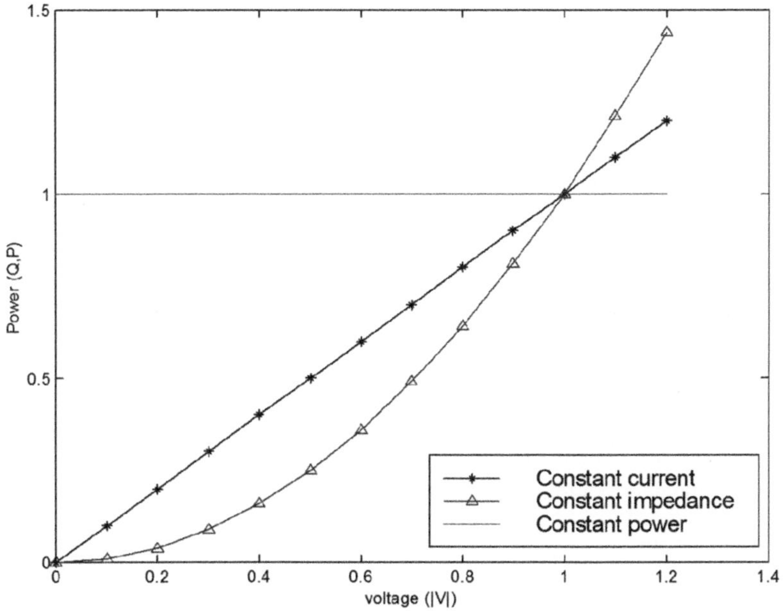

Fig. 5.2 Characteristics of different load models

5.2.2.2 Dynamic Model

A dynamic model expresses the active and reactive powers at any instant of time as functions of the bus voltage and frequency at past and present instants of time. The response of most composite loads to frequency and voltage changes is fast, and a steady state response is reached very quickly. The case of static models is justified in such conditions. There are, however, many components of power systems which respond relatively slowly. Studies of inter-area oscillations and long-term stability often require load dynamics to be modelled.

5.2.3 Generator Loading

In some extreme circumstances, the generator may be required to supply an isolated load and it is necessary to enhance the generator model to account for the load power characteristics of the local load. The most commonly accepted static model is to represent active power as constant current and reactive power as constant impedance [82]. Throughout the book, these models are used when analysing a single unit supplying an isolated load. As described earlier, most loads consist of a large quantity of diverse equipment and since we are interested in the short-term stability of the system, it is adequate to use the static load models of Eq. 5.9. Assuming the system voltage remains constant and using the constant current model, the active power can be expressed as

$$P_L = P_0 + P_0 D_{pf} \Delta f. \tag{5.10}$$

Equation 5.10 is developed for small signal analysis and neglecting the second-order terms yields Eq. 5.11 that separately represents the nonfrequency sensitive load change and frequency sensitive load change:

$$\Delta P_L = \Delta P_0 + D\Delta f. \tag{5.11}$$

The damping term $D = P_0 D_{pf}$ is proportional to the connected load and the load frequency sensitivity parameter. Thus, the damping will be close to zero if the unit is lightly loaded. Combining the load Eq. 5.11 with the generator representation of Eq. 5.7 yields a transfer function describing the mechanical motion of a single machine connected to a load with damping D:

$$\Delta \bar{P}_m - \Delta \bar{P}_e = (T_m s + \bar{D})\Delta\bar{\omega}. \tag{5.12}$$

5.3 Parallel Operation

In today's world, an isolated generator supplying its own load independently is very rare. The usual situation is that the generators are operating in parallel (synchronised) sharing the load of the power system. Operating is this manner increases the reliability of the power system because the failure of any generator will not lead to total power loss. Moreover, having a large number of generators in parallel allows a bigger load to be supplied and permits the shutdown of some generators for maintenance.

5.3.1 Electrical Coupling Between Generators

There is considerable electrical coupling between a generator connected in parallel to another generator or to a power system. To investigate this effect, consider an electrical system represented by two generators connected in parallel across a reactance X_T with their resistances neglected. Figure 5.3 illustrates the system representation. The reactance X_T consists of the sum of the armature reactance and the coil's self-inductance of each generator. The classical model of a generator as described in Sect. 5.2.1 is used in this analysis. Generator #1 operates at a voltage E_1 and rotor angle δ_1. Similarly, Generator #2 operates at a voltage E_2 and rotor angle δ_2. In steady state, the two generators will be operating at base frequency and any load imposed on the system will affect the frequency of both generators. Since

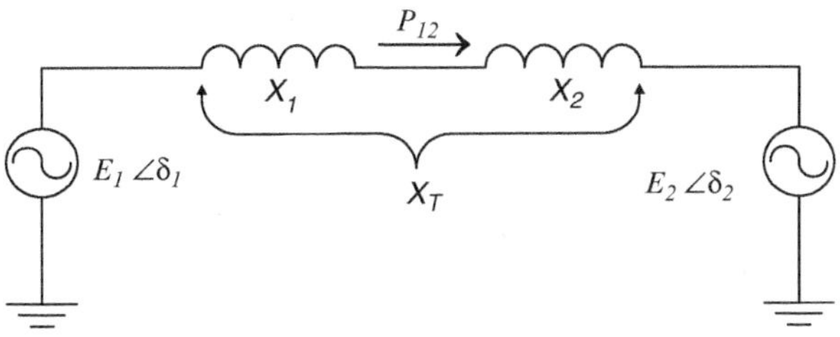

Fig. 5.3 Parallel generators model

the generators are operating in parallel, the power-angle expression of Eq. 5.13 is used to calculate the power exchange between the two units. This power will act as a load on machine #1 and as a generation to machine #2 [83]:

$$P_{12} = \frac{E_1 E_2}{X_T} \sin\delta \tag{5.13}$$

where P_{12} is the power exchange between the machines, E_1 and E_2 the generator voltage, X_T the total reactance between the two generators and δ the torque angle $(\delta_1 - \delta_2)$.

Since we are interested in small changes around the operating point, Eq. 5.13 can be linearised around the operating conditions represented by $\delta = \delta_0$ to yield

$$\Delta P_{12} = \frac{E_1 E_2}{X_T} \cos\delta_0(\Delta\delta). \tag{5.14}$$

The deviation in the torque angle $\Delta\delta$ is calculated by integrating the difference in frequency between the two machines. For a torque angle in radians, the frequency deviation is multiplied by $2\pi f_0$. Figure 5.4 shows the block diagram of two generators operating in parallel sharing a load with a damping coefficient (D). The model developed can be used for internal machine oscillation studies to describe a unit connected in parallel with the power system.

The strength of the electrical coupling depends on the synchronising coefficient T_s of the power transmission line, which is a measure of the incremental change in power resulting from an incremental change in the power angle:

$$T_s = \frac{\Delta P}{\Delta\delta} = \frac{E_1 E_2}{X_T} \cos\delta_0. \tag{5.15}$$

The strength of coupling approaches zero when operating close to the stability limits ($\delta = 90$). Therefore, it is not desirable to operate the line near to its power limit.

Substituting Eq. 5.15 in Eq. 5.14 yields

$$\Delta P_{12}(s) = \frac{2\pi T_s}{s} \left[\Delta F_1(s) - \Delta F_2(s)\right]. \tag{5.16}$$

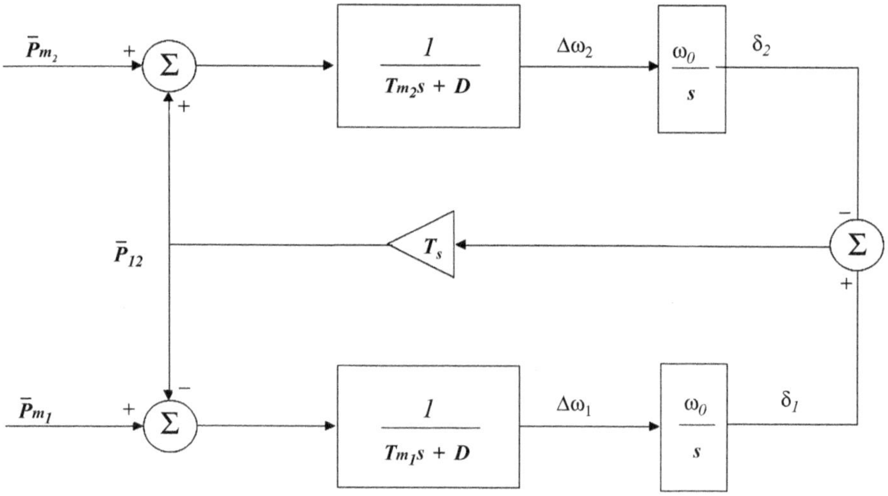

Fig. 5.4 Block diagram of two generators operating in parallel

Using the representation of Fig. 5.4, the characteristic equation of a generator operating in parallel can be written as Eq. 5.17:

$$s^2 + s\frac{D}{T_m} + \frac{T_s\omega_0}{T_m} = 0. \tag{5.17}$$

Comparing Eq. 5.17 with the general representation of a second-order system, the undamped natural frequency ω_n and the damping ratio ζ of the generator are given by

$$\omega_n = \sqrt{\frac{T_s\omega_0}{T_m}}$$

$$\zeta = \frac{D}{2\sqrt{T_m T_s \omega_0}}. \tag{5.18}$$

An increase in the machine inertia constant decreases both the natural frequency and the damping ratio. Conversely, increasing the synchronous torque coefficient increases the natural frequency and decreases the damping ratio. An increase in the damping coefficient increases the damping ratio without affecting the natural frequency. Figure 5.5 shows the behaviour of interconnected generators to a sudden load change in Generator #2 assuming no change in prime mover power (no governor action). Since the generators are operating in parallel, they will share the total load ΔP_L and the power exchange between the generators will be equal to half of the applied load. The final steady state deviations of frequency in both generators

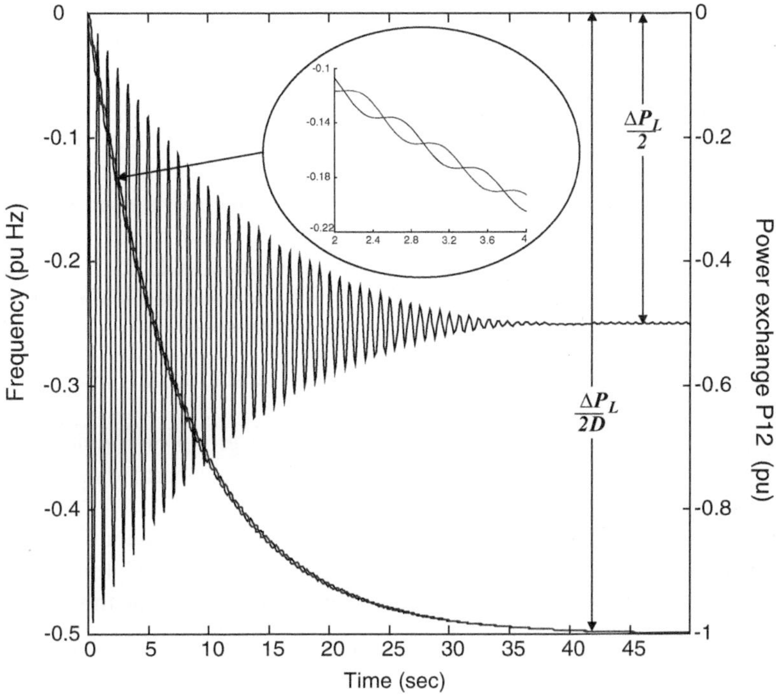

Fig. 5.5 The effect of loading interconnected generators

settle at a value of $\Delta P_L/2D$. The natural oscillation frequency of the speed and power exchange can be calculated using Eq. 5.18.

The change in electric torque of a synchronous machine following a perturbation can be resolved into two components, the synchronising torque component and the damping torque component. System stability depends on both components of torque of each synchronous machine. Lack of adequate synchronising torque results in instability through a periodic drift in rotor angle, the rotor angle continues to increase steadily until synchronisation is lost; this form of instability is referred to as first-swing instability.

5.4 Power System Model

A stable power system is one in which the synchronous generators, when perturbed, either return to their original states if there is no load change or modify the generation to match the load change. The perturbation may cause an oscillatory transient but, with a stable system, the oscillation will be damped. These oscillations appear as fluctuations in the power flow over the transmission line. A reliable power system maintains a constant voltage and frequency at all times. In practice, both voltage and frequency must be held within close tolerance so that the consumer's equipment operates satisfactorily.

Two types of stability analysis are associated with a power system. Recovery from a sudden large disturbance is referred to as *transient stability*. Typically, the disturbances are short circuits of different types which occur on the transmission lines and bus and transformer faults. The period of interest is usually limited to 3–10 s following the disturbance. On the other hand, *small signal stability* is the ability of the power system to maintain synchronism despite the occurrence of small disturbances. This type of stability is largely due to insufficient damping in the system and different types of oscillation can be identified:

- Local machine: the oscillations are localised to one station or a small part of the system.
- Tie line: this relates the effect of oscillations of a group of coupled machines at one part of the power system with machines at other parts, due to weak ties between different parts of the system.
- Control: here the oscillations are associated with the behaviour of the speed governors, the automatic voltage regulation and the HVDC link converters. Tuning the control system to an optimum setting will overcome this problem.

An electrical energy transmission system is characterised by its nominal frequency, voltage profile and load flow [84]. In any such system, the demanded real (P_L) and reactive (Q_L) powers change throughout the day. However, over short periods of typically a few minutes, they can be considered constants with superimposed first-order disturbances ΔP_L and ΔQ_L. The changes in the generated power, ΔP_G and ΔQ_G, must match the load disturbances if the exact nominal state is to be maintained. Due to the random nature of the load fluctuation, this cannot be achieved fully and the goal is then to regulate the system to within sufficiently small tolerances of the nominal. Therefore, controlling system behaviour can be divided into two more or less independent problems.

5.4.1 Megawatt-Frequency Control (P-F Control)

For satisfactory operation of the power system, the frequency should remain nearly constant. Relatively close control of frequency ensures constant speed of motors; the frequency of a power system is dependent on active power balance, as frequency is a common factor throughout the system. A change in the active power demand at one point is reflected throughout the system by a change in frequency. Thus, the frequency error is the most sensitive indicator that a real power mismatch has occurred – an increase or decrease in the real power generated is required in response.

5.4.2 Megavar-Voltage Control (Q-V Control)

The control of voltage level is accomplished by controlling the production, absorption and flow of the reactive power at all levels of the system. The generating units provide the basic means of voltage control. The synchronous generator can generate or absorb reactive power depending on its excitation. When overexcited they supply reactive power, and when underexcited they absorb reactive power. A change in voltage magnitude is the most sensitive indicator that reactive power mismatch has occurred; an increase or decrease in the reactive power generated is required in response.

Dinorwig is configured to operate as a power/frequency (P-F) controller; thus, it is sufficient to implement the frequency changes and ignore the voltage changes. In order to establish a model of the power system to be used in conjunction with the hydraulic system, consider a controlled region of the power system connected via a tie line to a neighbouring region. Assume that the controlled region experiences a real load change of magnitude ΔP_{L} and that a change ΔP_{G} in generated power occurs due to the action of the frequency control machine. The net power surplus (ΔP) is given by

$$\Delta P = \Delta P_{\mathrm{G}} - \Delta P_{\mathrm{L}}. \tag{5.19}$$

The surplus power can be absorbed by the system in one of three ways:

- Changes to the system kinetic energy, which is generally associated with a change in system frequency.
- Increased power export to the neighbouring area via the tie lines.
- Increased load consumption.

The sensitivity of a system load to frequency is expressed by the load damping factor D, which represents the change in the power for a given change in the frequency in an interconnected system. Typical values of D are 1–2%; a value of $D = 2$ implies that a 1% change in frequency would cause a 2% change in load. The smaller the changes in frequency for a given load change, the stiffer the system [33]. The per unit area transfer function using the total load as power base and 50 Hz as frequency base can now be represented as follows:

$$\frac{\Delta \bar{F}(s)}{\Delta \bar{P}_{\mathrm{G}}(s) - \Delta \bar{P}_{\mathrm{L}}(s) - \Delta \bar{P}_{\mathrm{tie}}(s)} = \frac{1}{\bar{D} + \bar{M}s}, \tag{5.20}$$

where $\bar{M}$ is the combined inertia constant of the local machine and the effective rotating inertia of all the other machines connected via the power system.

The inertia constant of the power system can be estimated using the measured transients of the frequency. Inoue et al. [85] have used the frequency transient responses which occur during events such as load rejection tests. A polynomial approximation was applied to the waveform of the transient in estimating the inertia constant. The result shows that the inertia constant in Japan is around 11–18 s. There is also a suggestion that there is a positive correlation between the inertia constant and the load damping D, which affects the value of the inertia for a given system loading. In the absence of a speed control governor (blocked governor) and the tie line, the inertia and the damping constants determine the system response to a load change, as the steady state frequency deviation is calculated as $\Delta f_{ss} = -\Delta P_L/D$.

5.5 Load Frequency Control

The characteristic of synchronous generators requires that they rotate at a fixed speed. If two or more generators are connected on the same electrical system (utility grid), they will operate as if they are on the same shaft. The real power generated depends on the prime mover torque, which is controlled in a hydraulic turbine by the guide vane position.

Two types of control schemes are used for most generator drives. The first type is called isochronous (constant speed) control where the governor continues to adjust the generator output until the measured frequency matches the set point precisely. An isochronous governor scheme works satisfactorily when a generator is supplying an isolated load or in a relatively small power system, where one generator only is used to respond to load changes. Isochronous governors are not used in multimachine systems because of the need for proper load sharing between the machines, since they will counteract each other in trying to control system frequency to their own setting. Isochronous speed control would cause an individual turbine to load up completely if its reference is slightly higher than the power system or to totally unload if its reference is slightly lower than the power system [86].

The second type is droop control (speed regulation) where the governor opens the guide vanes to a fixed position determined by the relationship between speed (frequency) of the power system and a speed reference. The speed droop operates as a steady state offset with regard to a constant frequency reference. The turbine speed cannot be changed when the generator is locked to a power system but it is possible to change the speed reference of the governor. Therefore, all the control machines will pick up load if the power system frequency falls and will drop load if the power system frequency rises. Whatever governor mechanism is present on a turbine, it will always be adjusted to provide a slightly drooping characteristic with increasing load. The value of the speed droop (R) determines the speed versus load characteristic of the generating unit as shown in Fig. 5.6.

The speed droop of a unit is defined by as follows:

$$R\% = \frac{\Delta f(\Delta w)}{\Delta P} = \frac{f_{nl} - f_{fl}}{f_0} \times 100, \tag{5.21}$$

where f_{nl} is the steady state frequency at no load, f_{fl} the steady state frequency at full load and f_0 the rated frequency.

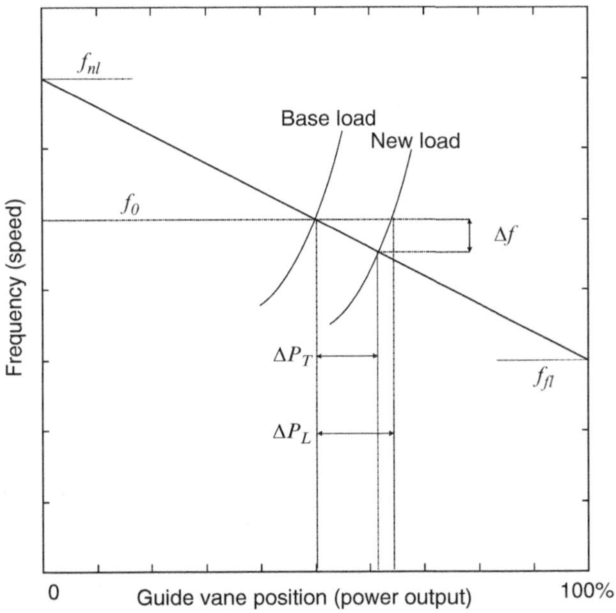

Fig. 5.6 Characteristics of governor with speed droop

A typical droop or 'speed regulation' characteristic is 4% in the UK, which means that a frequency deviation of 4% causes 100% change in the generator output.

The load frequency control characteristic of the power system depends on the collective effects of all the droops of the speed governors and on the frequency characteristics of the load damping D of Eq. 5.20. The steady state frequency deviation following a load change ΔP_{L} for a power system with units operating in speed droop mode is given by

$$\Delta f_{\mathrm{ss}} = \frac{-\Delta P_{\mathrm{L}}}{\left(\frac{1}{R_1} + \cdots + \frac{1}{R_n}\right) + D} = \frac{-\Delta P_{\mathrm{L}}}{\beta}, \tag{5.22}$$

where n is the number of units operating with speed governor.

The power system frequency response characteristic β, which is referred to as the stiffness, is normally expressed by MW/Hz. The physical significance of β can be stated as follows: If a power system was subject to a step load change, it would experience a static frequency drop inversely proportional to its stiffness. The smaller the changes in frequency for a given load change, the stiffer the system. Figure 5.6 illustrates the steady state relationship between load change, frequency change and the increase in power output provided by the governor action. The power output increase of each individual unit under governor control is given by

$$\Delta P_{\mathrm{T}} = -\frac{\Delta f_{\mathrm{ss}}}{R_{\mathrm{unit}}}. \tag{5.23}$$

The power system Eq. 5.20 is rewritten to include the stiffness β; thus, the transfer function of the power system which relates the change in frequency to a change in power is given by

$$\frac{\Delta \bar{F}(s)}{\Delta \bar{P}_{\mathrm{G}}(s) - \Delta \bar{P}_{\mathrm{L}}(s)} = \frac{1}{\bar{M}s + \bar{\beta}}. \tag{5.24}$$

The linearised equation (5.24) is used to represent the power system for load frequency control. This is justifiable because the study is based on analysing the effect of Dinorwig's operation on the power system and the amount of power that is involved is very small compared to the total power system size. Therefore, it is justifiable to assume that there is no change in the system operating point and the perturbations due to Dinorwig are around that point.

5.6 Conclusions

This chapter has presented the background to power system dynamics, as appropriate for use in a modelling study of a power system. The complexities of the models depend on the type of transient and the system being investigated. Since we are mostly concerned about the frequency control of the power system, more effort has been placed on the active power flow in the system.

Generally, the components of the power system that influence the electrical torque include the network state before and after the transient, the loads and their characteristics, the synchronous generator, the speed governor and other supplementary controls such as tie-line power flows. A linearised model representing the power system was developed for use in frequency control analysis of Dinorwig plant. The justification for using the linearised model in the simulation is that the effect of Dinorwig generation is small compared to the total power system loading; it is therefore possible to assume that the system variation is close to its operating point.

Chapter 6
Speed Governor

6.1 Introduction

Having described the dynamic characteristics of the hydraulic prime mover (penstock/turbine) and the electrical power system in Chaps. 4 and 5, this chapter proceeds to describe the next link in the chain, the primary speed/load control, the speed governor. Using the information from the system frequency and power demand, the governor regulates the flow of water through the penstock by controlling the turbine guide vanes in order to restore the frequency to its set value.

This chapter is dedicated to developing models of the governor and the guide vane dynamics, using an analytical approach. Subsequently, these models are verified and revised by system identification techniques. The Dinorwig governor model is obtained by frequency response measurement; this model includes the additional internal functions in the PLC diagram such as filters and extra operational loops. Finally, a transient analysis based on a step response test is carried out in order to obtain a guide vane approximate model [33].

6.2 The Three Term (PID) Controller

As in many other industrial applications, the PID (proportional-integral-derivative) type controller is one of the most widely used control laws in hydro power station governing. This controller can be implemented mechanically, pneumatically, electrically or, as is most common today, by a computer-based device. Computer-based controllers are easy to adjust and configure, in addition to providing possibilities of improvement in operation and control. The major advantage of using a digital controller is that it allows the governing system to be more sensitive and it has fast action. In addition, digital controllers are accurately reproducible, which means that individual controllers produce an identical response from all the units online in

G.A. Munoz-Hernandez et al., *Modelling and Controlling Hydropower Plants*, Advances in Industrial Control, DOI 10.1007/978-1-4471-2291-3_6,

contrast to the variability of mechanical or electrical governors. The PID parameters are separately adjustable; they can be tuned to meet the needs of the plant [87].

The proportional term of the controller produces a control signal proportional to the error in the system, so that

$$u(t) = K_{\mathrm{p}} e(t). \tag{6.1}$$

Typically, given a step change in set point, low values of K_{p} give rise to stable responses but large steady state errors. Higher values of K_{p} give better steady state performance, but worse transient response. Therefore, the proportional action is used to reduce the steady state error, although increasing the gain K_{p} decreases the system time constant and damping. In many systems, first and second order without time delay, the proportional action can never eliminate the steady state error, because some (small) errors must be present in order to produce a control output.

A common way of reducing the steady state error is by incorporating integral action into the controller. Here, the control signal generated is proportional to the integral of the error signal, that is:

$$u(t) = K_{\mathrm{i}} \int e(t)\,\mathrm{d}t, \tag{6.2}$$

where K_{i} is the integral gain.

While an error exists, the integrator tends to increase control action, thus driving the plant towards the demand output. Then, when the error disappears, the continuing integrator output can be used to maintain the control action necessary for steady state conditions. Although the steady state error may be reduced to zero, such performance is achieved at the expense of stability. This is because the integral term increases oscillation amplitude and settling time by introducing extra 90° of phase lag at all frequencies, thus reducing the gain and the phase margins. To reduce the oscillation a third term can be added, which gives a control signal proportional to the time derivative (rate of change) of the error signal,

$$u(t) = K_{\mathrm{d}} \frac{\mathrm{d}e(t)}{\mathrm{d}t}, \tag{6.3}$$

where K_{d} is the derivative gain.

Since the output signal responds only to the rate of change of error, it has no effect upon steady state operation (rate of change is zero). Pure derivative feedback is not practical to implement and it is usually used in conjunction with proportional and/or integral gains to increase the damping of the system. During a transient, the 90° phase lead introduced by the zero at the origin of the s-plane increases the system's phase margin and hence increases the damping of the system. This increase in the damping will allow higher values of K_{p} and K_{i} to be used than would otherwise be the case. Although there are many equations to represent the PID controller, the overall controller transfer function can be written as [24, 87]

$$G_c(s) = K_p + \frac{K_i}{s} + K_d s. \tag{6.4}$$

Often, the controller is represented in terms of the interactions created as

$$G_c(s) = K_p \left[1 + \frac{1}{sT_i} + sT_d \right], \tag{6.5}$$

where T_i is the integral action time (reset time) $= \frac{K_p}{K_i}$ and T_d the derivative action time (rate time) $= \frac{K_d}{K_p}$.

6.2.1 Digital PID Representation

The principle of PID control applies also to digital control, and the controller is transformed into discrete time using Tustin's method (bilinear approximation) to yield [87]:

$$G_c(z) = K_p + K_i \frac{T_{sa}(z+1)}{2(z-1)} + K_d \frac{2(z-1)}{T_{sa}(z+1)}, \tag{6.6}$$

where $z = e^{sT}$ and T_{sa} the sampling period.

The sampling period selection is usually limited by the speed of the computer and physical consideration of the system. The sampling frequency must be at least a little more than twice the value of the highest significant frequency in the signal [88]. Sampling rates used in practice are generally much higher and may be between 4 and 20 times the system bandwidth. For the specific case of Dinorwig hydropower station, it is not necessary to sample all parts of the process at the fastest rate and its PLC governor uses different sampling times as follows [89]:

50 ms	Frequency measurement, frequency control and test terminals
200 ms	Operating mode, selection of parameters and fault monitoring
2,000 ms	Auxiliary circuits

6.2.2 Dinorwig Governor Configuration

The unit governor has the general configuration shown in Fig. 6.1 where the speed and power control loops are integrated, with the power loop providing the permanent droop on the speed [33]. The system comprises two main control loops as follows:

- A frequency control loop, here the frequency deviation is fed to the controller whose output, is used to adjust the turbine guide vane position. An adjustable

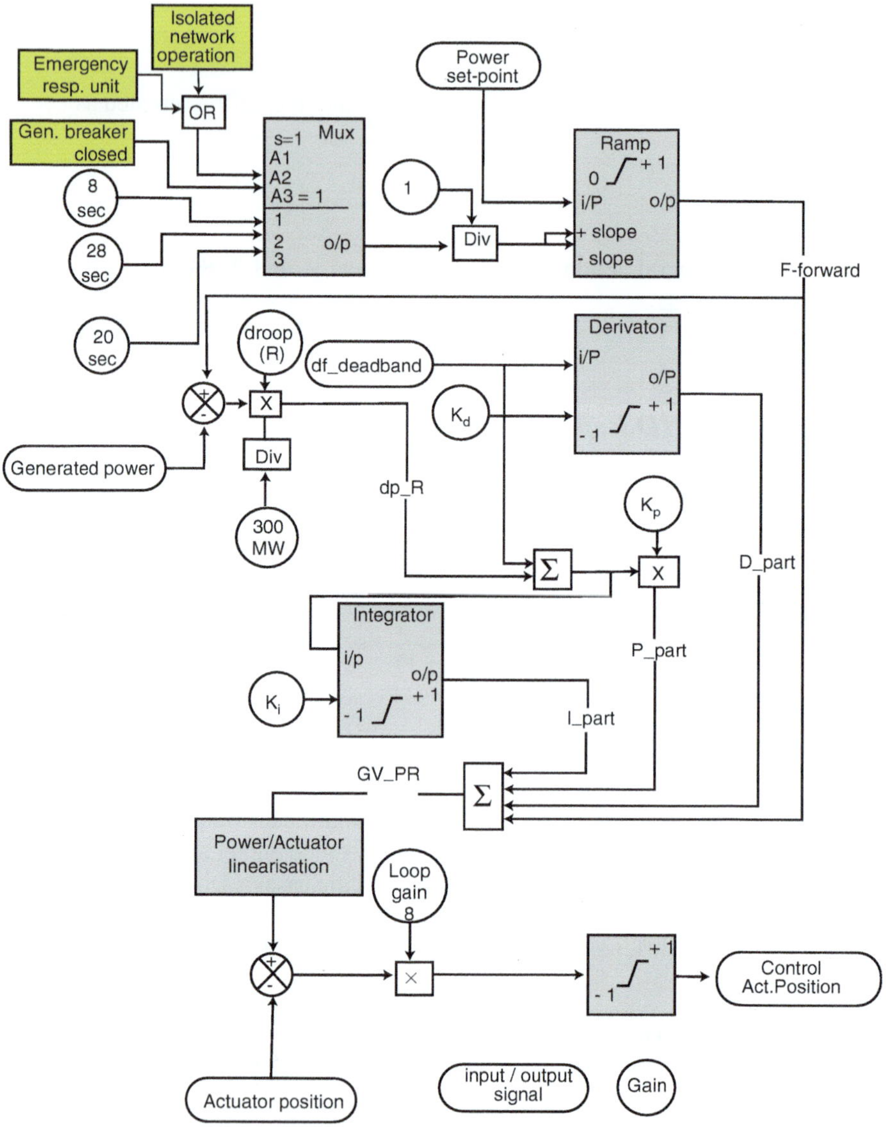

Fig. 6.1 Dinorwig PLC diagram

frequency dead-band, typically of 0.5 Hz, would be included in this loop to limit the response and prevent hunting.

- A power control loop in which the power output deviation is multiplied by the speed regulation droop (R) and fed into the proportional and integral sections of the governor thus creating a compensating signal for the frequency deviation. The derivative action is not used in this loop.

Due to its particular response characteristics, a hydraulic turbine requires a governor with transient droop characteristics for stable speed control response. Traditionally, this is implemented in a mechanical-hydraulic governor. The term transient droop implies that for fast transient conditions (deviations in frequency) the governor exhibits a high regulation (low droop gain), while for slow transients the governor has low regulation (high droop gain) [57]. At Dinorwig, this is implemented by setting the governor to operate with two droop settings: 1% for high regulation and 4% for low regulation. They are set on a control panel by the station operator at the request of the National Grid Company (NGC).

To set the reference position for guide vane opening, the power reference signal is used by injecting a feed-forward signal. This enables a rapid initial response to be achieved after which the governor acts to trim the guide vane position to obtain the required output power. In this manner, the response time from closed to fully open is 28 s in normal operation and 8 s under emergency conditions. Besides, the power reference signal is employed to define the set point for the unit when used in the part–load response mode for frequency regulation. The output signals from the P, I parts and D are summed with the feed-forward signal to produce the guide vane position reference signal (GV_PR), which can be written as

$$\mathrm{GV\_PR} = \left[(\mathrm{d}f + \mathrm{d}p \times R)\left(K_\mathrm{p} + \frac{K_\mathrm{i}}{s}\right) + \mathrm{d}f \times K_\mathrm{d}s + \mathrm{feedforward} \right]. \quad (6.7)$$

The governor normally operates with a power feedback loop; hence, it is necessary to compensate for the nonlinear relationship between guide vane opening and power (torque). This compensation is brought about by using a linearisation function in the PLC's programme. The function is defined by 3 coordinates for different net head (490–520–546 m), 10 coordinates for power and 30 coordinates for resulting actuator position. The values between these coordinates are interpolated linearly [33]. These characteristics can be treated using advanced controllers; some of them will be discussed in Chaps. 11, 12 and 13.

The reference guide vane position is then compared with the position signal from the actuator and the deviation signal is converted into a hydraulic signal using an E/H (electrical to hydraulic) transducer. The hydraulic signal drives the actuator servomotors to adjust the guide vane position. The deviation signal influences the E/H transducer so that it deflects in proportion to the amplitude of the signal from neutral position. The deflection of the E/H convertor causes the actuator servomotor to move in the closing or opening direction depending on the direction of the deflection. The speed of the servomotor is proportional to the deflection of the E/H convertor. The servo loop gain parameter is adjusted during the commissioning period and a suitable response to step disturbances in the actuator loop is obtained by setting the gain to a value of 8 [33].

6.3 System Identification

Since even the best theoretical models are only approximations to the real system, system identification is an important step to verify the theoretical model with experimental data. The identification process fits a model to the recorded experimental data by assigning suitable numerical values to its parameters. System identification can be grouped into *non-parametric* methods and *parametric* methods [90].

The time and the frequency response of a linear time invariant system can be identified using nonparametric methods; these methods do not need prior knowledge of the model structure of a system. For this reason they are often used as a preliminary tool to obtain an initial estimate of the model structure of a system with unknown dynamics. The different types of experimental data for generating a model can be classified as transient analysis (step or an impulse response), frequency analysis, spectral analysis (pseudo-random binary sequence 'PRBS') and correlation analysis; all these methods are easy to apply but often sensitive to noise.

On the other hand, parametric methods can be characterised as a mapping from recorded data to an estimated parameter vector (curve-fitting techniques). These methods can be applied to offline or online estimation. One of the most common types of linear system model is the *ARMAX* model (autoregressive moving average exogenous variable). Estimation of the parameters of this model is usually done by the least-squares method, in which a certain quadratic criterion, based upon the system output measurement and the parameters to be estimated, is optimised to minimise the fit error [90, 91].

The design of experiments and data collection are the critical first steps in system identification. Making an experiment on an industrial plant always involves a large economic risk. Thus, the experiment must be planned carefully and executed with caution to ensure that normal operation is undisturbed while obtaining the data. An efficient experiment is possible if the process and the disturbances are known *a priori* very well. Figure 6.2 summarises the principal steps in a practical identification study.

Collecting the data (observing the system): Collecting observations of the process variables (inputs and outputs) is the first step in identification. The input–output data are recorded during a specifically designed experiment, where the user may determine which signal to measure and when to measure. The objective of the experimental design is to make these choices so that, subject to practical constraints, the data contain the maximum information.

Selecting a model set: On the basis of observations a specific model is selected; this is probably the most important and most difficult choice within the system identification procedure. It is here that *a priori* knowledge and engineering intuition and insight have to be combined with formal properties of models.

Choosing a selection criterion: Any model is only an approximation to the behaviour of the real system. So the intended use of the model will largely determine the selection criteria in order to optimise its desirable features.

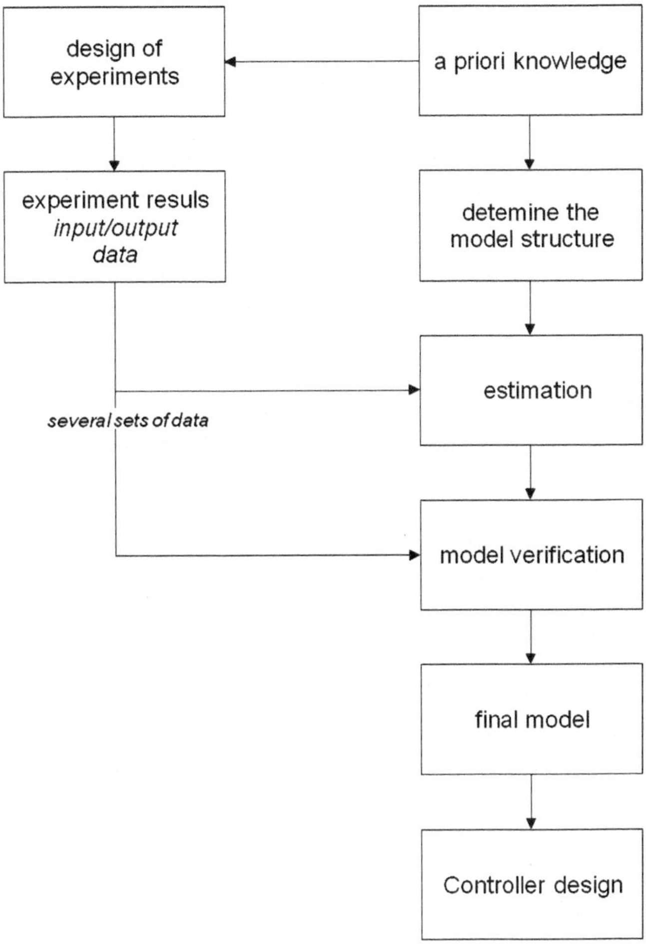

Fig. 6.2 System identification process

Computing model parameters: The evaluation of a model's quality is typically based on how it performs when attempting to reproduce the measured data. The computation of the model parameters can be seen as an optimisation problem (the selection of the 'best' model).

Model validation: Validation is a process where the model is tested in order to determine if it is 'good enough', that is, if it is valid for its purpose. This involves various procedures to quantify how the model relates to observed data, to prior knowledge and its intended use. Deficient model behaviour in these respects causes the model to be rejected, while good performances will develop a certain confidence in the model.

6.3.1 Dinorwig Governor Frequency Response Test

An open loop frequency test was conducted on the spare governor at Dinorwig, using a Solartron 1253 gain-phase analyser [92], in order to verify the mathematical model of the governor described in Eq. 6.1. A sinusoidal signal of varying frequency is applied to the system and the output is measured in terms of both magnitude and phase relative to the input. The system gain is then calculated at each frequency as the ratio of output to input magnitudes. The results are then plotted as a Bode diagram, from which a system transfer function can be deduced.

The governor used in the test is a digital ABB-HPC 640 controller as utilised at Dinorwig for all six operating units. The input signal to the governor is the power system frequency. The signal is in the shape of successive pulses of constant amplitude with time intervals dependent upon the value of the frequency. The frequency is sampled every 50 ms and the output signal is measured using the mean average value over ten cycles. The generator output of the gain-phase analyser is a voltage signal; consequently, it was necessary to convert it to a pulsed frequency and this was achieved by using an AD537 V/F converter [93]. The converter is a controlled oscillator whose output signal has a frequency proportional to the input voltage.

The governor output to the actuator is a voltage signal whose magnitude varies by ± 10 V, which represents the actuator position from fully open to fully close. No signal conversion is required here as the analyser compares the two voltage signals. Figure 6.3 shows the hardware connections for the system under test, where the voltage generator is connected to the V/F converter and its output signal used to supply the frequency input to the governor. One analyser channel is used to monitor the voltage input to the V/F converter and the other monitors the actuator signal of the governor. The gain-phase analyser generator was set to a -5.14 V bias, which is a steady state DC offset corresponding to 50 Hz output of the V/F converter, which is the nominal frequency of the power system. A frequency error signal ($\pm \Delta f$) was superimposed on the 50 Hz and this was implemented by setting the amplitude (rms) of the AC component of the generator output to 20 mV equivalent to ± 0.3 Hz. The frequency range chosen was from 0.02 to 1.26 Hz swept at 18 points from the low to high frequency.

6.3.1.1 Test Results

The swept frequency test was carried out using gain settings for isolated operation mode, $K_\mathrm{p} = 2.5$, $K_\mathrm{i} = 0.8$ and $K_\mathrm{d} = 4$, while the droop was set to zero to disconnect the power feedback loop. A system transfer function was obtained by fitting asymptotic approximations into the data to determine the corner frequencies and hence the pole-zero locations.

MATLAB® software [94] was used to write a program to plot the frequency response characteristics of the governor using the data collected from the analyser

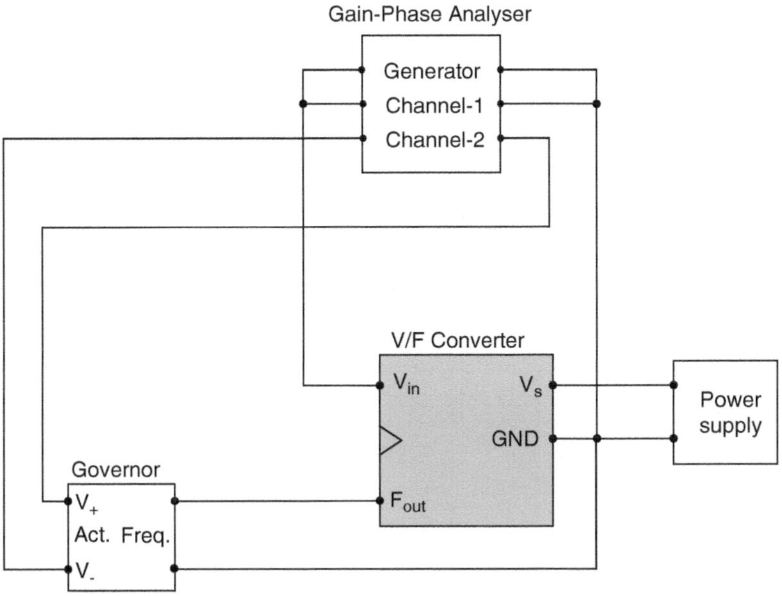

Fig. 6.3 Test connection

as shown in Fig. 6.4. Initially the Bode gain plot has a slope of −20 dB/decade and the phase plot shows a constant phase shift of −90°, so that the system has a pole at $s = 0$, which represents the integral action of the controller.

The system proportional gain is represented by the horizontal line and it is apparent that the value of the gain is ≈27 dB, which is higher than expected (7.95 dB corresponds to $K_p = 2.5$). The source of the extra gain is the internal servo loop gain of 8 (18 dB) implemented in the governor software as shown in Fig. 6.1. Three corner frequencies can be determined at frequencies $\omega_1 = 1/T_i \cong 0.325$ rad/s, $\omega_2 = 1/T_d \cong 0.625$ rad/s and $\omega_3 \cong 0.325$rad/s. The break point due to the integral effect occurs at ω_1, which represents the ratio between the integral and proportional gains.

It can be seen that there is a decrease in the phase below ω_1, and to prevent this from affecting the overall system's phase margin it is necessary to locate ω_1 below the crossover frequency of the system. Meanwhile, the next break point is due to the derivative effect and occurs at ω_2, which is the ratio between the proportional and the derivative gains. Here, the phase is increasing above ω_2; therefore, it is desirable to locate ω_2 so that the increase in the phase occurs near the crossover frequency to insure high-phase margin. This causes a stabilising effect on the system response and reduces the overshoot. The purpose of the derivative action is to extend the crossover frequency beyond the constraints imposed on the PI type controller.

At frequency ω_3, there is an additional 20 dB per decade roll-off, indicating the presence of a second filter term. Investigation revealed a low pass filter with time constant $T_f = 0.3$ implemented to limit the derivative action. While the

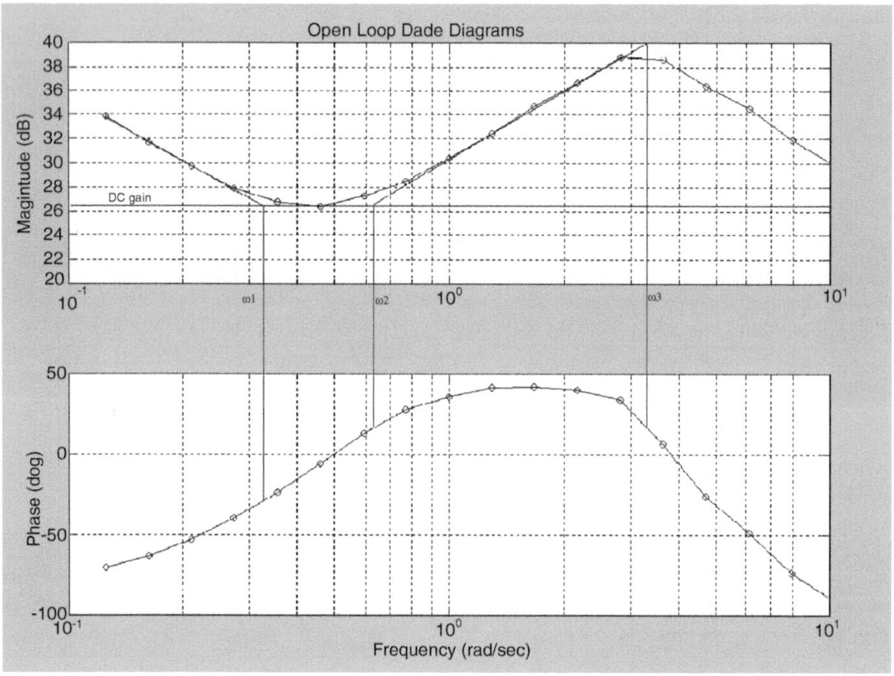

Fig. 6.4 Test frequency response of the governor

derivative term causes the gain to increase with frequency, thereby improving the phase margin, it also has the undesirable effect of increasing the high frequency gain, making the system more noise sensitive and encouraging undesirable natural resonance effects. For this reason, the derivative action is usually cut off at some point between the system bandwidth and the frequency of first resonance.

It was also necessary to model the signal conditioning included by the control system manufacturer that takes a ten cycle running average of the measured power system frequency. This was represented by the transfer function:

$$\frac{\Delta F_a(z)}{\Delta F_f(z)} = \frac{\sum_{j=0}^{9} z^j}{10z^9}, \tag{6.8}$$

where F_a is the average frequency and F_f the power system frequency.

To verify the revised governor model, its frequency response was compared with the experimental results. The Bode plot was obtained by combining the discrete transfer function of the PID compensator (including the derivative limiter filter) as represented by Eq. 6.9, with the averaging filter of Eq. 6.8:

$$G_c(z) = \left.\frac{15.7z^2 - 31z + 15.3}{z^2 - 1.85z + 0.848}\right|_{Tsa=0.05}. \tag{6.9}$$

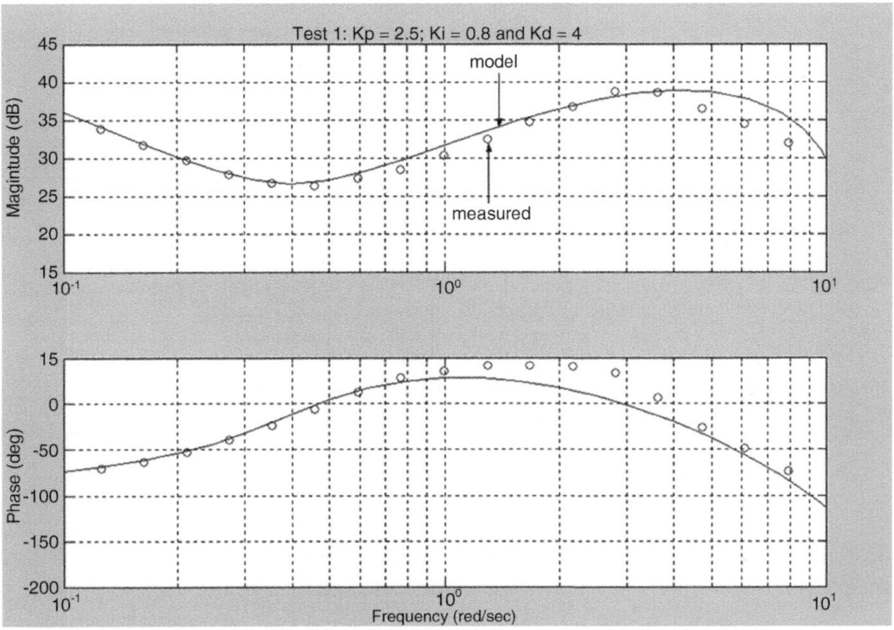

Fig. 6.5 Frequency response of the governor

Equation 6.9 is derived using the same PID gains as the experiment and a sampling period of 50 ms. The results, shown in Fig. 6.5, exhibit a good agreement between the revised model and the experimental results in both magnitude and phase. As was expected some discrepancies remain between the two responses, especially at high frequencies $\sim 5 +$ rad/s. This is thought to be due to the noise present during the test and the saturation of the guide vane position signal. The test was repeated using different PID gain parameters and the results when compared with the model were found to exhibit similar agreement to the previous test. The verification procedure was considered complete at this point and the revised governor model shown in Fig. 6.6 is used in subsequent work.

6.3.2 Guide Vane Modelling

The guide vane controls the water flow into the turbine; this position depends upon the control signal from the governor. Two-stage actuator, with an internal feedback loop, arrangement is used to represent the guide vane dynamics. The governor electrical output is converted by an E/H transducer to drive the actuator servomotors and therefore to adjust the guide vane's position. The transfer function described by Eq. 6.10 can represent, in general, the guide vane dynamics; this equation relates the

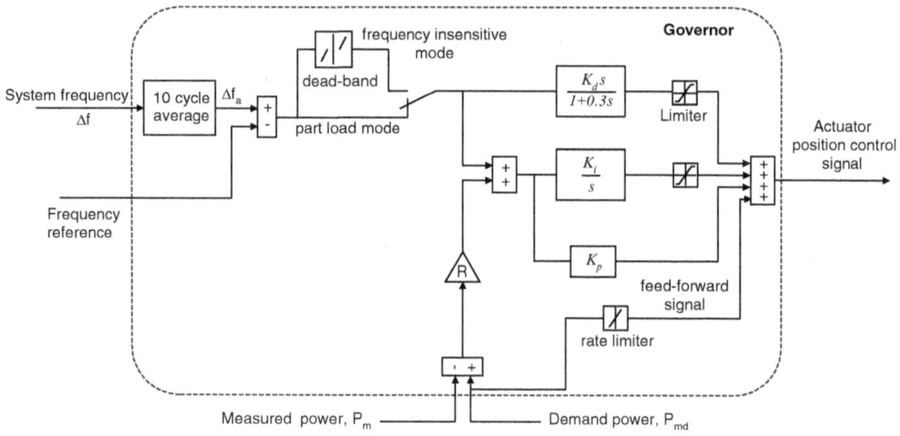

Fig. 6.6 Control loop model for the governor

desired and the actual positions. The two time constants T_1 and T_2 are determined by the pressure/flow characteristics of the guide vane and its actuator servomotors [95]:

$$\frac{Y(s)}{U(s)} = \frac{1}{(T_1 s + 1)(T_2 s + 1)}. \tag{6.10}$$

6.3.2.1 Step Test

Using a step response test, it was identified a guide vane low-order approximation model of the two-stage servomotors from an off-line input/output data test record. Due to this method being quick and relatively easy to implement, it was chosen; besides all the signals required for performing the test are available in the governor cubicle. Information about the steady state and transient response is obtained by this test. The test is performed by using a step-input signal as a guide vane position reference and then recording both the actuator and the guide vane position. At the same time the actuator feedback loop is used to adjust the guide vane position. Figure 6.7a shows the actuator step response, which is smooth and monotonic. This can be represented by a sum of exponentials as shown in Eq. 6.11 [96]:

$$y(t) = y(\infty) + Ae^{-\alpha t} + Be^{-\beta t} + \cdots \tag{6.11}$$

Assuming that $-\alpha$ is the slowest pole in the system, it is possible to write

$$y(t) - y(\infty) \cong Ae^{-\alpha t}. \tag{6.12}$$

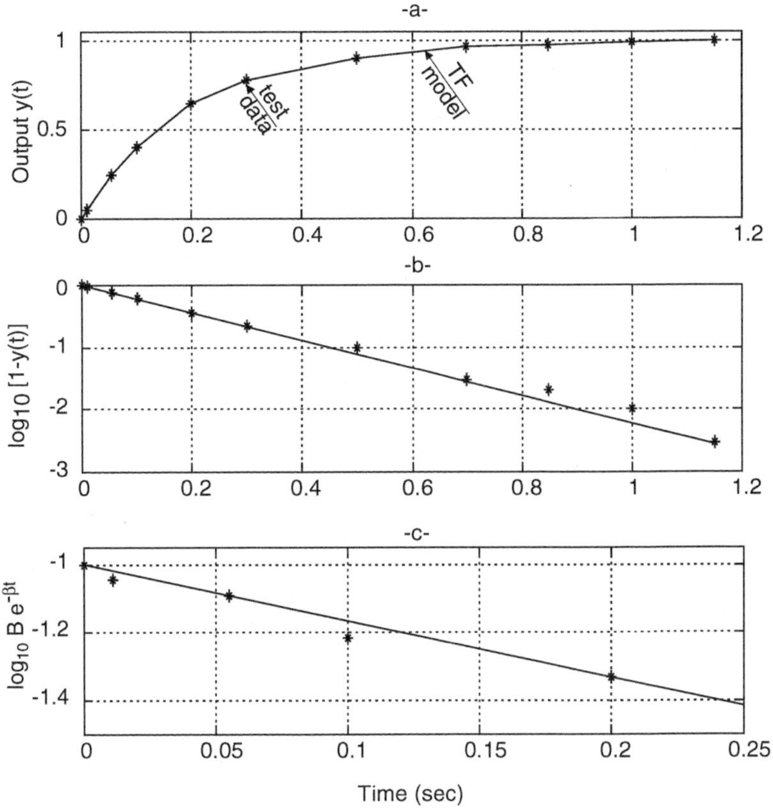

Fig. 6.7 Actuator response

Because $y(\infty) > y$ as shown in Fig. 6.7a, the equation can be solved as

$$\log_{10}[y(\infty) - y(t)] = \log_{10}(-A) - \alpha t \log_{10}(e). \tag{6.13}$$

Figure 6.7b shows the plot of $\log_{10}[y(\infty) - y(t)]$, which is an equation of a line whose slope determines α and intercept determines A. Using this method, the parameter values were estimated to be $A = -1.1$ and $\alpha = 5.1$. The next pole is found by subtracting $1 - Ae^{-\alpha t}$ from the data, and then $\log_{10}$ of the result is plotted as shown in Fig. 6.7c. Following the same procedure, the parameter values are estimated to be $B = 0.1$ and $\beta = 3.8$. Combining the results, an expression for the response is estimated as

$$y(t) = 1 - 1.1e^{-5.1t} + 0.1e^{-3.8t}. \tag{6.14}$$

Equation 6.14 is plotted as a solid line in Fig. 6.7a and shows a reasonable fit to the data and no further terms were calculated. From $y(t)$ we compute,

$$Y(s) = \frac{1}{s} - \frac{1.1}{(s+5.1)} + \frac{0.1}{(s+3.8)}$$

$$Y(s) = \frac{(s+5.1)(s+3.8) - 1.1(s+3.8) + 0.1(s+5.1)}{s(s+5.1)(s+3.8)}.$$

Solving will result in a low-order transfer function:

$$G(s) = \frac{1}{(0.19s+1)}. \tag{6.15}$$

The process was repeated to determine the transfer function of the second stage servomotor, which was found to be a first-order system with a time constant of $T_2 = 0.4$; then the guide vane subsystem dynamics can be represented by the transfer function:

$$\frac{G(s)}{U(s)} = \frac{1}{(0.19s+1)(0.4s+1)}, \tag{6.16}$$

where $G(s)$ is the guide vane position and $U(s)$ the control signal.

6.3.2.2 Nonlinearity

Due to its movement limitation, the guide vane dynamics contain a nonlinearity that imposes upper and lower bounds on the corresponding model variable. When the input signal is within the range specified by the *lower limit* (fully closed) and *upper limit* (fully open) parameters, the input signal passes through unchanged. When the input signal is outside these bounds, the signal is clipped to the upper or lower bound. This behaviour can be modelled by saturation element. The model of the guide vane dynamics used in the simulation including the required saturation blocks is shown in Fig. 6.8.

To verify the guide vane model, a feed-forward signal was applied to the servo loop. This signal was ramped for 7 s to fully open the guide vane; then after 8 s, another signal was applied to close the guide vane. The results are shown in Fig. 6.9a, where it can be seen that the actuator and the guide vane start opening at the same time and the actuator is opening at a faster rate than the guide vane. The actuator is fully opened in 7 s while the guide vane response trails by 3 s, and to compensate for this effect in the model a rate limiter is introduced between the actuator and the guide vane loop as shown in Fig. 6.8.

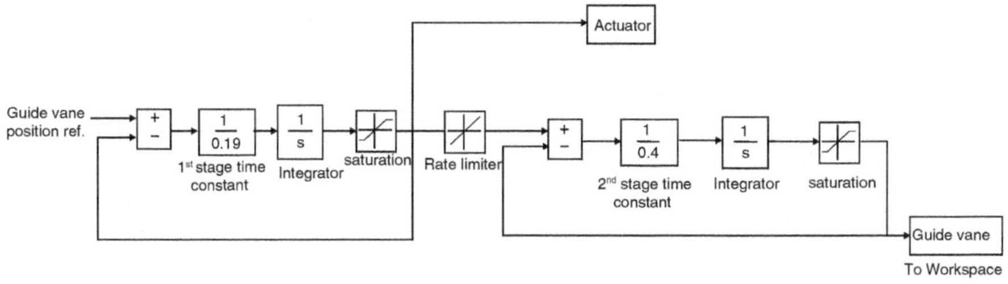

Fig. 6.8 Simulation model of the guide vane

Fig. 6.9 Guide vane test

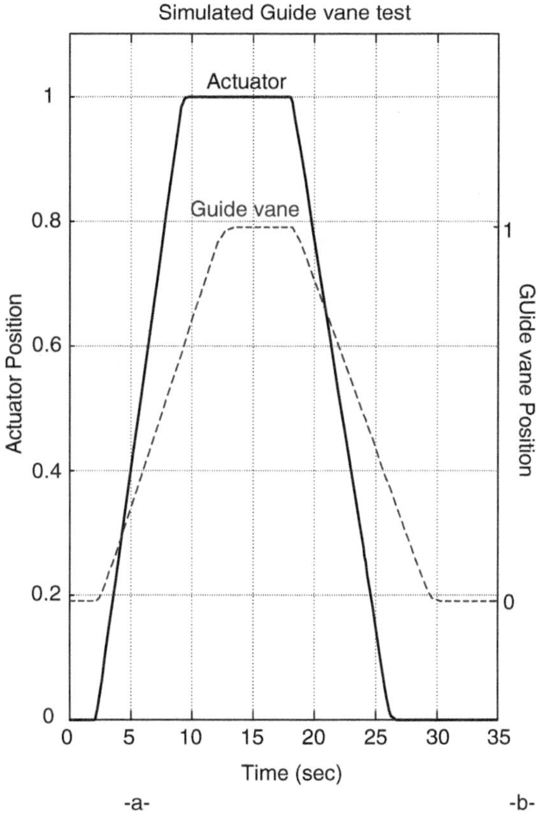

A similar test was conducted on the model, and the results of the simulation run are presented in Fig. 6.9b. The only discrepancy between the two responses occurs during closing down because the guide vane slows down before it is fully closed in order to limit the pressure in the system. This effect is not incorporated in the model of Fig. 6.8, which is used to represent the guide vane dynamics in the studies.

6.4 Parameters Specification

As was stated before, the role of a hydroelectric station in frequency control mode is to provide timely and accurate supply of its target power contribution to the power system. The actual form of the power demand is related to grid frequency variation; however, it can be specified in terms of step, ramp and random input signals [41, 97]. In this section, the step and ramp response specifications for single unit operation, which were proposed by Jones et al. [97], are described.

The intention here is to introduce, for research purposes, a more challenging set of specifications than are currently in use for commercial purposes. The specifications in this section represent a balance between a significant improvement in speed and accuracy of response but not so demanding that they result in unrealistic control activity and tunnel pressures. The specifications themselves are for single unit operation, but it is implicit that any controller that achieves them also maintains an acceptable response over the remainder of the operating envelope [63, 97].

6.4.1 Step Response

The step response specification for single unit operation is expressed in Fig. 6.10 and Table 6.1 (these are not valid for commercial purposes). The most important criterion is usually Test P1 for the *primary response*, which requires that the station,

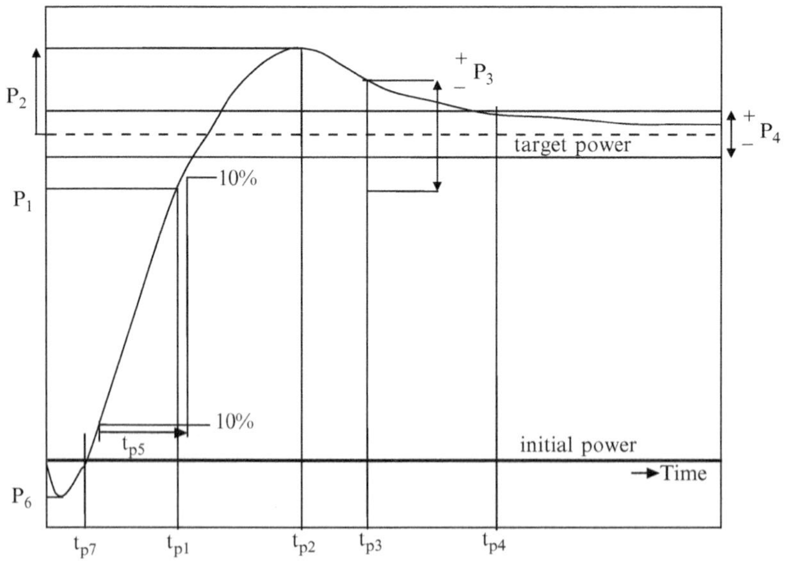

Fig. 6.10 Specifications for a response to a step change in demanded power

Table 6.1 Specification of step response for advanced control design at Dinorwig

Test	Specification for single unit operation	Single unit response with current governor
P_1	$P_1 \geq 90\%$ at $t_{p1} = 10$ s	81% at 10 s, 90% at 13.7 s
P_2	$P_2 \leq 5\%$ and $t_{p2} \leq 20$ s	No overshoot
P_3	$t_{p3} = 25$ s for $P_3 \leq 1\%$	25.9 s
P_4	$t_{p4} = 60$ s for $P_4 \leq 0.5\%$	29.2 s
P_5	$t_{p5} = 8$ s	12.1 s
P_6	$P_6 = 2\%$	1.75%
P_7	$t_{p7} = 1.5$ s	0.88 s

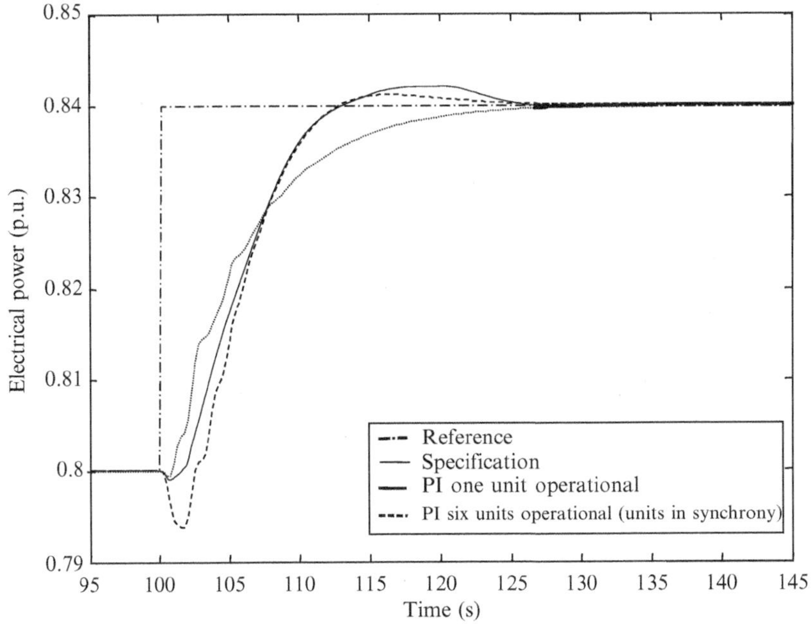

Fig. 6.11 Step response of the hydroelectric plant under a PI controller

under defined conditions, achieves at least 90% of the demanded step power change within 10 s of initiation. Table 6.1 also shows that the overshoot P2 must not exceed 5% and the initial negative excursion P6 (undershoot), associated with the nonminimum phase response, must not exceed 2%.

Figure 6.11 shows the small-step response of the nonlinear elastic model of Dinorwig. In order to illustrate the operational envelope at Dinorwig, one and six units operational, with all units in synchrony, are included. Also, an artificial response that complies with all step response specifications is shown. It can be seen that the response in the case of one-unit operational is slower than the specification, although it has a shorter undershoot. The six-unit operational response is faster than the specification, but has a larger undershoot. If the control parameters are tuned

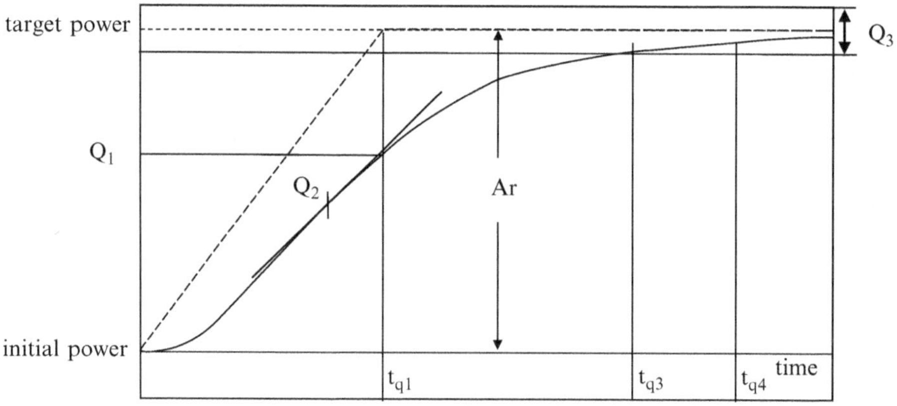

Fig. 6.12 Specification for a ramp input power target

Table 6.2 Specification of ramp response for advanced control design at Dinorwig

Test	Specification for a single unit operation	Single unit response with current PI control
Q_1	$Q_1 \geq 90\%$ at $t_{q1} = 15$ s	14.7
Q_2	$Q_2 = 90\%$ of 6 MWs^{-1}	1.8 MWs^{-1}
Q_3	$t_{q3} = 30$ s for $Q_3 \leq 1\%$	27
Q_4	None specified	E(RMS) = 3.09 MW for $t_{q4} = 50$ s

to increase the speed response in the one-unit operational case, the response of the six-unit operational case will be faster but its undershoot and overshoot will fail to comply with the specification.

6.4.2 *Ramp Response*

The ramp response specification for single unit operation is expressed in Fig. 6.12 and Table 6.2. Again, the most important criterion is usually Test Q1 for the *primary response* (t_{q1}), which requires that the station, under defined conditions, achieves at least 90% of the demanded power change, ramp amplitude (A_r), within 15 s of initiation. Table 6.2 also shows that the maximum rate Q2 must not be less than 90% of the ramp rate and the steady state accuracy Q3 must not be longer than 30 s. Test Q4 shows the effective under-delivery of power over the period of the ramp [97]. The ramp response of the nonlinear elastic model of Dinorwig is shown in Fig. 6.13.

Figure 6.13 shows the ramp response of the nonlinear elastic model of Dinorwig. Again, the one- and six-unit operational cases, with all units in synchrony, are presented, plus an artificial response that complies with all ramp response specifications. As can be seen, in the case of one-unit operational, the response is slower than

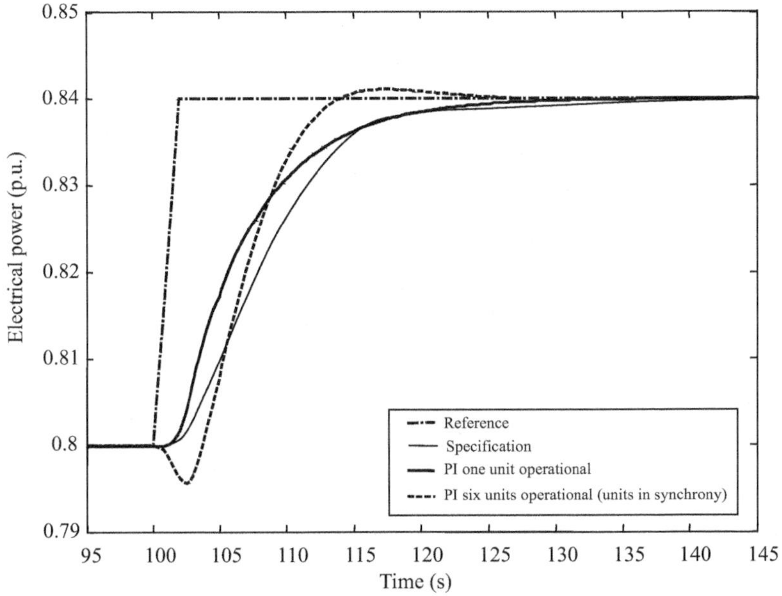

Fig. 6.13 Ramp response of the hydroelectric plant under a PI controller

the specification and has no undershoot. The six-unit operational response is as fast as the specification, and it has a lower undershoot. As for the small-step response, if the response of the one-unit operational case is accelerated, the response of the six-unit operational case will be faster but its undershoot and overshoot will exceed the specification.

The comparison of the model's response with a PI controller with the specification has shown that several criteria are not satisfied. The application of a small-step demand in the six-unit operational results in overshoot because the hydraulic coupling has increased the effective water starting time. The standard PI controller setting is a compromise between one- and six-unit operational. Increasing the loop gain would improve the one-unit response but would make the six-unit response even worse. The aim of the improved controller considered in this book should therefore be to satisfy the single unit specification and also achieve fast, well-damped and low-interaction responses during multiunit operation. This is not possible with a fixed-parameter PI controller [63, 98, 99].

6.5 Closed Loop Analysis

In order to increase the understanding of the relationship between parameters of the hydraulic system response and the parameters of the P and PI controllers, using an unit step as an input signal, several simulations were carried out, using MATLAB®

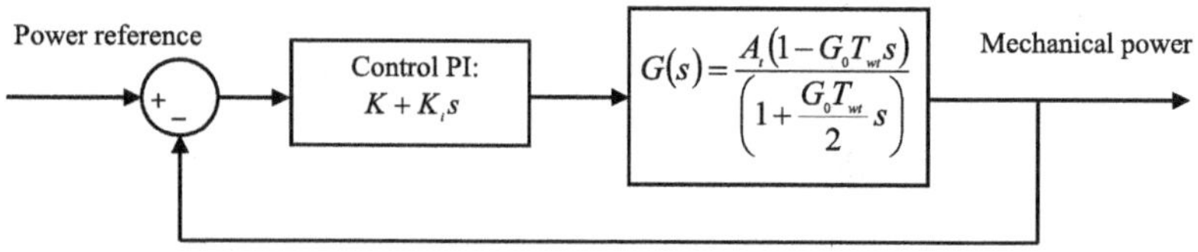

Fig. 6.14 Closed loop hydraulic system

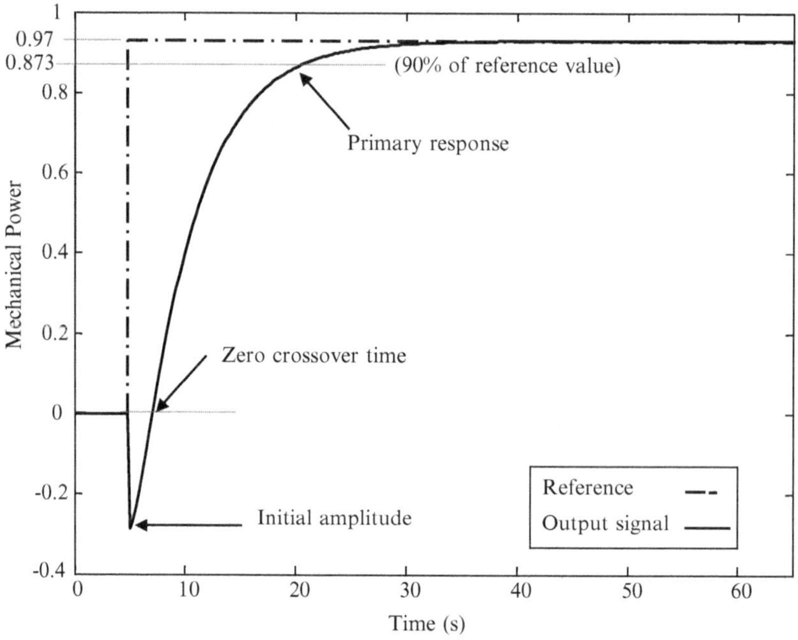

Fig. 6.15 Closed loop step response with a PI control

SIMULINK® [94] combined with theoretical analysis. The study can be seen as a visualisation approach to control tuning [100]. These methods are designed to visualise the system's dynamic behaviour across variations in system and control parameters. Heuristic rules, based on *a priori* knowledge, are employed to select the tests required to evaluate the system performance. It is necessary to generate several performance tables and graphics to visualise a system adequately. This method was selected not only with the intention of not repeating the analysis developed by Mansoor [33], but also to offer another perspective which could provide a better understanding of the tuning problem in Dinorwig.

Figure 6.14 shows the most basic system model in closed loop control. The parameters under study were primary response, undershoot and zero crossover time. As is shown in Fig. 6.15, the initial amplitude (undershoot) is the output signal value when the time is zero. The zero crossover time is the time when the output signal reaches zero after the nonminimum phase behaviour.

Table 6.3 Variations of zero crossover time and T_{wt}

Units under operation	T_{wt}	Zero crossover time
1	0.70	0.31
2	1.00	0.45
3	1.31	0.59
4	1.62	0.73
5	1.93	0.86
6	2.23	1.00

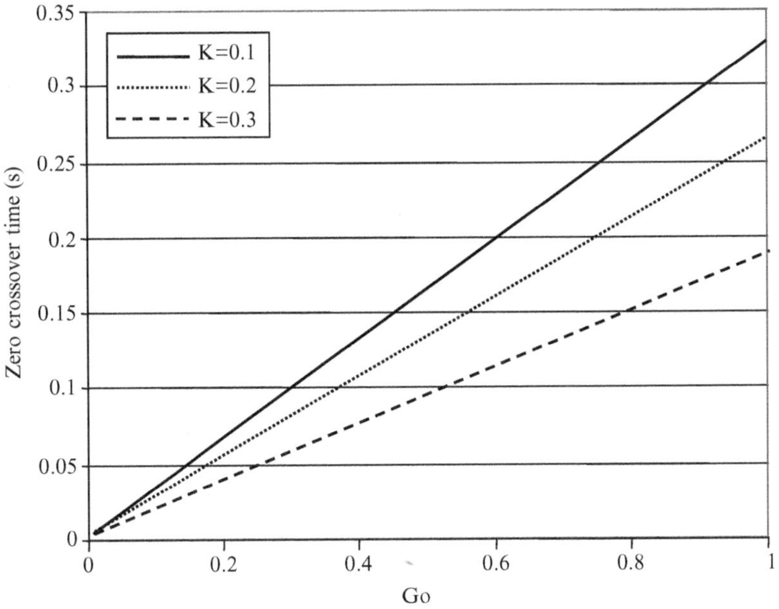

Fig. 6.16 Zero crossover time under P control with different values of K and G_o

The effects of the control parameters on zero crossover time and the relationships between it and other parameters were investigated. Table 6.3 shows that T_{wt} depends on the number of units under operation and that there is a linear relation between zero crossover time and T_{wt}, the zero crossover time increasing by 0.14 s per unit in operation. As is shown in Fig. 6.16, variations in the operating point, G_o, produce a steady increase in the zero crossover time, while the slope of this relationship depends inversely on K.

When the plant parameters are fixed at $T_{wt} = 0.7$ and $G_o = 0.95$, the zero crossover time decreases as K increases (see Fig. 6.17). On the other hand, its value grows as K_i increases for a fixed K (see Fig. 6.18). Its value has an inverse dependence on K and a direct dependence on K_i.

The relationship between the initial amplitude and the control parameters was examined. It was found that the initial amplitude (undershoot) depends only on K,

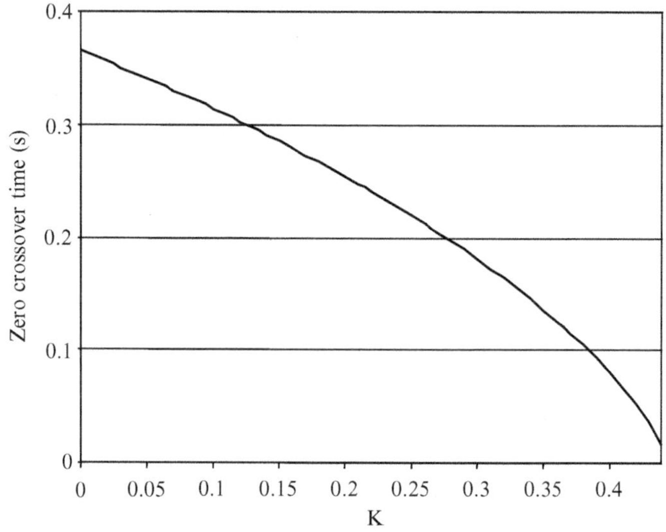

Fig. 6.17 Zero crossover time under P control

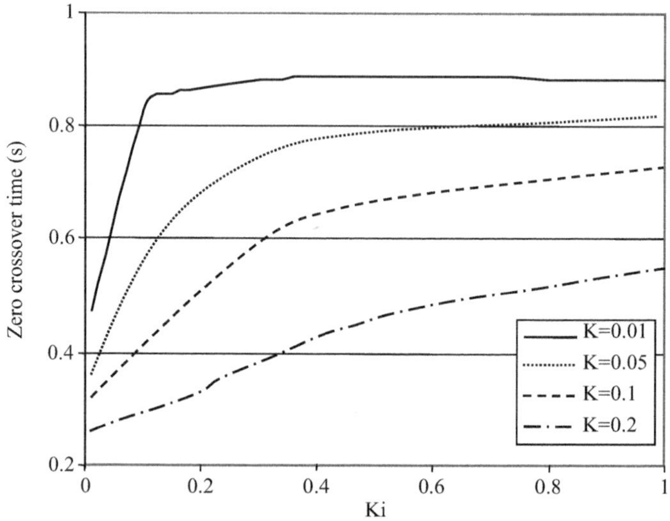

Fig. 6.18 Zero crossover time under PI control

even when PI control is used. According to Fig. 6.19, its value decreases rapidly as K increases. With values of $K > 0.23$, this parameter is outside the saturation value.

The effect of variation of the control parameters over the primary response was studied. As is shown in Fig. 6.20, variations in lower values of K_i ($K_i \leq 0.4$) decrease dramatically the primary response; variations of K_i over 0.4 show a gradual drop in primary response.

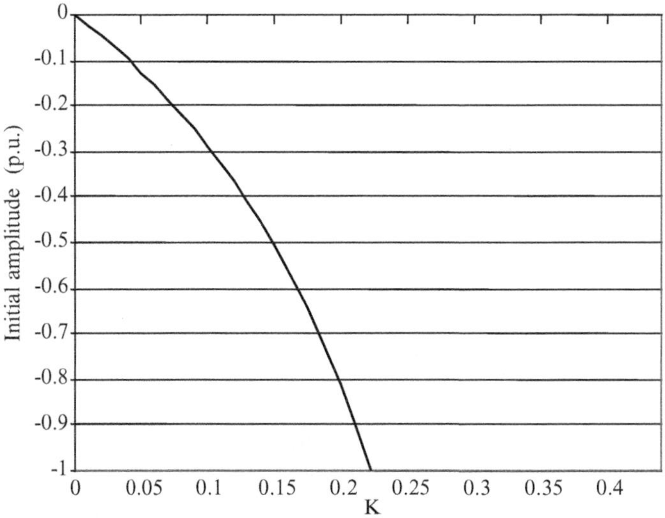

Fig. 6.19 Initial amplitude under P and PI controls

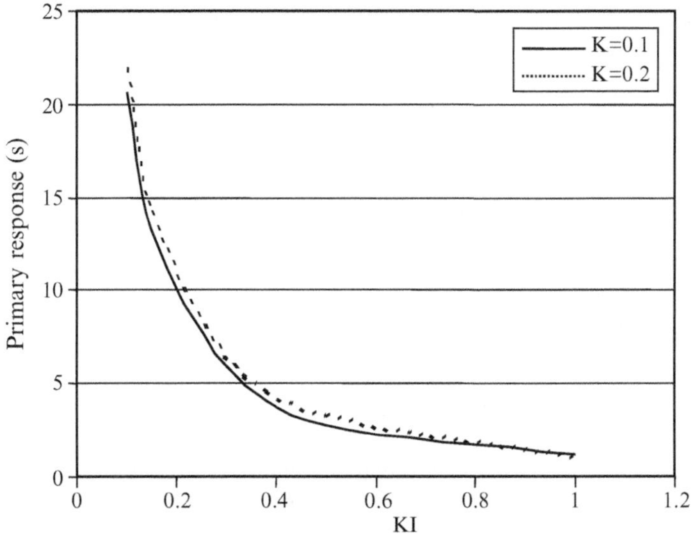

Fig. 6.20 Primary response under PI control

Finally, the connection between primary response and zero crossover time was investigated. As shown in Fig. 6.21, there is an inverse relationship between these parameters: it is not possible to decrease one of them without increasing the other. Variations of K_i from 0.3 to 0.7, when K is fixed to 0.1 and 0.2, are shown in Fig. 6.21. These values were selected because lower values of K produce a long zero crossover time and a very long primary response (Figs. 6.19 and 6.21).

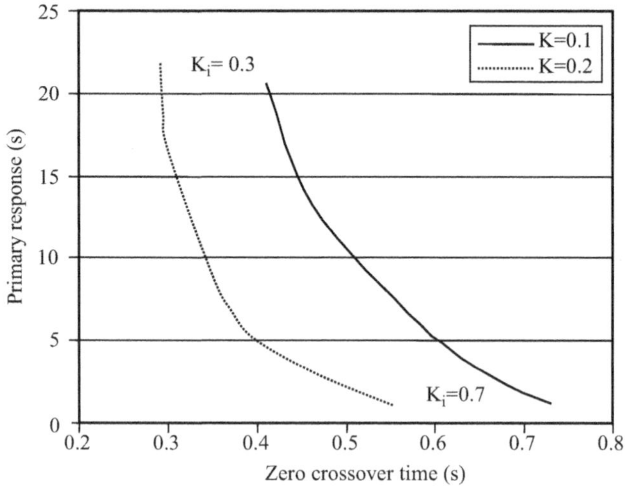

Fig. 6.21 Zero crossover time and primary response with K fixed and K_i variable

Considering the points discussed in this section as suitable, the range for the parameters of a PI controller can be selected for K and K_i as follows: $0.23 \geq K \geq 0.1$ and $0.5 > K_i > 0.1$. Similar analyses have been conducted with classical control techniques [28, 33] and the results are congruent with those stated here. The conclusion that there is a close relationship between zero crossover and primary response is valid despite the structure of control applied to the system. However, this and other effects can be attenuated if features such as multivariable behaviour and nonlinear dynamics are considered in the controller.

6.6 Conclusions

The first part of the chapter concentrated on extracting the salient dynamic features (those relevant to the simulation) from the complex PLC programme. This was achieved by implying system identification techniques to establish accurate models of the governor and the guide vanes. The results demonstrate that it is essential to verify the system mathematical models because the uncertainty about exactly how the program functions and what the parameter values are can lead to errors.

In the case of the governor, a frequency analysis method was chosen, although it was more time consuming. The test results in identifying an averaging filter due to frequency measurement and the derivative action limiter filter which were not included in the original mathematical models.

A step response test was chosen to identify the guide vane dynamics. First, the system has a low order and the step response results in an adequate approximation. Secondly, the frequency response analysis could not be considered for commercial reasons as the length of the test would have resulted in loss of revenue due to lost generation. Thirdly, the power oscillation in the system due to the guide vane's continuous movement would have been unacceptable to the NGC.

In the last part of the chapter, the general features of speed governors were discussed and the configuration of the Dinorwig speed governor was presented. This closed loop analysis offers guidance for selecting the parameters for standard PI control. It has been shown that, because of the relationship between the dynamical parameters of the system, it is not possible to improve one of them without deterioration in another parameter.

Chapter 7
Models Verification

7.1 Introduction

This chapter presents a systematic integration of the models derived in previous chapters into a complete model of the system. The simulation is made with MATLAB®/SIMULINK® tools which are easy to integrate with dSPACE® tool to perform online or hardware-in-the-loop simulation in Chap. 8. In Sect. 7.2.1, a model of a unit supplied from a single penstock is developed. A review of linearised dynamic models generally used for predicting hydropower station responses illustrates their shortcomings when applied to a fast-response station. This is followed by constructing a nonlinear model by adding the water hammer effect, the control system and the dynamics of the electrical power system. Section 7.2.2 extends the integration of the model by including the hydraulic coupling effects due to the common tunnel and the surge tank to complete model of Dinorwig connected to the power system. Section 7.3 presents the steps taken to verify the model behaviour which includes comparison with an independent model and real-plant responses. Section 7.4 lists the main parameters used in the SIMULINK® models, while Sect. 7.5 shows a comparative study of these models.

7.2 Model Integration

7.2.1 Single Penstock Plant

Traditionally, models of hydropower stations have concentrated on the behaviour of low to medium head base load stations connected to an isolated load [19, 28]. They have tended to use relatively simple representations of the hydraulic and mechanical systems, such as shown in Fig. 7.1. In this model [32], the turbine/penstock combination is represented by the classical transfer function, Chap. 4, relating

G.A. Munoz-Hernandez et al., *Modelling and Controlling Hydropower Plants*, Advances in Industrial Control, DOI 10.1007/978-1-4471-2291-3_7,

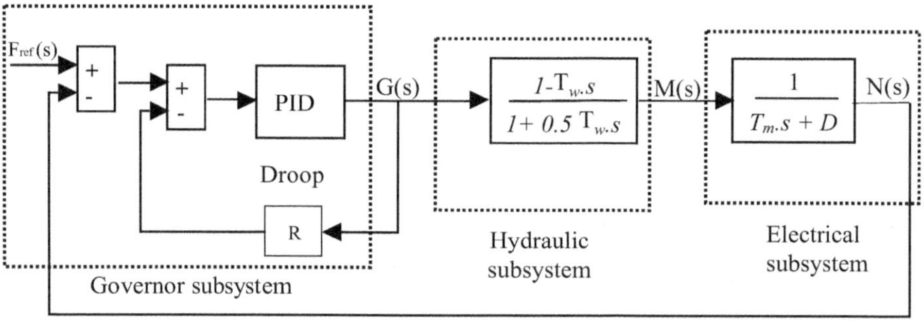

Fig. 7.1 Block diagram of dynamics suitable for a base load hydropower station

changes in the mechanical power to changes in the guide vane opening which assumes an ideal turbine operated at rated flow and head and uses the water starting time (T_w) to represent the transient behaviour of the hydraulics.

The speed governor is modelled as a standard PID controller as described in Chap. 6 and the generator dynamics are represented by the 'swing' equation, which relates the rotating machine inertia (T_m) to acceleration torque. The load is represented by the constant current model of equation where the loading condition depends upon the value of the load damping D [19].

7.2.1.1 Limitation of the Linearised Model

Although this model has been used for many studies, the inherent approximations hold only for low-medium head conditions and it is not suitable for many fast response stations, which tend to operate with a high head. A simple illustration of the limitations of this type of model is shown in Fig. 7.2, where the unit was subject to an initial 30% (0.3 p.u.) increase in load demand, followed a short time later by a further 50% (0.5 p.u.) increase.

The initial part of the curves in Fig. 7.2 shows that the station acts to pick up the 30% load to restore the system frequency back to its nominal value. However, when the response to the subsequent 50% increase in demand load is considered, the inaccuracy of the model becomes clear as the guide vane opening exceeds 1.0 p.u., (which is physically impossible) before settling down to a final load figure of 80%.

The water starting time for the penstock T_w is calculated using the equation from Chap. 4, which, in the linear representation, is calculated from the initial value of flow. Thus for simulation studies, the linearised penstock model requires different values for T_w to reflect changes in water flow rate [33, 63, 101]. The linear model is valid only as long as the unit's load does not deviate greatly from the initial load. For the nonlinear model, the value of T_w is calculated from the value of the flow at rated conditions. The change in the effective time constant is implicit within the nonlinear expressions and the model is valid for all operation conditions.

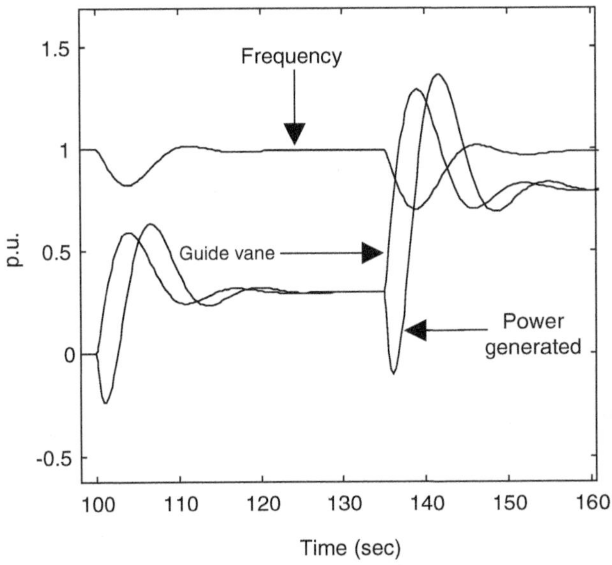

Fig. 7.2 Unit response of a hydropower station to step changes in demand load

A model needs accurately to reflect the true system; this implies the inclusion of factors such as the influence of the power system, limits on guide vane opening and the water hammer phenomenon resulting from the guide vane movement, especially where the penstock may be considered elastic. That yields to a nonlinear model.

7.2.1.2 Enhanced Nonlinear Model

The approach to modelling a system requires an ability to break the problem down into smaller elements that can be addressed individually. Here, this is apparent in the way the hydraulic, electrical and control systems were initially modelled as separate subsystems in Chaps. 4, 5 and 6, respectively. However, having modelled each of these individually, they must then be integrated together to form a model of the whole system. The system integration was implemented by collecting the low-level SIMULINK® blocks into three different subsystems as follows:

- Hydraulic subsystem: consisting of the water column, the surge tank (if utilised) and the guide vane dynamics.
- Electrical subsystem: consisting of the power system, the generator, the power transferred between the two and the load perturbation.
- Control subsystem: consisting of the speed governor.

Stations operating with high head and long penstocks experience pressure changes in the penstock due to the water hammer phenomenon [33]. In Chap. 4, these effects were considered in detail and a conclusion was made to include them into the turbine-penstock dynamics. As a result, a nonlinear hydraulic system was assembled as described in Chap. 4, where the turbine power is a function of head across the turbine and the guide vane opening. In the model, the travelling wave effects are represented by the equation:

$$\frac{H(s)}{Q(s)} = -\frac{T_{\mathrm{w}}}{T_{\mathrm{e}}}\tanh\left(T_{\mathrm{e}}s + F\right). \tag{7.1}$$

The guide vane dynamics were integrated using the model shown in Chap. 6, which is developed from an experimental data. In the meantime, an improved speed governor representation based on experimental data is used where the feed-forward signal is added into the control loop.

The primary purpose of development of the plant model is to analyse its response to load/frequency control; hence, the influence of the power system size and the effect of other plants operating in frequency control mode should be included in the model. This is achieved by using an equation that relates the system frequency change with stiffness. Generally, in the analysis of load-frequency controls, the intermachine oscillations are not considered as the interest is based on the collective performance of all generators in the system. However, if we are interested in a specific plant performance, then the electrical coupling between the machine and a power system could be included.

At steady state the unit and the power system will be operating at base frequency; hence, if load appears it will affect the frequency of the system. It is shown in Chap. 6 that the power exchange between the unit and the power system $P_{\mathrm{Gi}}(t)$ is equal to instantaneous shaft angle difference between the machine and the power system. Once all these enhancements are applied, a generic model of single tunnel hydropower plant is assembled as shown in Fig. 7.3. The parameter B_{c} represents the MW base conversion between the unit and the power system.

The function $\tanh(T_{\mathrm{e}}s)$ in Eq. 7.1 can be simulated using the relationship:

$$\tanh(T_{\mathrm{e}}s) = \frac{1 - e^{-2T_{\mathrm{e}}s}}{1 + e^{-2T_{\mathrm{e}}s}}. \tag{7.2}$$

The term $e^{-2T_{\mathrm{e}}s}$ represents a time delay which can be implemented in SIMULINK® by means of a transport delay block. However, using this representation causes algebraic loops constraints which stop the simulation. Therefore, another solution was required and a Padé approximation to the time delay $e^{-2T_{\mathrm{e}}s}$ was considered. The results of the approximation were compared with the frequency response of $\tanh(T_{\mathrm{e}}s)$ as shown in Fig. 7.4. It can be seen that increasing the order of the approximation improves the match to the frequency but unfortunately it has the disadvantage of causing time-domain oscillations. This is demonstrated in Fig. 7.5 where the step response of $\tanh(T_{\mathrm{e}}s)$ is compared with that of the sixth-order Padé approximation where large high-frequency oscillations are evident. This causes the simulation to become unstable. An alternative is to use a second-order approximation but its frequency response exhibits a shift of 0.2 rad/s in the resonant

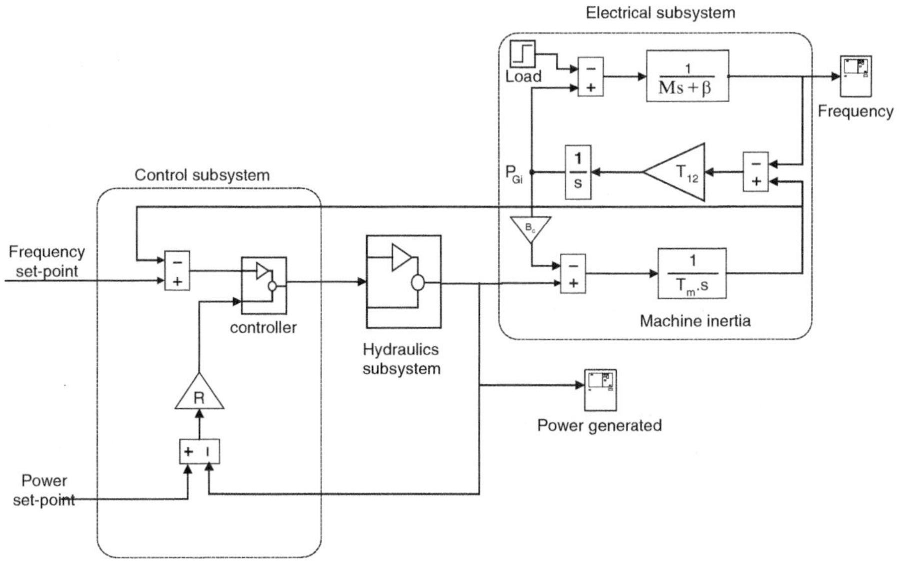

Fig. 7.3 Single tunnel hydropower plant connected to the power system

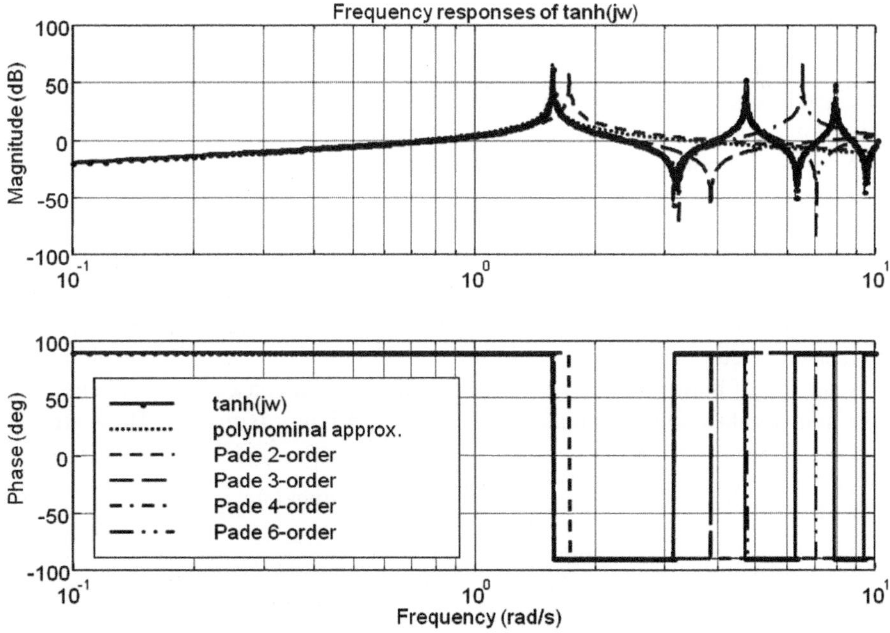

Fig. 7.4 Frequency responses of tanh $(T_e s)$

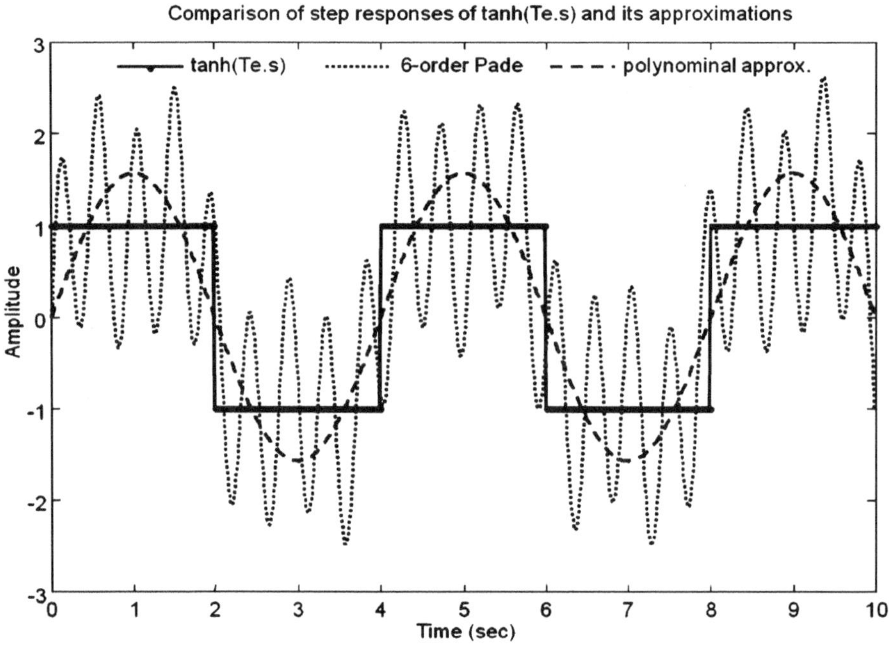

Fig. 7.5 Step responses of tanh $(T_e s)$

peak and it is accurate only for frequencies below 1.3 rad/s. Consequently, it was decided not to use this method of approximation and a third approach was considered.

A low-order polynomial approximation was found by converting Eq. 7.2 to its equivalent z-transform, which was then used to obtain a continuous representation of the system. Using this method, a second-order approximation was obtained which gives a better response than the Padé approximation as demonstrated by its frequency and step responses of Figs. 7.4 and 7.5, respectively. The function of Eq. 7.2 is replaced in the simulation by the low-order polynomial approximation and gives a satisfactory representation of the travelling wave effect while avoiding the introduction of high-frequency oscillatory artefacts and consequent problems with poor numerical stability associated with a high-order approximation. These are:

$$\tanh(T_e s) = \frac{3.59s}{s^2 + 5.21} \quad \text{for the main tunnel } (T_e = 0.642)$$

$$\tanh(T_e s) = \frac{16.67s}{s^2 + 112.6} \quad \text{for the } j\text{th penstock } (T_e = 0.146)$$

To illustrate the improvements in the model structure a simulation test was carried out, where the unit is connected to a 40 GW power system with load damping $D = 1$. The unit is configured to be the only frequency regulator in the system

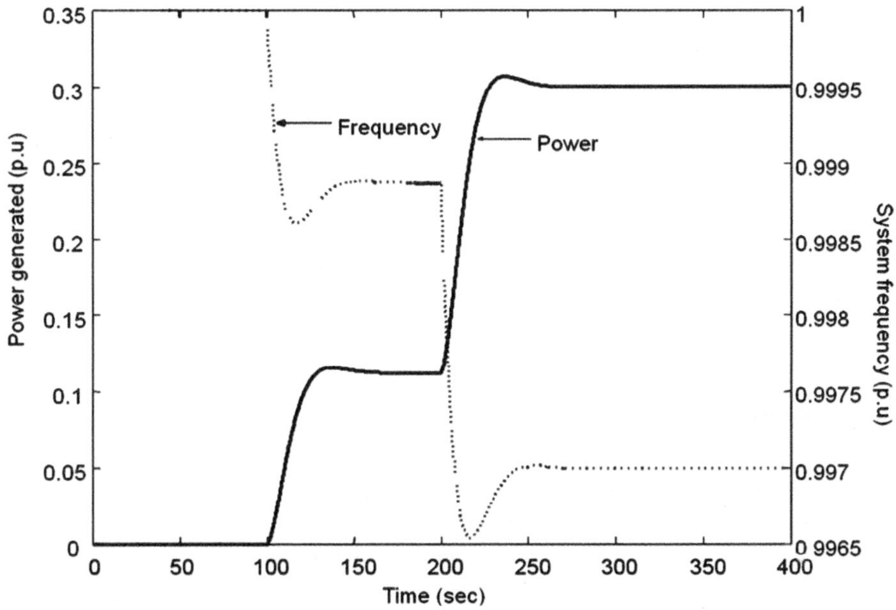

Fig. 7.6 Unit response to step changes in system demand load

operating with 1% droop. A step load of 90 MW is applied at $t = 100$ s, followed by an additional 120 MW step load at $t = 200$ s. The unit responses are shown in Fig. 7.6, where the unit is initially at nominal frequency (1 p.u. Hz); at $t = 100$ s, the load insertion causes an initial drop in the frequency (causes the unit to slow down). Subsequently, the unit governor responds by opening the guide vane increasing the output until it reaches a new common operating frequency (0.9987 p.u. Hz). The amount of load picked up by the unit depends on its droop characteristic; in this case the unit loaded to 0.11 p.u. (MW). Similar responses are obtained at $t = 200$ s, where the unit picks up extra load to reach a new operating frequency.

The steady state frequency deviation following a load change is given by Eq. 5.22 where the power picked up by the unit can be calculated using Eq. 5.23. Applying the equations on the system scenario results in steady state frequency deviation of 0.0013 p.u. (Hz) after the first load and 0.0017 p.u. (Hz) after the second load, while the total load picked up by the unit is equal to 0.3 p.u. (MW) of its rating; these values agree with the simulation result.

7.2.2 *Multiple Penstocks Plant*

The preceding discussion represents the initial stage of model development. However, as explained in Chap. 4, the Dinorwig station has a single tunnel with surge chamber, which splits into six separate penstocks supplying individual turbines. The approach used to integrate the system is similar to the one used for the single penstock model, where the model was broken down into small elements (subsystems). However, the system complexity prevents the hydraulic system from being represented as a single subsystem and this applies to the electrical system too. Therefore, it was decided to create a subsystem for each of the following elements:

- Unit individual penstock
- Low-pressure tunnel and the surge tank
- High-pressure tunnel
- Unit controller
- Electrical coupling between the unit and the power system
- The power system

Combining all subsystems together results in a model of the Dinorwig plant as shown in Fig. 7.7. In the SIMULINK® model, the individual penstocks and the high-pressure tunnel are represented by the elastic water column developed in Sect. 4.3.4.2, where the effects of the hydraulic coupling and water hammer are included.

Meanwhile, the low-pressure tunnel and the surge tank subsystem is based on the model of Fig. 4.15 which describes the low-pressure tunnel as an inelastic water column, because the surge tank alleviates the travelling wave effects. The unit controller subsystem considered here is also based on the Dinorwig governor of Fig. 6.6, where two feedback loops (power and frequency) are included. This subsystem also includes the guide vane dynamics shown in Fig. 6.8. Consequently, the subsystem output signal is used to adjust the guide vane position of the penstock subsystem.

As before the electrical coupling subsystem is implemented by comparing the power system frequency with the unit frequency, which is then multiplied by the integral of the synchronising coefficient to yield the power exchange between the unit and the power system.

7.3 Model Verification

Before using the simulation to investigate plant behaviour and hence to predict the effect of control parameter changes, it is necessary to confirm that its response agrees with the real plant. However, for financial and regulatory reasons there are tight constraints on the field tests which can be performed on the working plant. Thus, it was not feasible to apply system identification techniques such as those described before on the plant and alternative methods are considered based on simulation analysis to verify the model.

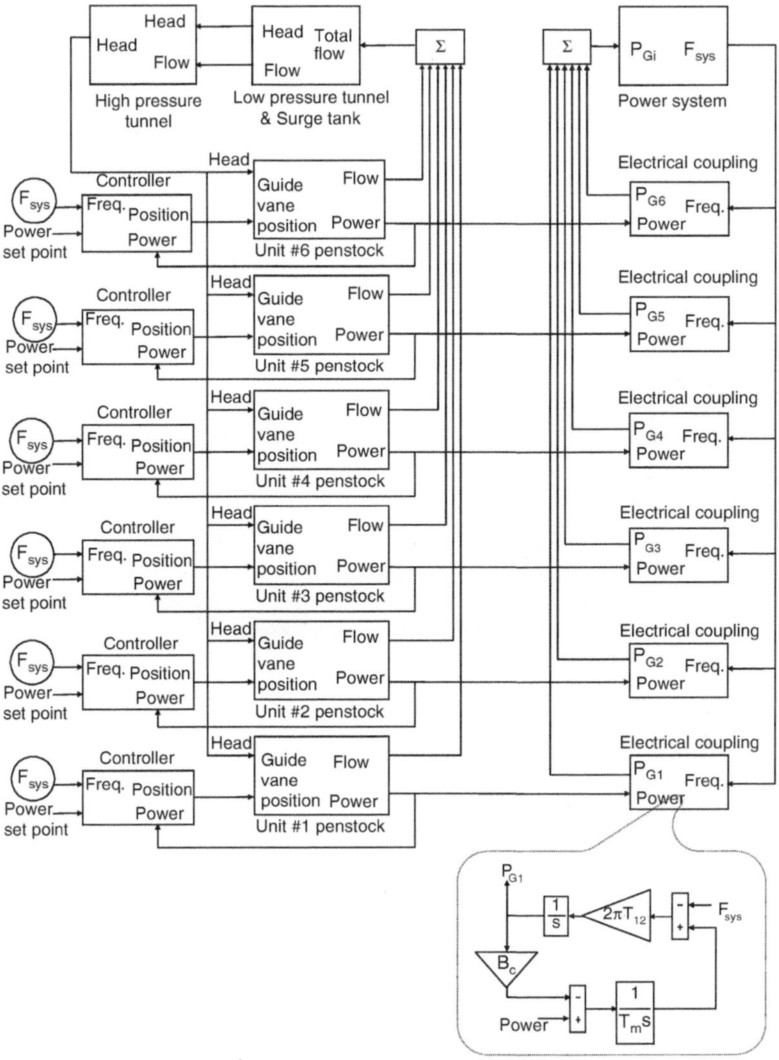

Fig. 7.7 The overall nonlinear multiple penstock offline model structure

7.3.1 Comparison with Linear Response

Initially, the response of the simulation was compared with that of the linear model to check that its qualitative behaviour and time scale were correct. For comparison purposes the governor representation was deactivated in both models; both no-load flow and speed deviations were set to zero in the travelling wave model. Figure 7.8 shows the simulation results when the models are subject to a large 4 p.u. step in guide vane position. The power increase predicted by both models has a similar

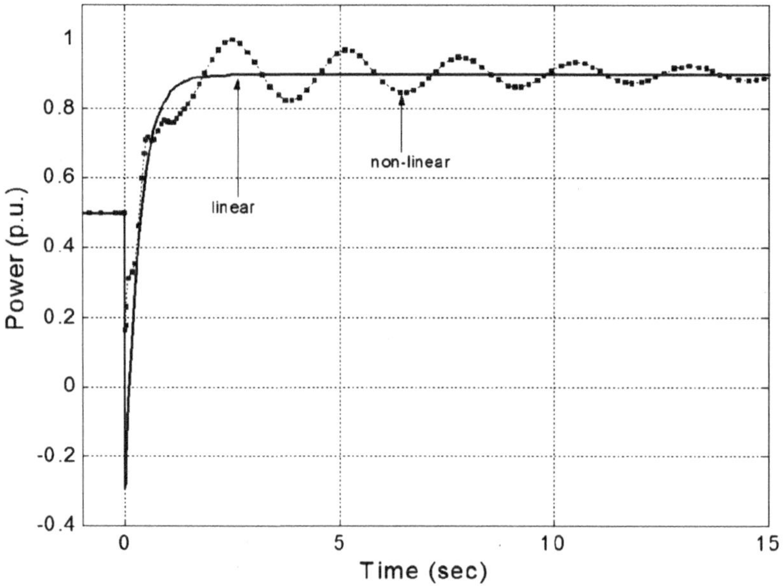

Fig. 7.8 Comparison between linear and nonlinear model responses

form and time constant (note that nonminimum phase behaviour is exhibited). The nonlinear model, however, predicts an output power oscillation as expected due to the travelling wave effects in the water column, which are entirely absent in the linearised model.

7.3.2 Simulation of Hydraulic Coupling Between Units

The effect of hydraulic coupling in the supply tunnel is confirmed by simulating a sequence of four units brought online in turn, with their governors blocked. The turbines head (pressure) variations are shown in Fig. 7.9, where 20 s ramp openings are used, as would be the case in practice. As the guide vane of the unit coming online is opened, its inlet pressure is seen to decrease as the flow increases. With the guide vanes on the other units held stationary, the flow to the remaining units is reduced because the total flow in the tunnel cannot change immediately; the consequent drop in pressure at those units is clear in Fig. 7.9. Note that the pressure drop for the units coming online will be greater than for the remaining units and the hydraulic stiffness is increased because of greater flow in the common tunnel when there are more units online. This test confirms that the simulation exhibits the expected nonlinear and multivariable behaviour. Although this does not constitute a systematic verification of the model, it does provide good evidence for its authenticity [102].

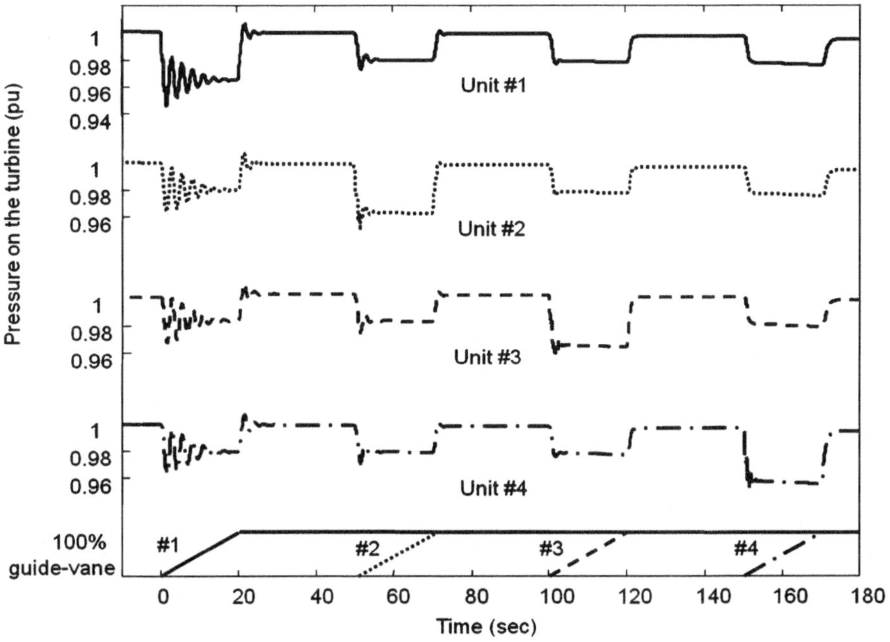

Fig. 7.9 Simulation result showing the hydraulic coupling effect

7.3.3 Comparison with an Independent Model

The third element of model verification is comparison with responses predicted by an independent computer model based on an entirely different method of solution and separately developed code. The computer programme called PTRAN was used to investigate the hydraulic system transient responses [103]. The programme is based on the method of characteristics (Sect. 4.7) which converts the quasi-linear, partial differential equations of motion and continuity, describing the unsteady flow of a fluid in any internal flow system, into ordinary differential equations relating pressure and flow at 89 nodes in the Dinorwig system.

An open loop response was calculated for several well-defined conditions within the plant's working envelope to emphasise the hydraulic effects between two units. The passive unit is generating 288 MW at full head under steady state conditions, while the active unit is initially loaded to 100 MW. Figure 7.10 presents the transient responses obtained using the PTRAN programme. Once the active unit loading increases to 150 MW, due to its guide vane opening in 2.626 s, the passive unit generation initially falls by 10 MW because of pressure drop and after ~4 s the unit returns to its steady state condition. The same scenario was applied to the simulation model and again the results are shown in Fig. 7.10. There is very good agreement between the two responses for both the unit being brought online and

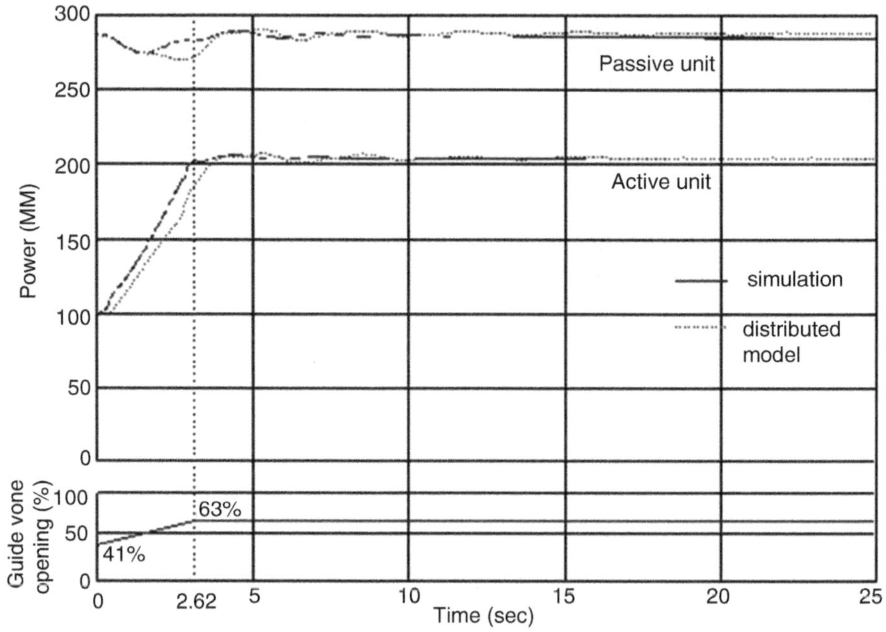

Fig. 7.10 Comparison of simulation results with an independent distributed parameter model

the unit which is perturbed as a consequence. The major discrepancies are that the distributed parameter model predicts a rather longer delay in the response of the active unit and a more pronounced dip in the power output of the passive unit than the lumped parameter simulation.

7.3.4 *Comparison with Measured Response*

The final verification measure was to compare the simulation with the results of a test on the plant, which was possible because it is occasionally necessary for commercial purposes. This is a dynamic test usually carried out by NGC on Dinorwig plant to ensure compliance with the Grid Code and to establish the plant frequency response capability for commercial payment purposes. The tests involve injection of a simulated frequency error signal into the governor. In addition to the step as shown in Fig. 7.11, ramp injections are also carried out together with dead-band tests to check the robustness of the unit control system. Since the response characteristic of a unit varies depending on its loading level, tests are repeated at various critical loading points across the whole loading range.

The results shown (Test No. 4) represent Unit #2 loaded with 220 MW and operating at specific governor setting of $K_p = 20$, $K_i = 3$, $K_d = 4$ and droop of 1%,

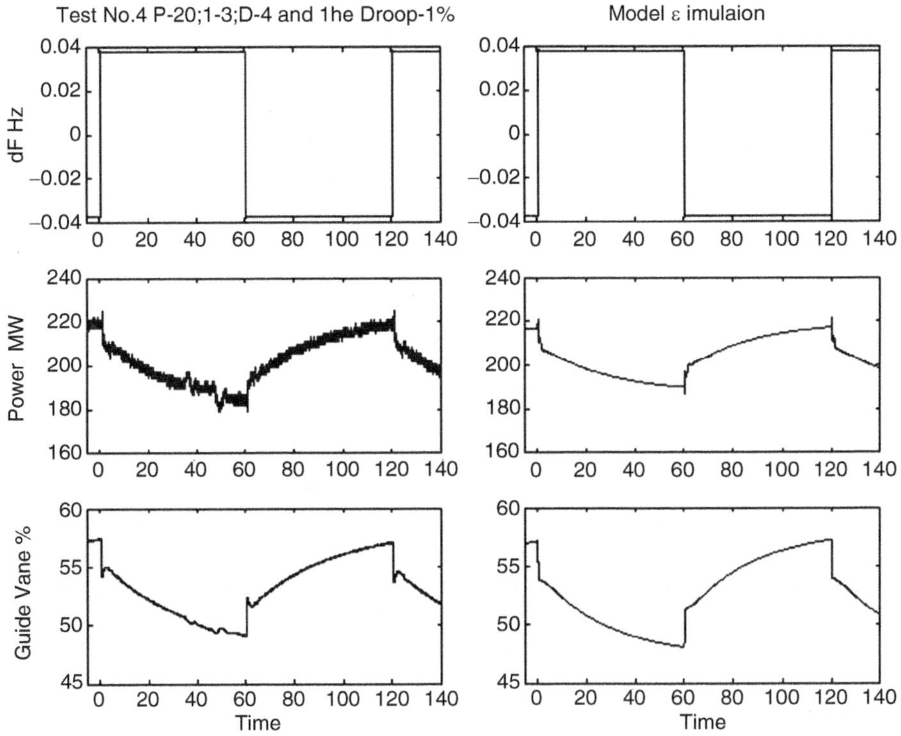

Fig. 7.11 Frequency response test

responding to an injection of ±0.035 Hz step error signal cycled every 60 s. The unit responds to the positive signal representing loss of load in the power system by shedding load of 30 MW.

The automatic reduction in the power output is termed high frequency response [20]. It is important that the generation output is reduced increasingly within 10 s, to minimise the initial frequency peak. Meanwhile during the negative frequency cycle, which represents loss of generation in the power system, the initial short-term automatic power output increase is termed primary response. It is important that the primary response from the synchronised generation is increased with time through automatic governor action in the period of 0–10 s after the incident and sustained for a further 20 s; this is critical to minimise the initial frequency dip. The automatic positive power response in the subsequent frequency stabilisation phase beyond 30 s after the incident is termed secondary response. The unit primary response increases the power by 18 MW and the secondary response is stabilised with extra generation of 30 MW.

An identical test setting was applied to the simulation and the results are shown in Fig. 7.11. There is very good agreement for both the generated power and the guide vane opening, although close inspection shows that the simulation output is rather better damped than the measured response during the period immediately after the step is applied. The slight perturbation to be seen in the measured response in the interval 35–50 s is due to a change in the operational conditions of the plant (another unit being brought online at this time), this was not included in the simulation. Further tests, using different sizes and types of frequency error signals, have concluded that while a systematic comparison with measured system responses over the complete working envelope would have been preferable, the measures taken provide sound evidence for the model's authenticity and the verification procedure has established a good degree of confidence in its use for further investigation [33].

7.4 Models for Simulation

As stated in Sect. 7.1, a software tool for SIMULINK® was developed to facilitate studies of the power plant under different governors. The tool has libraries of special functions (blocks) and the power plant models were constructed by connecting these functions to the standard SIMULINK® functions [94]. Using a dialog box, the parameters of a specific block can be adjusted. For example, the operating point of linear models may be changed. These models can represent the power plant as SISO or MIMO system and linear or nonlinear behaviour may be selected. Figure 7.12 shows a schematic of the SIMULINK® power plant model. The values of constants and parameters of Dinorwig working at busbar were obtained from the work of Mansoor [33], as given in Table 7.1.

As was already explained, the full hydroelectric station model is constructed combining the subsystems: Guide vane dynamics, hydraulic subsystem and turbine/generator (including sensor filters). Each block is part of the SIMULINK® library developed for this study; they can be selected to represent a diversity of modes of operation. For example, there are three models available to simulate

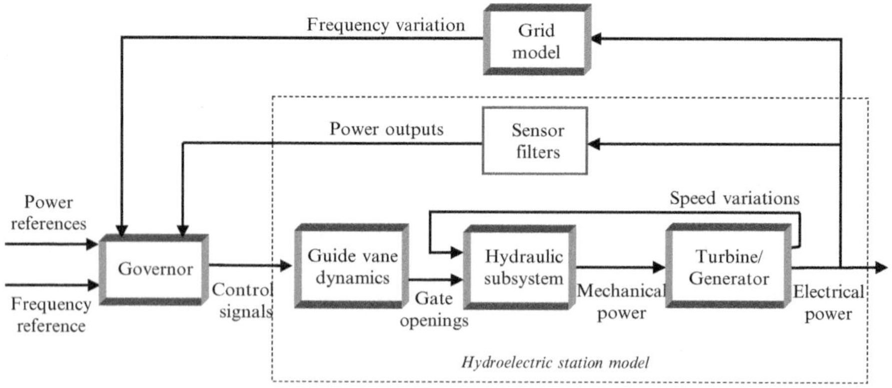

Fig. 7.12 Schematic of the SIMULINK® model developed

Table 7.1 Dinorwig parameters used in simulations

Symbol	Name	Subsystem	Value
T_{wt}	Water starting time of the main tunnel	Hydraulic	0.388 s
T_w	Water starting time of a single penstock	Hydraulic	0.3066 s
T_{em}	Wave travel (propagation) time in the main tunnel	Hydraulic	0.642 s
T_e	Wave travel (propagation) time in one penstock	Hydraulic	0.148 s
A_t	Turbine gain	Hydraulic	Max 1.18 Avr. 1.12 Min 1.05
f_{pt}	Head loss coefficient in main tunnel	Hydraulic	0.00002873 m/(m^3/s)^2
f_{pn}	Head loss coefficient in penstock	Hydraulic	0.00052 m/(m^3/s)^2
Z_{0T}	Surge impedance main tunnel	Hydraulic	0.6044
Z_0	Surge impedance single penstock	Hydraulic	2.1
D_n	Turbine-damping coefficient	Hydraulic-electric	0.5
T_m	Machine starting time	Hydraulic-electric	7.99 s
K_D	The per-unit coefficient of damping torque	Electric	8.38
H	Turbine/generator inertia constant	Electric	3.995 J nm^2/MVA
K_s	Synchronising torque coefficient	Electric	0.7071
ω_0	Base rotor electrical speed	Electric	314.1592 rad/s

the hydraulic subsystem: linear, nonlinear nonelastic and nonlinear elastic. The guide vane dynamics can be selected with or without rate limitation and saturation. The sensor filters is a fixed block. The grid model can be adjusted to represent different conditions of the national grid. Through the governor block classical and advanced controllers can be selected, as will be discussed in subsequent chapters.

7.5 Evaluation of the SIMULINK® Models

As shown in previous sections, several simulations were carried out in order to establish consistency between different models, over the operational envelope. In this section, the evaluation is concentrated mostly on the hydraulic subsystem. Again, the nonlinear elastic model was taken as a basis for comparison [33, 104]. The hydraulic models evaluated were the linear model, which was discussed in Chap. 4, and the nonlinear nonelastic model, also discussed in Chap. 4.

Figure 7.13 shows the open loop step response of the linear and nonlinear hydraulic models when only one unit is in operation. A step of 0.76 (p.u.) was applied to all models to fix the operating point, and then a 0.04 step was applied at 100 s of simulation. The nonlinear nonelastic and linear models have a very similar

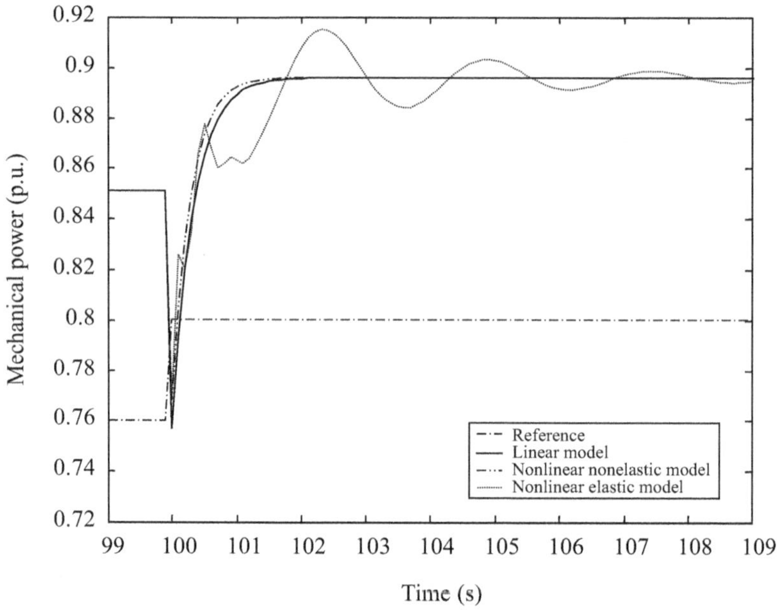

Fig. 7.13 Step response of the hydraulic models with one unit active

response, although the linear model produces a more pronounced NMP response. The response of the nonlinear elastic model has a marked oscillation, which is not present in the other models; it is attributable to the elastic characteristics of the fluid included in this model.

As can be seen from Fig. 7.14, when six units are active the open loop responses of the nonlinear nonelastic and linear model have some differences. The nonlinear model also has oscillation but it is less significant than in the one unit operational case.

Figure 7.15 shows again the open loop case with six units operational but at a different operating point (0.6). The nonlinear nonelastic and linear models now have marked differences. The nonlinear models react to this change; however, the linear model depends on this parameter.

Figure 7.16 shows the direct and cross-coupling open loop step responses of the hydraulics plus guide vane models. In the upper graph is shown the response of the unit one and in the lower graph is shown the response of the units 2–6 in synchrony. In this simulation, an operating point of 0.76 p.u. was first established in all units. Then a step of 0.04 was applied to unit one at 100 s of simulation, followed by a step of the same amplitude applied to units 2–6 at 110 s.

As can be seen from Fig. 7.16, the responses of the models are comparable. In both graphics, the response of the nonlinear elastic and linear models has a good

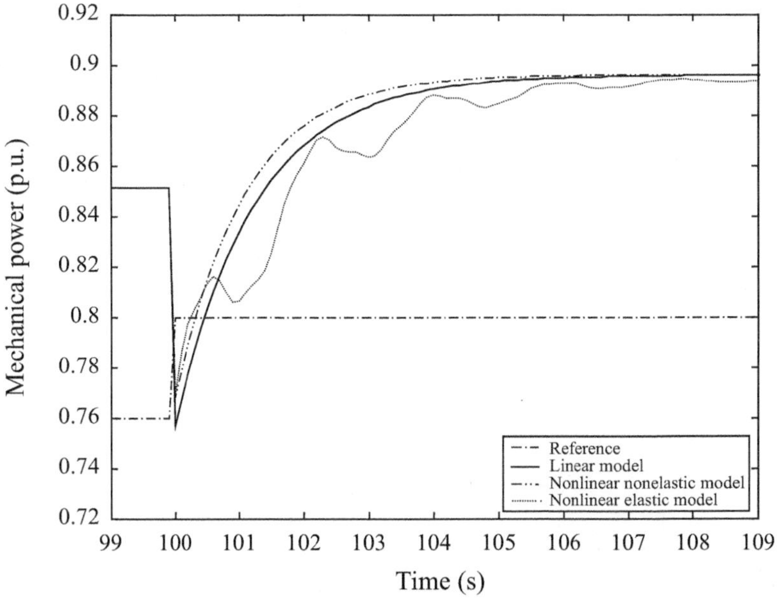

Fig. 7.14 Step response of the hydraulic models with six units active

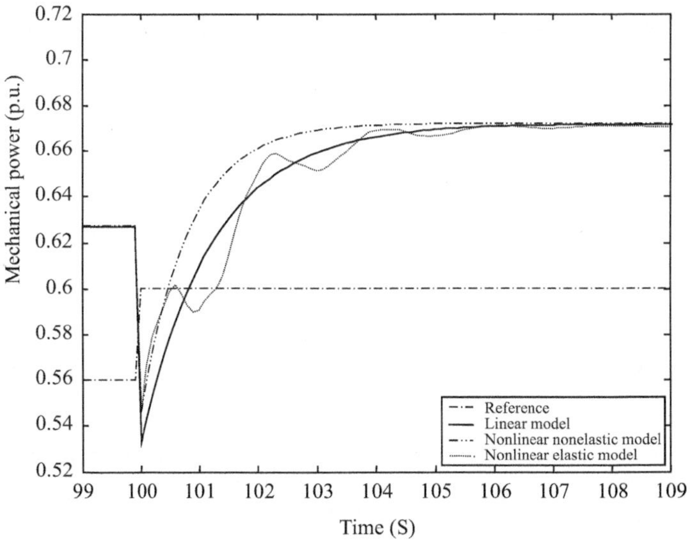

Fig. 7.15 Step response of the hydraulic models with six units active, operating point at 0.6

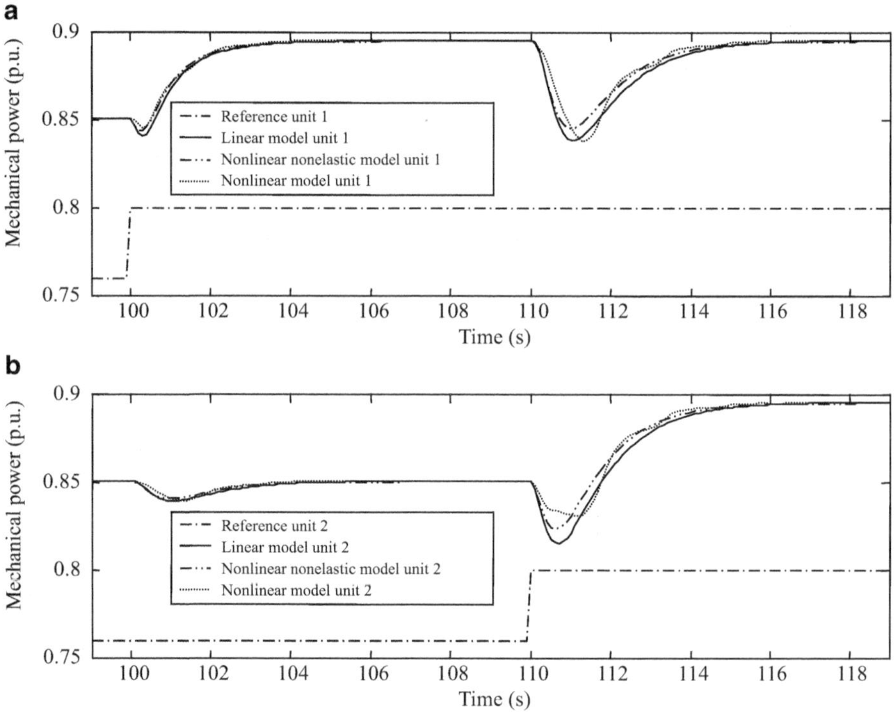

Fig. 7.16 Responses of integrated models to 0.04 p.u. step power demand: (**a**) Unit 1 and (**b**) Units 2–6

approximation to the response of the nonlinear nonelastic model. The guide vane dynamics acts as a filter for the oscillation of the elastic nonlinear model; other effects are the reduction of the NMP response and an increase in the time constant of all models. The linear model seems to be a good approximation to both nonlinear models and is less computationally demanding.

7.6 Conclusions

This chapter has developed the integrated model of a pump storage station and applied it to the Dinorwig power station. The model includes representations of the hydrodynamics, control system and electrical power system. The real benefit of the simulation is that it gives better insight and improved understanding of the plant dynamics whose complex features cannot be fully understood in terms of linear models. Different verification measures such as comparison with real plant responses and an independent model were applied to establish a good degree of confidence in the veracity of the model. The models will be used in subsequent chapters to compare the performance of the hydroelectric station under different controls. The linear model will be used in time domain simulation when small perturbations are analysed, while the nonlinear elastic model will be employed in simulations when small and large amplitude signals are fed to the power plant.

While the model now includes the majority of the factors affecting the plant's dynamics, there are other secondary factors which could be incorporated at the expense of additional complexity. For instance, studies of interactions between turbine hydraulics including draft tube pulsation. In such cases, the model must correspond as closely as possible to the actual turbine and controls that exist at the plant.

Chapter 8
Hardware-in-the-Loop Simulation

8.1 Introduction

Up to now, simulation of the models described in preceding chapters has not required synchronism between the simulation rate and the rate of progress of physical time; in other words, these simulations were not *real time*. In fact, it is desirable that these 'offline' simulations run as quickly as possible, preferably much faster than the physical process itself, and particularly so if the simulation is used for iterative optimisation. On the other hand, if accuracy or limited computational power (or a combination of the two) dictates that the simulation runs slower than the physical process, then this presents no particular difficulty except perhaps a test of the researcher's patience.

This relatively relaxed approach to the simulation rate is acceptable for 'offline' computation, where the entire system is represented in software. A typical use of this mode of simulation would be designing and 'tuning' prototype control software while assessing and optimising its effect on system performance. The model executes as an autonomous whole, generating logical decisions and control signals internally and solely as a function of the model's state. All events are therefore synchronous with the rest of the simulation.

However, there are circumstances where it is necessary for the simulation to maintain a strict relationship with physical time, preserving synchronism with an external clock, or executing in *real time*. This usually happens when part of the model must be implemented externally to the simulation computer. Several examples of HIL simulation for hydroelectric plant are described in the open literature [105–108]. Broadly, two situations can arise:

- 'Software-in-the-loop'
- 'Hardware-in-the-loop'

The former is the conventional case. It commonly occurs when the prototype control software is converted to a real-time digital implementation and interfaced electronically with the physical plant, the plant simulation being discarded. This is

G.A. Munoz-Hernandez et al., *Modelling and Controlling Hydropower Plants*, Advances in Industrial Control, DOI 10.1007/978-1-4471-2291-3_8,

often known as 'software-in-the-loop'. It is a valuable step in the design process because the real plant introduces additional factors, such as noise and unmodelled dynamics, which may require further development of the controller.

Note that this scenario assumes that it is acceptable to connect the prototype software (as designed offline and quite possibly running on the same computer) directly to the plant. Clearly, it would be essential to employ adequate safety measures to limit damage to the plant, should the prototype software fail. Also, a risk assessment would need to ensure that the probability of failure is low and that the situation would be recoverable, even if all the safety mechanisms were defeated. This approach could be quite satisfactory to, say, bench-testing a small robot in a laboratory but is not appropriate for high-value plant in continuous operation.

Once the prototype control software is considered satisfactory, it can be transferred to its production form. This may require recoding into a prescribed computer language, using a particular real-time operating system (RTOS) and porting to specific hardware, such as a PLC. Unfortunately, this procedure can be error-prone. For instance, the production controller may be the responsibility of a different team to the prototype developers, raising the likelihood that specification and coding errors will occur. Another round of tests must be undertaken, this time to verify the production controller on the physical plant.

In contrast to the above scenario, a 'hardware-in-the-loop' (HIL) simulation combines a simulated plant model with a version of the controller whose hardware and software approach its production form. It is often used as a tool for systems integration where the proper function of several components (possibly from different suppliers) must be tested both individually and collectively. Here, the inputs from the individual components converge on the simulated plant, so that it may be verified they interact correctly. The technology is widely used for systems integration and testing in the automotive and aerospace industries, sometimes relying on multiple high-speed processors and specialised high-bandwidth interface devices to achieve the required speed and accuracy.

Another situation that favours the use of HIL simulation is when modifying or upgrading a production controller (its hardware or software or both). For high-value plant, the overriding priority is to minimise any risk of damage; this is certainly the case when commissioning controller upgrades for hydroelectric plant. The financial loss, physical damage or even injuries that could result from a controller malfunction make it essential to verify its correct operation before it is connected to the physical plant. Hence, a new controller is 'dry-commissioned' on the HIL simulation before being 'wet-commissioned' on the physical plant. A popular technique is first to connect the current controller to the HIL simulation, recording its responses over a range of operational conditions. The new controller is then adjusted to reproduce these responses, thus establishing a high level of confidence that it will replicate the current controller when attached to the real plant. Further software modifications can then be added incrementally from this known safe condition.

An additional benefit of using a HIL simulation is that it decreases the plant down-time needed for commissioning. For example, if it is assumed that the wholesale cost of electricity is £50 per MWh, the loss of 100 MW of generation for a single 8-h day will cost £40,000. Although simplistic, this reckoning illustrates that if a HIL simulator yields even a small reduction in the commissioning period, all its development costs will be recovered rapidly.

HIL simulation is also useful for operator training [109, 110]. The simulator will usually become part of a larger system which incorporates operator level procedures and methods of communication. The software must then be capable of reacting in real time to asynchronous external events and be interfaced to extra hardware such as video displays, control panels and digital communications channels.

In common with other industries, real-time simulation is an invaluable aid to the efficient installation and operation of hydroelectric plant. The core idea is illustrated in the remainder of this chapter by reference to a simplified HIL simulation of Dinorwig power station, consisting of an operational governor controlling one of the models described in Chap. 6. The next section outlines the basic concepts and terminology of real-time systems. Subsequent sections present the details of the simulator and the comparisons undertaken with offline computations and measured plant data in order to verify the responses. Some of the features of a more recent version of the simulator are then described to show how it can be extended to improve accuracy and realism.

8.2 Real-Time Systems

A general definition of a real-time system [111–113] is one in which the correctness of the system depends not only on the logical results of computation but also on the time at which the results are produced.

Real-time systems range from simple microcontrollers to complex distributed systems and directly control much of the world's industrial and commercial infrastructure. They have two distinctive characteristics:

- They are *reactive*, being aware of and reacting directly to events that occur in their environment.
- They are *manipulative*, affecting directly the course of events in their environment.

In these statements, the term 'directly' can be interpreted as 'without a human as an intermediary', although, of course, a human may well be a component of the overall system.

A digital controller that is interacting with a physical plant makes decisions based on measurements made by sensors and influences the plant by sending signals to actuators. It is the archetypal real-time system.

It is possible to distinguish two types of process with which the control computer must interact:

- *Periodic processes*, where the computer samples data and executes a control algorithm at regular, known intervals.

Table 8.1 Comparison of soft and hard real-time characteristics

Soft	Hard
Poor predictability (can only be expressed as a probability of meeting deadlines)	Good predictability (worst-case timing constraints are guaranteed to be met)
Dynamic scheduling (task preemption) leads to some tasks taking longer to execute than expected	Fixed scheduling
Performance degrades at peak workload	Constant performance
Efficient use of resources (high average CPU utilisation, dynamic memory allocation)	Inefficient use of resources (low average CPU utilisation, fixed memory allocation)

- *Aperiodic processes*, where the computer must react to external events that occur at irregular and unknown intervals (although there may be some information available about the probability of these events occurring).

In either case, temporal correctness requires that the computation time satisfies a prescribed time constraint.

A computational task in a real-time system can be broadly categorised according to the strictness of its associated time constraint (deadline):

- *Hard real time* deems that there is no value to executing a task after its deadline has been missed.
- *Soft real time* deems that there is some diminished value to executing a task after its deadline has passed.

In practice, the process of classifying a task into the hard or soft category according to this rule is not always straightforward; in any case, real-time systems tend to be a mixture of hard and soft tasks. Table 8.1 briefly contrasts the characteristics of soft and hard real-time systems.

Another important concept for real-time systems is *predictability*, which is the idea expressed by Stankovic [111], that:

> If a real-time system can be shown to be able to meet its deadlines (using a worst-case rather than an average-case analysis), it is said to be predictable.

Despite this apparently simple rule, detailed consideration of real-time systems reveals that predictability is a difficult term to pin down, either theoretically or in practice. Stankovic [111] suggests the following elaboration:

> Predictability means that when a task or set of tasks is activated, it should be possible to determine its completion time, subject to failure assumptions. This must be done taking into account the state of the system (including the state of the operating system and the state of the resources controlled by it) and the tasks' resource needs.

In terms of a HIL simulator, the primary requirement is for the model code to complete execution within one sample time – this is a 'hard' timing constraint. It has the following implications:

- A fixed step integration method must be used.
- Because the time taken to execute a simulation step may vary (due to data-dependence or the occasional need for communication with a data monitoring

task), the integration step size must exceed its worst-case value plus some safety margin.
- No iterative procedures (such as the Newton iteration often used to 'break' algebraic loops in offline simulation) are allowed.
- 'Stiff' parts of the model (parts that have short time-constants) may have to be executed with a smaller step size.
- Using virtual memory is not permitted because it could cause page-faults.
- If the simulator CPU is unable to meet the time constraint, then it may be necessary to reduce model complexity (but accept diminished accuracy), invest in a faster CPU, partition the model so that it can be computed in parallel on several CPUs or implement part of the model in specialised hardware, e.g. a field programmable gate array (FPGA).

It should also be remembered that a HIL simulation forms a digital control loop. This could lead to a conflict between the desire to lengthen step size, in order to accommodate a complex model, and to shorten it in order to satisfy feedback stability criteria.

8.3 HIL Simulator for Dinorwig Power Station

8.3.1 Hardware and Software for the Development System

An early implementation of a real-time simulator for a hydropower plant is described by Throckmorton and Wozniak [105]. Their simulation was based on two linear models and was primarily intended to evaluate the performance of a generator speed control system. They used a TMS320C30 DSP to compute the plant model but substituted an IBM PC in place of the operational governor to compute the control. Their work highlights the need to choose appropriate values of sampling time and signal quantisation if accurate simulated responses are to be produced.

The architecture of the HIL simulator built for Dinorwig is basically the same as that of Throckmorton and Wozniak but implemented in SIMULINK® in the dSPACE® real-time environment. This provides advanced tools for:

- Creating the system models in a block diagram editor under Windows
- Specifying sample rates for various parts of the model
- Creating virtual instrument panels to display simulation variables and allow parameter changes to be made while the simulation is running
- Generating and compiling the real-time code automatically from the block diagram
- Providing device driver blocks for the I/O interface cards
- Running the code in a RTOS on a separate DSP
- Forming a communication link between the DSP and the host PC

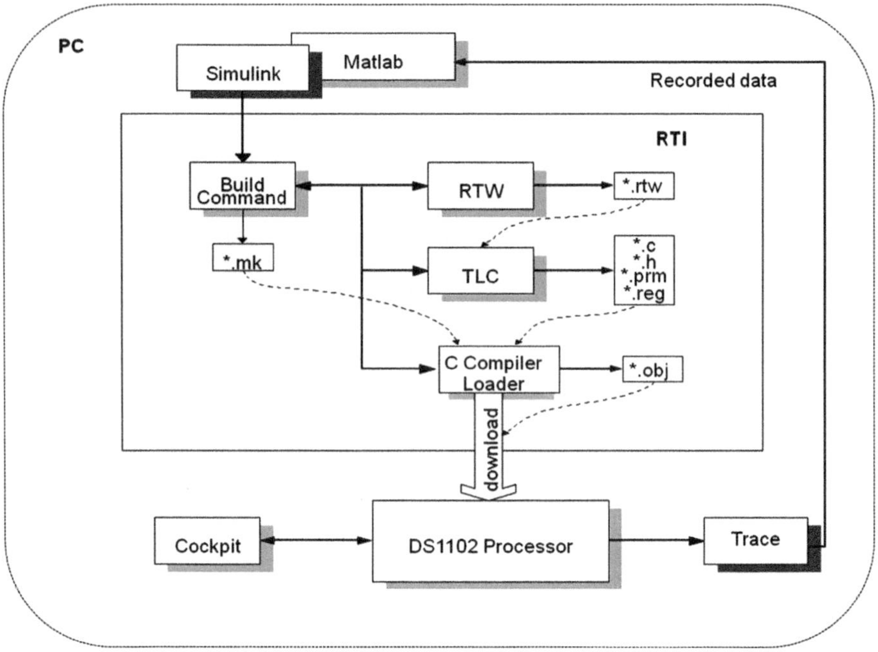

Fig. 8.1 Block diagram of the HIL simulator

Figure 8.1 is a block diagram of the HIL simulator showing the relationship of the dSPACE system [114] and the MATLAB® development software [94].

The simulator is based on the dSPACE DS1102 plug-in board mounted on the PC's ISA bus. This gives a fast communication channel between the target processor on the DS1102 and the host PC which plays a multiple role as:

- A development system
- An interface for display and user input
- A data-logger

The target processor on the DS1102 is a Texas Instruments TMS320C31 floating-point DSP. The board also has a two-channel 16-bit ADC, a two-channel 12-bit ADC and 16 DIO (digital input/output).

As shown in Fig. 8.1, the workflow for producing a real-time simulation consists of the following stages:

1. Express the dynamic model for the plant in SIMULINK®.
2. Add blocks for the ADC, DAC and DIO channels which invoke the device drivers and allow scaling and sample times to be selected.
3. Make the hardware connections between the DS1102 and the governor PLC. It is convenient to do this by means of a 'break-out' box so that an oscilloscope can be used to monitor the signals and, if necessary, allowing selected connections to be switched out as a diagnostic aid.

4. Automatically generate ANSI/ISO C code from the SIMULINK® model using real-time workshop (RTW). The first stage of the code generation analyses the source model file (*.mdl) to produce a description file (*.rtw) which contains a hierarchical structure of records describing the systems and blocks in the model and how they are connected. In the second stage, a target language compiler (TLC) interprets the intermediate model description and generates code (usually C) to express the model as a programme (*.c, *.h).
5. Compile and link the generated code for the model with code for the real-time environment. In this stage, the generated C programme is compiled to object code (*.obj) and the dSpace real-time interface tool (RTI) links in the run-time support files to provide services such as:

   - Timers
   - Monitoring code for selected parameters and signals
   - Interrupt service routines
   - Hardware I/O drivers

   The outcome is an executable code file (*.exe).
6. Use the Cockpit tool to design a virtual instrument panel. This software allows the user to:

   - Create displays such as time-plots, gauges, bar-graphs, etc., for selected simulation variables
   - Create controls such as switches, dials or buttons, etc., for selected simulation variables
   - Create textual editing panels for selected simulation parameters

7. Download and run the executable code on the target processor.
8. Monitor and control the simulation in real time from the PC host. This is facilitated by the TRACE software which tracks all the programme variables on the dSPACE processor with minimal overhead on the real-time application.

In practice, stages 4, 5 and 7 are fully automated and running through the development cycle requires no more than a few mouse clicks.

The equipment setup is shown in Fig. 8.2 with the governor PLC at lower left. A touch-sensitive screen control panel is directly connected to the governor. This is identical to those installed in the station control room and allows the operator to make changes in a transparent manner to:

- The demanded values of power and frequency
- Parameters such as the speed droop setting
- The operational mode by adding/removing the deadband nonlinearity in the control loop.

The control panel also displays variables such as the system frequency, unit power generated, guide vane opening and system operating limits, in a form that is familiar to the operator.

Fig. 8.2 (*Left*) Photograph of the HIL simulator, (*right upper inset*) the operator's touch-panel and (*right lower inset*) the system response displayed on the PC monitor

8.3.2 Preliminary Real-Time Implementation

As a preliminary to proper HIL simulation, real-time closed loop control of a single generating unit supplying an isolated load was simulated. Here the governor, as well as the plant, is represented by a SIMULINK® model. This is done in order to evaluate the effect of the computer-sampling rate on the response, as discussed in Sect. 8.2.

The model for the turbine/generator is based on the nonlinear representation of the water column described in Sect. 4.3.3.2, and standard PID control with the gains set for isolated operation at $K_p = 2.5$, $K_i = 0.8$ and $K_d = 4$.

The only system variables transferred between the governor and unit models are the system frequency and the guide vane signal. A small Cockpit model was built for the experiment, as shown in Fig. 8.3, so that three virtual control knobs can be used to adjust the individual PID gains as the simulation proceeds. A single numerical display shows the power generated by the unit.

As stated in Sect. 8.2, it is necessary to use a fixed integration step in real-time simulation and the first choice was a first-order (Euler) integration algorithm with a step size of 100 ms. The real-time response to a 30 MW step load change is compared in Fig. 8.4a to the 'continuous' result computed using the default SIMULINK® variable step size integration method set to 100 μs. It is clear that there are discrepancies between the two responses.

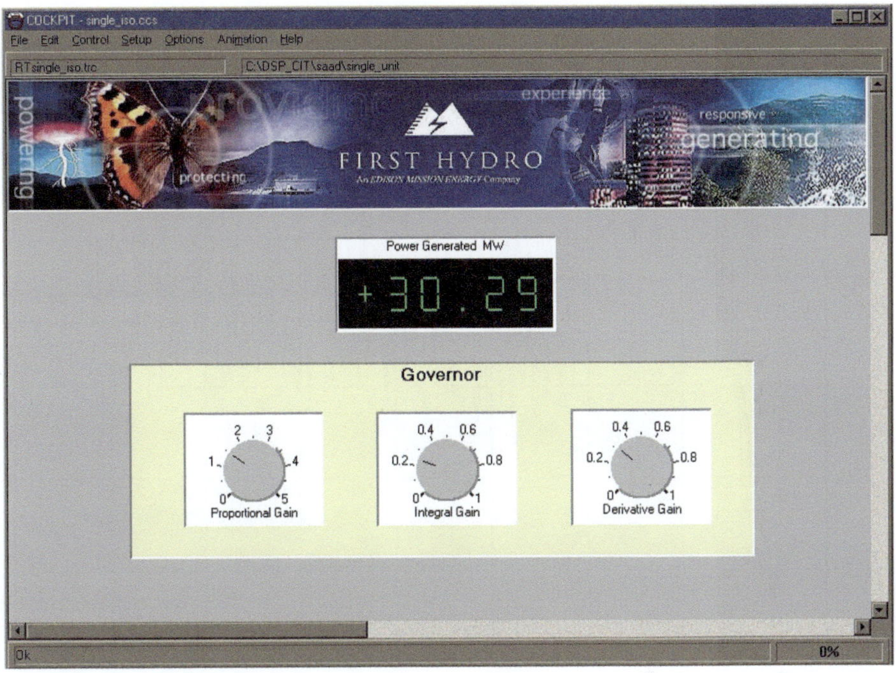

Fig. 8.3 Cockpit instrument panel to adjust PID control gains

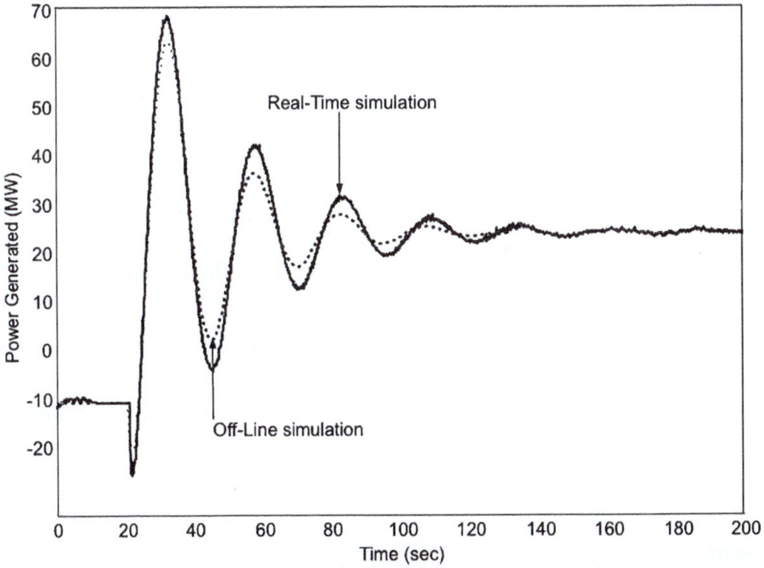

Fig. 8.4 Response of a single unit connected to an isolated load

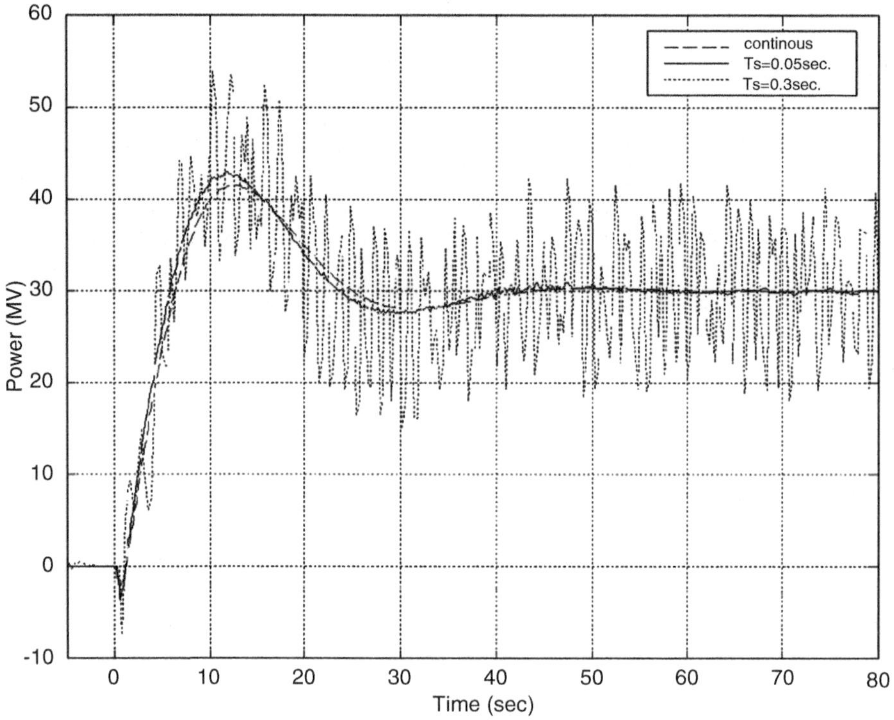

Fig. 8.5 The effect of simulation step size on the unit response

Different integration step sizes were tried and Fig. 8.5 shows that a value of 50 ms gives good agreement with the offline simulation. However, sampling with periods as long as 300 ms leads to a large numerical oscillation superimposed on the correct response.

8.3.3 Connecting the Real Governor to the Plant Model

The next stage was to connect the simulation to a real governor. Dinorwig power station always maintains a fully functional spare governor PLC which can be used as a rapid replacement if an operational unit fails. The hardware of the spare governor (type ABB-HPC 640) is identical to the operational units and programmed with the latest version of the control software [89]. Figure 8.6 shows the system setup. The simulator produces three analogue signals:

- Network frequency
- Total power generated by the unit
- Guide vane position

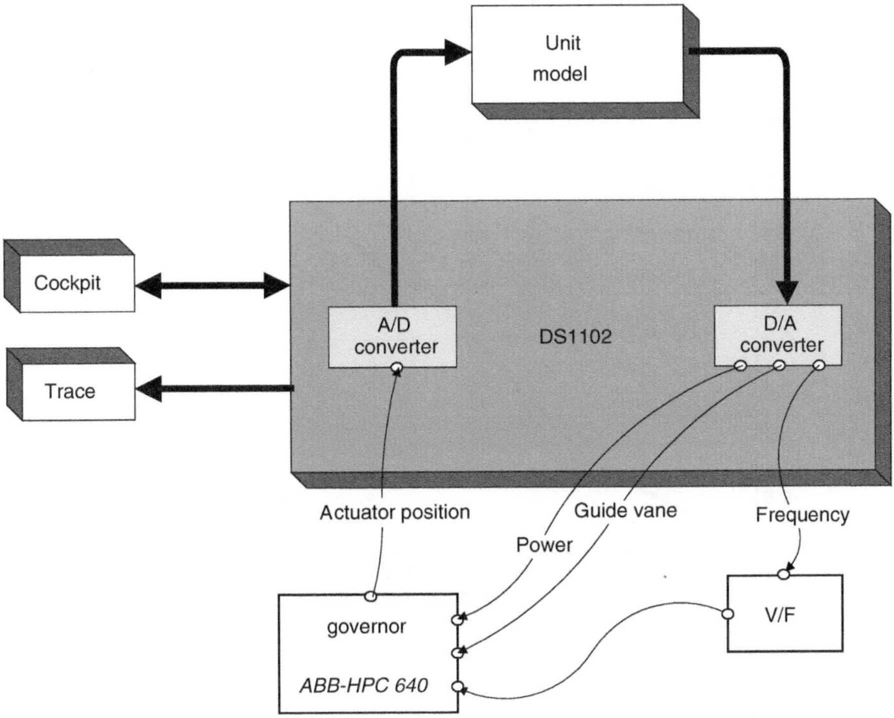

Fig. 8.6 HIL simulator control configuration

In return, the governor produces a signal representing the demanded actuator position which is fed into the simulator and used to control the simulated turbine/generator.

The governor I/O is of course intended for interfacing with the plant rather than the simulator. The governor expects a nominal 50 Hz sine wave on its network frequency input and this was synthesised by driving an external voltage/frequency converter circuit from one of the simulator's analogue outputs. Most other signals only required appropriate offset and scaling to produce the per-unit signals used in the model (± 1 p.u. $= \pm 10$ V analogue). However, interfacing the actuator position input from the governor required additional blocks to more accurately represent the guide vane actuator.

The governor output is a ± 10 V analogue voltage but, in practice, this is later converted to a ± 60 mA current signal that controls the actuator servo valve. This signal actually controls the guide vane opening and closing rates, with the actuator position being measured and fed back to the governor to form a position control loop. The steady state servo valve characteristics were measured on one of the operational units during an outage. Fixed level voltages were injected at the servo valve input and its opening and closing rates observed as shown in Fig. 8.7.

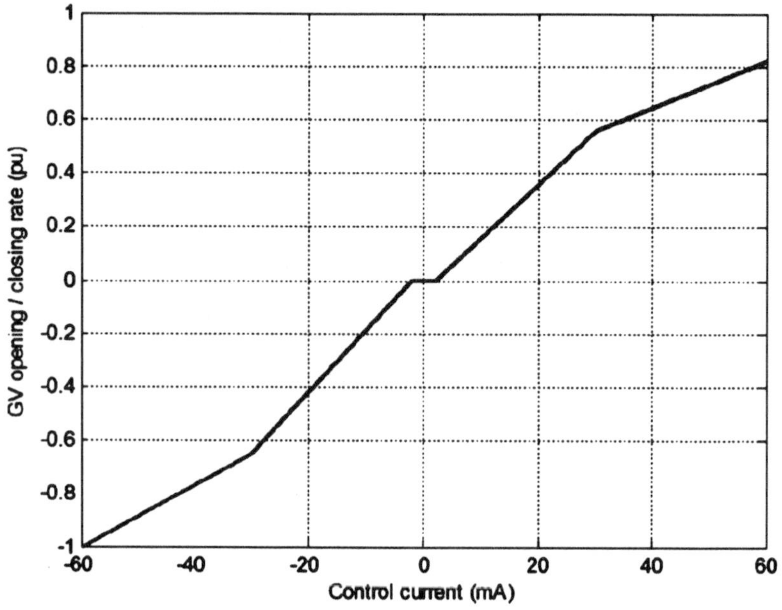

Fig. 8.7 Measured servo valve characteristics

Note that the closing rate is faster than the opening rate and when the current signal is within ±2 mA of zero, the actuator moves very slowly (near standstill). It was necessary to include these nonlinear characteristics as a look-up table in the SIMULINK® block called 'actuator signal conversion' in Fig. 8.8.

8.3.4 *Test Results*

Tests on the HIL simulator were done using the plant model with nonlinear representation of the water column and multiple penstocks, as described in Chap. 4. The investigation was based on three units connected to the power system. Two were set to generate near full load with deadband on and speed droop of 1%. The third was used as a test unit whose response to parameter changes could be observed over a variety of operating conditions.

8.3.4.1 Test 1: Frequency Insensitive (Deadband) Mode

The power system's total load capacity was set to a typical value for Great Britain of 35 GW at 50 Hz and a load sensitivity of 1% for every 1% change in the frequency ($D = 1$ per unit or 700 MW/Hz). The system has 1,000 MW of spinning reserve.

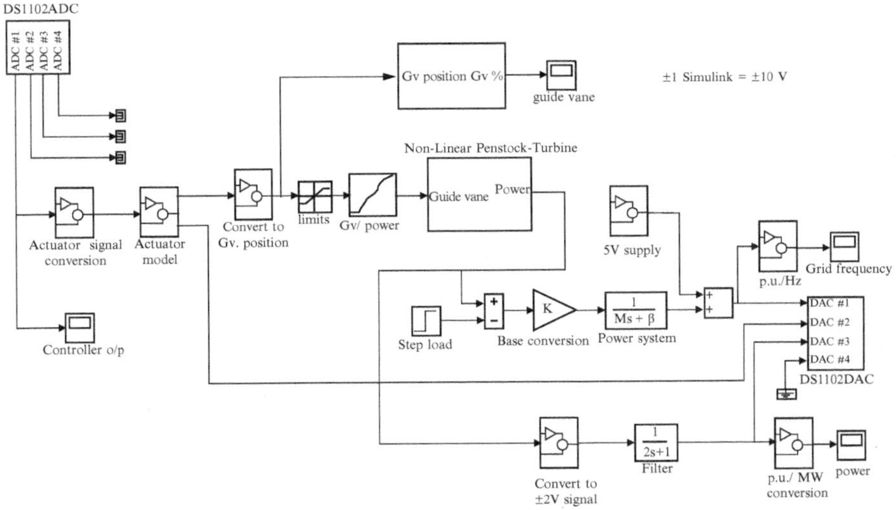

Fig. 8.8 HIL simulation – SIMULINK® block diagram

Table 8.2 Parameters for the test unit

	Droop (%)	K_p	K_i	K_d
Case 1	4	7	3	4
Case 2	1	10	8	2

Only 10% of the total generation is from units operating as frequency regulators, with their droop set at 4%; the rest of the generators are operating with valves wide open. Operating with this configuration, the generation contributing to the regulation is equal to 3,600 MW and the system composite frequency response characteristic (stiffness β) is equal to 2,500 MW/Hz. This normalises to 3.57 per unit on a base of 35,000 MW and 50 Hz. The total power system inertia constant (M) is set to 10 s and the hydraulic system parameters are as quoted in Chap. 4.

The first test studies the deadband response of the unit when instructed to increase generation from 150 to 288 MW. Several runs were performed according to the gain schedule in Table 8.2.

Figure 8.9 shows that using the higher droop gain with deadband causes a lively response, even though the governor PID gains are lower than for Case 2. The benefit is that both rise time and settling time are shorter but the drawback is the high overshoot which, in certain operating conditions, may lead to instability (as discussed in Chap. 4). In both cases, the frequency increases by 0.054 Hz, which agrees with the value calculated from the stiffness (β).

Figure 8.10 shows the measured response for Dinorwig Unit #1 operating at 4% droop with deadband and initially generating 150 MW. At time 06:28:30, the demanded generation is increased to 288 MW. The response rise time is about 9 s and the time taken to reach the maximum overshoot $T_p \approx 12$ s while the ±2% settling time is about 25 s. Comparison with the simulated response of Fig. 8.9

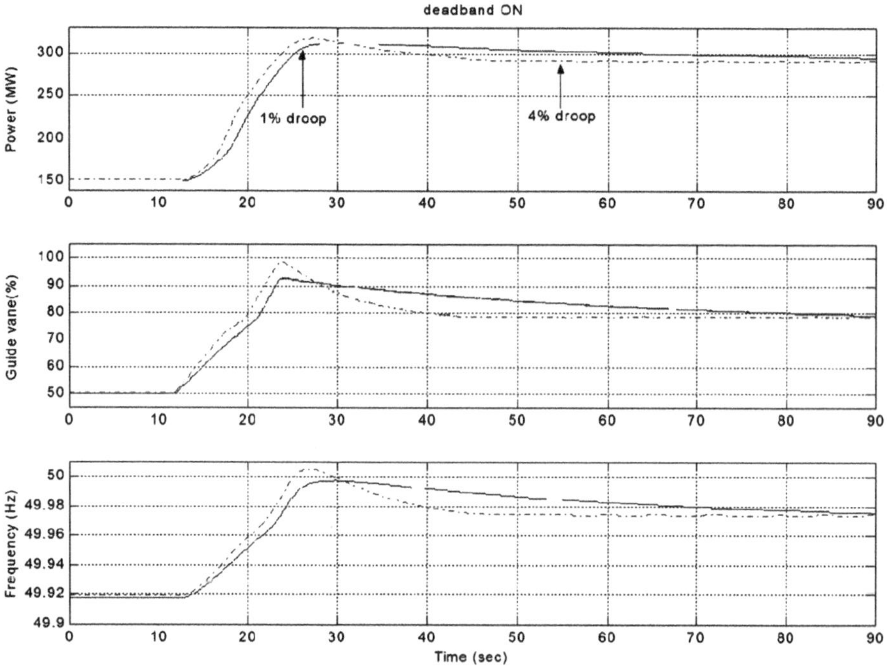

Fig. 8.9 Unit responses with deadband on

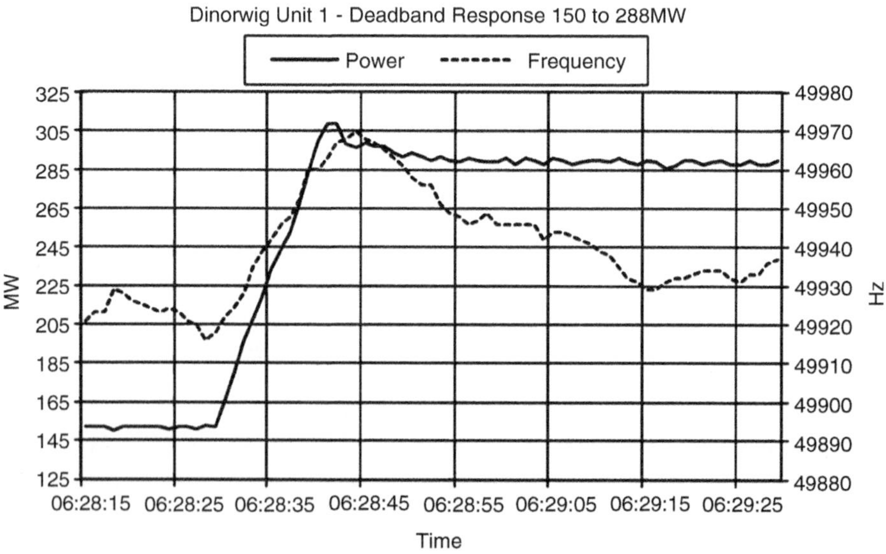

Fig. 8.10 Dinorwig Unit #1 measured deadband response, 150–288 MW

shows the strong similarity between the two. However, some discrepancies do exist, especially in relation to the size of the overshoot. In the measured response, the power overshoots by 12 MW while the simulated response overshoots by 20 MW. The network frequency increases by about 0.05 Hz in both cases, which indicates that the power system parameters used in the simulation are a good approximation to the real system conditions. However, following the initial increase, the network frequency in Fig. 8.10 is seen to fall away. This is due to a change in the load condition of the power system, which is not included in the simulation.

8.3.4.2 Test 2: Frequency Sensitive Mode

Here the HIL simulation is used to investigate the response to a step frequency error signal of −0.1 Hz inserted into the governor frequency-control loop, as described in Sect. 7.3.4. The unit under test is initially loaded at 230 MW and operating with 4% droop but now the deadband on the grid frequency measurement is *not* included.

The results are shown in Fig. 8.11. They exhibit very good agreement between the measured and simulated responses. Both have the initial nonminimum phase drop in generated power as the guide vane begins to open, although the measured drop is more pronounced than predicted by the simulation. This is followed by a load pickup of 13 MW occurring over the next 25 s, while the guide vane opens from 58.5% to 63.8%. Note the noisy fluctuation on the guide vane position signal which is absent in the offline simulation (Fig. 7.11). This is partly due to chattering in the actuator's deadband (see Fig. 8.7) and partly due to noise in the electronic hardware. This noise occurs in practice too although this is not evident in Fig. 8.11 because the measured trace has been sampled (effectively low pass filtered).

Other tests were conducted to verify the HIL simulation over a larger range of operating conditions. However, the two tests described above are sufficient to demonstrate that it is capable of computing the dynamic response of the plant in real time and in conjunction with external hardware.

8.4 Extending the Simulator

Having established the principle of HIL simulation with the case study presented in the previous sections, this chapter is closed by outlining some of the features that have already been included in a more recent version of the simulator.

We saw in Sect. 4.7 that the method of characteristics is a powerful method for modelling the hydraulic system. The dynamic responses more closely resemble the measured plant responses than those produced by any of the lumped-parameter models, even when the hydraulic system is represented by a small number of nodes. It is important to remember that there is a strict mathematical constraint that must be satisfied when using the method of characteristics. The spatial distance (ΔL) between solution points on the conduits must be related to the sample period (ΔT)

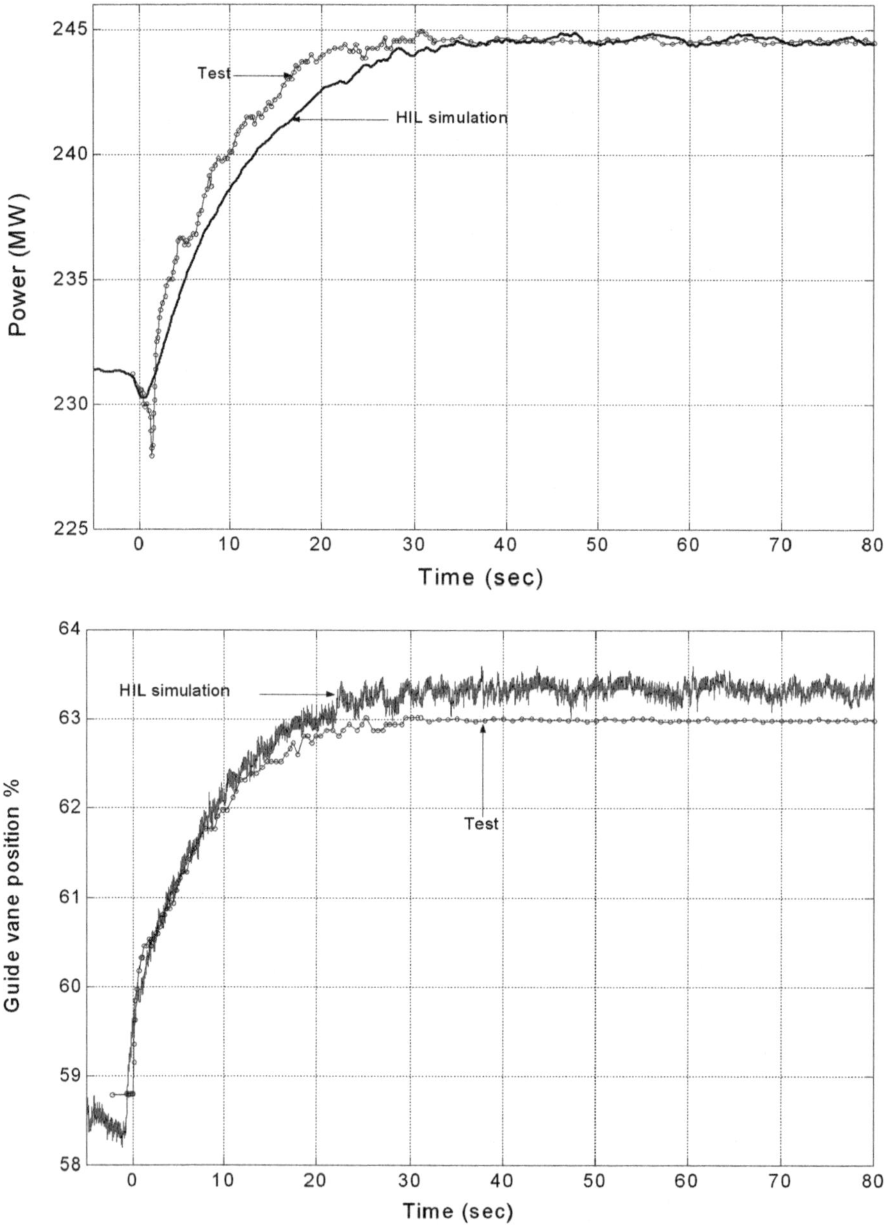

Fig. 8.11 Test unit in frequency-sensitive mode showing response to a −0.1 Hz step frequency deviation

by the wave velocity (a) by $\Delta L = \pm a\Delta T$. For real-time applications, it is usual to fix the sample period at a convenient value and this will determine the value of ΔL. In most cases, adjusting the lengths of the conduit sections to be integer multiples of ΔL introduces negligible error provided ΔT is small, as it must be for real-time computation.

One of the boundary conditions in a distributed-parameter model occurs at the inlet to the turbine where it is necessary to specify a functional relationship between the head (H_u) and the flow (Q) as a function of the guide vane opening (G). This nonlinear relationship is embedded into the characteristic equations for the hydrodynamics and the usual offline technique is to compute an iterative solution. However, this is not an option for real-time computation.

One possibility is to assume that the head and flow are related by a simple quadratic (which reduces the turbine characteristic to the basic orifice equation). Then

$$H_u = A_t'\left(\frac{Q}{GA}\right)^2, \tag{8.1}$$

where, $0 < G < 1$ is a linear modulation of the inlet area, A the area of the inlet and A_t' the turbine coefficient.

It may be shown that this reduces the characteristic equation to quadratic form that can be solved noniteratively by the usual formula. Avoiding an iterative solution becomes more difficult when a general nonlinear expression is used. If the static turbine characteristics (see fig. 4.5) that relate unit flow to unit speed, unit head and guide vane opening are used, it may be necessary to represent them as linearised regions of operation that are interpolated from a look-up table.

It is far easier to embed the other static turbine characteristic that relates unit torque to unit speed, unit head and guide vane opening. This simply requires a 2D interpolation table to approximate the unit torque curves and computation of the standard similitude expressions, which can be done by means of SIMULINK® blocks or a custom S-function. Combining a distributed-parameter model with the full turbine characteristics produces more realistic dynamic responses and steady state values than lumped parameter models. Another advantage of this approach is that it largely incorporates the pumping mode of operation, because driving the pump-turbine (as a motor) in reverse at synchronous speed selects an operating point in the third quadrant of the static characteristics.

The drawback of these relatively complex models is that the simulation becomes much more computationally intensive. Nevertheless, by coding the method in C and embedding it as an S-function in a SIMULINK® model, we have been able to run in real time at a sample rate of 1 kHz on a dSpace DS1006 board (AMD Opteron 64-bit 2.6 GHz processor).

A major part of commissioning a new governor involves testing the sequence of actions which take place as the mode of operation of a unit is changed. This is more complex at a pumped storage facility than a conventional hydroelectric station. The mode changes at Dinorwig are summarised in the finite state diagram

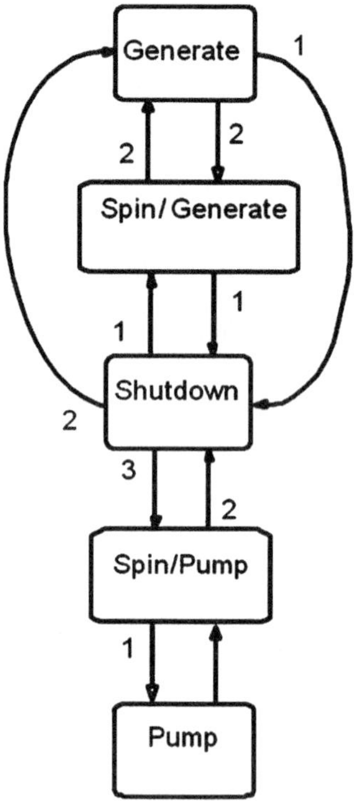

Fig. 8.12 Top-level state diagram for operational modes

of Fig. 8.12. Note that the transition between shutdown and pump modes is made via the spin/pump mode.

The mode changes are initiated by a control room operator. Each of the transitions shown in Fig. 8.12 involves a lengthy series of logical actions and interlocks that ensure the correct order and timing of the sequence. This is the responsibility of a separate sequencer PLC which coordinates the mode change throughout the plant and this necessarily includes communicating and interacting with the governor. Commissioning a new governor therefore requires the sequence logic to be represented.

A possible approach is to put the sequencer hardware 'in-the-loop' along with the governor. However, even if 'spare' hardware was readily available, it would be necessary to reprogram it to eliminate large parts of the sequence which are irrelevant to the governor. Instead, we chose to regard the sequencer as part of the plant and therefore a component which had to be modelled. The sequencer was modelled as a finite state machine, again coded in C and embedded in SIMULINK® as an S-function. This allowed the logical signals required to trigger the governor's mode-change software to be synthesised.

Another feature of the HIL simulation is the inclusion of simple models for ancillary plant. In some cases, such as the auxiliary plant or the main inlet valve (MIV), these models were substantially delays, modelled as integrators with threshold conditions set on various values of the output variable. During a mode change, these delays were used to determine the timing of transitions in the sequence. Other components were slightly more elaborate, such as:

- *De-watering of the runner*: In some modes of operation, the MIV is closed and compressed air blown into the spiral casing to remove the water from the runner. This causes a significant reduction of friction on the runner allowing it to be run at synchronous speed and drawing just sufficient electrical power from the grid to overcome windage, iron and copper losses, plus mechanical friction.
- *Variable frequency drive*: The unit is run as a motor by a variable frequency drive (VFD) drawing electrical power from the grid. This is used to raise its speed up to slightly super-synchronous in the reverse direction during a transition from shutdown to either Pump or SpinPump modes.
- *Auto-synchroniser*: The function of this component is to close the main relay between the power network and the generator as it runs up to synchronous speed. It uses the slip and load angle (δ) as its inputs and when these are between prescribed limits, switches are operated within the simulation to connect the generated output power to the grid input. From this point, the generator model reverts to the conventional small-signal linearised model, as described in previous chapters.

Another feature of the simulator is a customised user interface. The control room interface remains the touch-panel shown in Fig. 8.2 but the interface presented to the modeller provides much more information and control over the simulation:

- Time-plot displays of the primary simulation variables in real time
- Graphical displays of the progress of de-watering and MIV opening
- Graphical display of load angle to visualise proximity to auto-synchronisation
- Textual display of current simulation state and sequencer messages
- A facility for starting the HIL simulation
- A facility to control leaving/rejoining the HIL simulation for display and data acquisition purposes
- Edit buttons that allow grid external power and reservoir head to be changed online
- Buttons that mimic the operator's panel for:
  - Mode select and accept
  - Low-frequency relay (LF relay) setpoint entry
  - Arm part-load response (PLR) mode
- A pop-up panel for editing specified simulator parameters in real time.

The modeller interface is programmed as a MATLAB® function which (after initialisation) runs on the development laptop in a continuous loop and is asynchronous to the HIL simulation. Communication between the user interface and the simulation is implemented by means of calls to the dSpace real-time library. This allows current parameter and variable values to be read from the simulation and commands and modified parameter values to be sent back via an Ethernet link.

8.5 Conclusions

This work has demonstrated that it is possible to build quite complex models of hydroelectric plant and to execute them in real time, achieving responses that match well to those measured on the actual plant. Further, the model of the hydraulics and pump-turbine has itself become a component within an extended model that includes several different modes of operation.

In common with the experience of many other industries, HIL simulation is proving to be an invaluable tool for the safe and effective operation of hydroelectric plant. It provides a low-risk incremental route to commissioning and financial benefit in terms of shorter outage periods for governor debugging and adjustment.

The current state of the simulator does not in any way represent an end-point. There is ample scope for further development, some possibilities being:

- Modelling the individual characteristics of all six pump-turbine units, allowing maintenance schedules and outage decisions to be optimised
- Extended model of the electrical system, allowing studies such as the effect on the power network of interacting with renewable energy sources
- Inclusion of more components in the hydraulics model to help with system stability studies
- Improved model of the electro-hydraulic guide vane servo, allowing advanced control strategies to be studied
- More accurate grid model, allowing the investigation of advanced control strategies such as adapting the controller according to current grid characteristics

Beyond its utility as an engineering tool, it is possible that the simulator could be used for other purposes such as:

- Operator training
- Linking to trading simulations as an aid to tactical decision making
- Implementing advanced methods for fault diagnosis

In fact, it is foreseeable that in future virtually every aspect of station operation will be influenced by simulation, as part of the day to day decision-making process.

Part III
Controlling the Power Plant

Chapter 9
Classical Approach

9.1 Introduction

Classical controllers have been used, for many years, to regulate hydropower schemes; these controllers are normally tuned based on linear transfer functions. However, hydropower plants are, in fact, highly nonlinear, time variant and multivariable so any law designed on the basis of a linearised representation is a compromise. Advanced control techniques will be addressed in next chapters; however, the actual governors, for many hydropower stations, are classical controllers; then it is important to show the studies for optimising the performance of PID controllers. Therefore, in this chapter, the governor tuning is examined in terms of plant stability and dynamic performance. The primary objective is to improve the accuracy and speed of response of power plants to short-term power system load perturbations. Optimisation of plant performance will enhance the profits of the hydropower stations.

Various methods of governor tuning are introduced which not only indicate stability but also provide information on the adjustments needed to obtain a given specification of performance. First, the stability of a hydro unit configured to supply an isolated load is analysed using the Routh–Hurwitz criterion. The effect of the derivative gain is analysed using the root locus method and it is shown that the derivative gain limits can be extended further than recommended by Hagihara [32]. The work proceeds to investigate the stability of Dinorwig when operating as a frequency regulator and the results show that the power system size influences the plant response. However, when operating with a deadband (frequency insensitive mode), the response is mainly influenced by the droop setting. The stability limits were calculated using the gain and phase margins of the system [115].

G.A. Munoz-Hernandez et al., *Modelling and Controlling Hydropower Plants*, Advances in Industrial Control, DOI 10.1007/978-1-4471-2291-3_9,

9.2 Stability of the Unit in Isolated Operation

Usually, electrical generators are connected to power networks and the dynamics of the network must be considered during frequency (speed) control design. In the case of a small isolated network, the unit must act to maintain the system frequency. Most units at some time may be required to supply isolated loads either deliberately or by accident. For instance, a black-start operation may be required or the power system may split into smaller networks. Even when isolated, the dynamic characteristics of the load will vary and must be considered in control design [61]. In the derivation of the transfer function of the system for stability studies, the following statements are assumed:

- The hydro system operates with small load perturbations.
- The water is incompressible and no elastic effect in the penstock is considered. Moreover, the water time constant is held constant at full rated load value, not modified for part gate operation.
- The hydraulic coupling is ignored.
- The turbine is operating at a point on the power gate curve where a small change in gate position gives proportionate change in steady state torque. This assumes a linear relation between the guide vane opening and the power generated.
- The guide vane dynamics are assumed sufficiently fast to be omitted.

9.2.1 System Representation

Figure 9.1 shows the block diagram that represents the plant used for steady state analysis. Using the solid mass approximation for short/medium head penstock, the linearised representation of the hydraulic system (Chap. 4) is used in the analysis to represent the penstock-turbine dynamics. The generator is represented by its mechanical starting time, Chap. 5. Due to the primary interest in the frequency variation with power rather than voltage, the load dynamics are modelled by the constant power (MVA) model represented by Eq. 9.1 as described in Chap. 5. Depending on the load characteristic, the relationship between power and frequency can be represented by

$$P = P_0(1 + D_{\mathrm{pf}}\Delta f). \tag{9.1}$$

Linearising around the operating point to represent a small perturbation and neglecting the second-order terms yields

$$\Delta P = \Delta P_0 + P_0 D_{\mathrm{pf}}\Delta f, \tag{9.2}$$

where the load damping $D = P_0 D_{\mathrm{pf}}$ is proportional to the connected load and the load's frequency sensitivity.

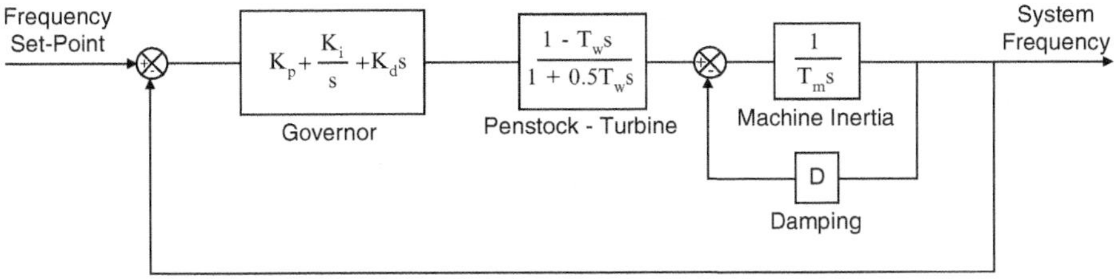

Fig. 9.1 Block diagram of a hydraulic turbine generating system

Then, the system transfer function can be written as

$$G(s) = \frac{(K_p s + K_i + K_d s^2)(1 - T_w s)}{s(1 + 0.5T_w s)(T_m s + D)}. \tag{9.3}$$

9.2.2 *Routh–Hurwitz Stability Criterion*

As is known, the transient response of a linear closed-loop system is determined by the position of the closed-loop poles on the complex plane; the system becomes unstable as soon as one of the closed-loop poles is located in the right-hand half of the complex plane [65]. The closed-loop poles are found by solving the characteristic equation (9.4), where $G(s)$ is the plant transfer function and $H(s)$ is the function of the feedback path:

$$1 + G(s)H(s) = 0. \tag{9.4}$$

Routh's method determines the number of roots located in the right-hand half of the complex plane using only the coefficients of the characteristic equation without solving the equation for the roots themselves [116]. The characteristic equation can be written in polynomial form as

$$A(s) = A_0 s^n + A_1 s^{n-1} + A_2 s^{n-2} + A_3 s^{n-3} + \cdots + A_{n-1} s + A_n = 0. \tag{9.5}$$

A necessary condition for stability of the system is that all the roots in Eq. 9.4 have a negative real part, which in turn requires that all $\{A_i\}$ coefficients will be positive in Eq. 9.5. Then, if any of the coefficients are negative or zero, it may be shown that there is either a closed-loop pole in the right-hand half of the s-plane or there is one or more closed-loop poles on the imaginary axis. This is not, however, a sufficient test for stability [115, 116].

The well-known Routh–Hurwitz test [96] requires the computation of a triangular array that is a function of $\{A_i\}$, a necessary and sufficient condition for stability is that all the elements in the first column of this array are positive. Using this condition an analytical expression for the system's stability boundaries, in terms of the plant and controller parameter values, is obtained. Combining the transfer function of the governor with that of the controlled system shown in Fig. 9.1, the characteristic equation can be written as

$$\left[K_{\mathrm{p}}+K_{\mathrm{d}}s+\frac{K_{\mathrm{i}}}{s}\right]\left[\frac{1-T_{\mathrm{w}}s}{1+0.5T_{\mathrm{w}}s}\right]\left[\frac{1}{T_{\mathrm{m}}s+D}\right]+1=0 \tag{9.6}$$

in the form of (9.5) is

$$\begin{aligned}A(s)&=s^{3}(0.5-X_{3})+s^{2}(X_{3}-X_{1}+1+0.5X_{4})\\&\quad+s(X_{1}-X_{1}X_{2}+X_{4})+X_{1}X_{2}=0\end{aligned} \tag{9.7}$$

where $X_1 = K_{\mathrm{p}}T_{\mathrm{w}}/T_{\mathrm{m}}$; $X_2 = K_{\mathrm{i}}T_{\mathrm{w}}/K_{\mathrm{p}}$; $X_3 = K_{\mathrm{d}}/T_{\mathrm{m}}$; and the system regulation $X_4 = D\,(T_{\mathrm{w}}/T_{\mathrm{m}})$

9.2.2.1 Routh's Test

The coefficients of the characteristic Eq. 9.7 are arranged in two rows in order to determine the Routh array, beginning with the first and the second coefficients and followed by the even-numbered and odd-numbered coefficients. Hence, the array for the system can be constructed as:

$$\begin{array}{lll}
s^3: & A_0 & A_2\\
s^2: & A_1 & A_3\\
s: & B_1=-\dfrac{\det\begin{bmatrix}A_0 & A_2\\ A_1 & A_3\end{bmatrix}}{A_1} & 0\\
s^0: & C_1=-\dfrac{\det\begin{bmatrix}A_1 & A_3\\ B_1 & 0\end{bmatrix}}{B_1} &
\end{array}$$

To ensure a stable response all the elements of the array A_0, A_1,A_2,A_3,B_1,C_1 must be positive, then the following criteria must be fulfilled:

- A_0 will be positive if $0.5-X_3>0$, which means that $0.5>X_3$; therefore, the derivative gain must be set to $K_{\mathrm{d}}<0.5T_{\mathrm{m}}$.
- A_1 will be positive if $1-X_1+X_3+0.5X_4>0$, which results in $X_1<1+X_3+0.5X_4$.
- A_2 will be positive if $X_1-X_1X_2+X_4>0$, which results in $1+(X_4/X_1)>X_2$.
- A_3 (C_1) will be positive if $X_1X_2>0$.
- B_1 will be positive if $A_1A_2-A_0A_3>0$, which results in:

$$(1-X_1+X_3+0.5X_4)(X_1-X_1X_2+X_4)-(0.5-X_3)(X_1X_2)>0. \tag{9.8}$$

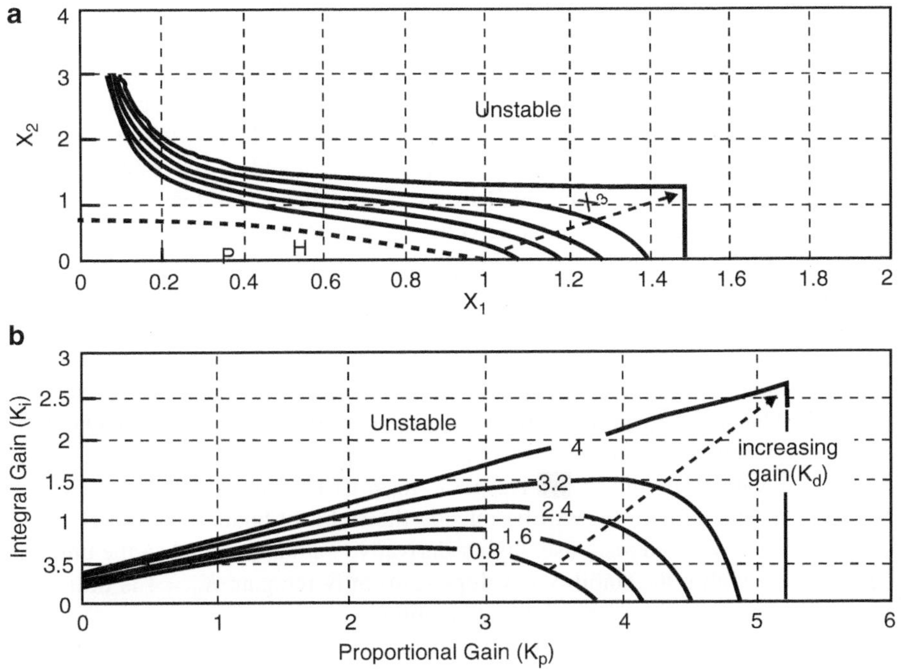

Fig. 9.2 Plots of stability boundaries for $X_4 = 0.2$

If all the former conditions are satisfied, the stability boundaries of the system can be established by setting Eq. 9.8 to be equal to zero which results in

$$X_1^2(X_2 - 1) + X_1(1 - 1.5X_2 + X_3 - 0.5X_4 - 0.5X_4X_2) + (X_4 + X_3X_4 + 0.5X_4^2) = 0. \tag{9.9}$$

9.2.2.2 Stability Limits

MATLAB® [94] was used to solve Eq. 9.9 and plot curves for various parameter values to establish the stability boundaries of the system. Figure 9.2a shows X_1 versus X_2 with changing X_3 from 0 to 0.4 in 0.1 steps, while setting X_4 to a fixed value of 0.2. All points lying within the region bounded by the curve and the axis are stable, and all points external to the region are unstable. The effect of X_3 is illustrated as the stability region is expanded by increasing the value of X_3 towards its stability limit, which is defined in condition 1 of the Routh–Hurwitz criterion, i.e. that $X_3 < 0.5$.

The defining of stability boundaries in terms of X_1 and X_2 is important because it determines the possibility of modifying the system behaviour modifying the values of them. In this context, Hovey [117, 118] has considered the stability boundaries for a system governed by a PI type controller and operating with regulation X_4 set to zero. The self-regulation of the turbine and connected load act to reduce the speed transient and to increase the stability of the unit. Nonetheless, since the grid amount of self-regulation is not always precisely known, the assumption of a purely resistive load for an isolated system will result in a conservative evaluation of the regulating capabilities of a particular hydroelectric generating unit and its governor.

The governor setting chosen by Hovey is conservative and does not utilise the maximum capability of the generating system; it can be seen that in Fig. 9.2a, where the dotted line represents Hovey's stability curve, there the operating point (H) for optimum transient response was chosen as $X_1 = 0.5$ and $X_2 = 0.25$. In the same way Paynter [69] has recommended an operating point (P) for optimum transient response at $X_1 = 0.4$ and $X_2 = 0.169$. For the particular case of Dinorwig, the stability boundaries of a unit in terms of the PID gains of the governor are shown in Fig. 9.2b. The operating point 'o' represents the PI gains currently used at Dinorwig for isolated network operation (synchronisation process) $K_p = 2.5$ and $K_i = 0.8$. Using low values for the derivative gain ($K_d < 1.6$) yields an unstable response because the operating point will be outside the stability boundaries. Therefore to ensure a stable damped response, higher derivative gain is required so the unit will be operating within the stability envelope. Currently the gain $K_d = 4$ is utilised in the governor which gives a maximum operating boundaries.

9.2.3 Root Locus Method

The root locus method is used to examine the effect of the governor gain parameters on the system dynamic behaviour of a unit supplying an isolated load. The root locus shows how changes in the system's feedback gain or other system parameters influence the pole location [116]. As it is known closed-loop system poles that lie in the right-half of the plane represent terms in the solution which will grow with time and thus indicate an unstable system. Whereas closed-loop system poles that lie in the left-half plane represent terms that die out and the further they are to the left, the faster these terms will disappear. Then, to improve the transient response of the system, it is necessary to move the closed-loop system poles to the left, thus increasing the damping. The distance of the complex roots from the real axis indicates the frequency of the oscillatory terms in the transient response. Those roots near the real axis have a low frequency while those further away will have a higher frequency [119].

In this study, the system representation of Fig. 9.1 was used. The open-loop system transfer function (9.3) can be rewritten relating to the parameters X_1, X_2, X_3 and X_4, Eq. 9.10:

$$G(s) = \frac{-X_3 s^3 + s^2(X_3 - X_1) + s(X_1 - X_1 X_2) + X_1 X_2}{s(1 + 0.5s)(s + X_4)}. \tag{9.10}$$

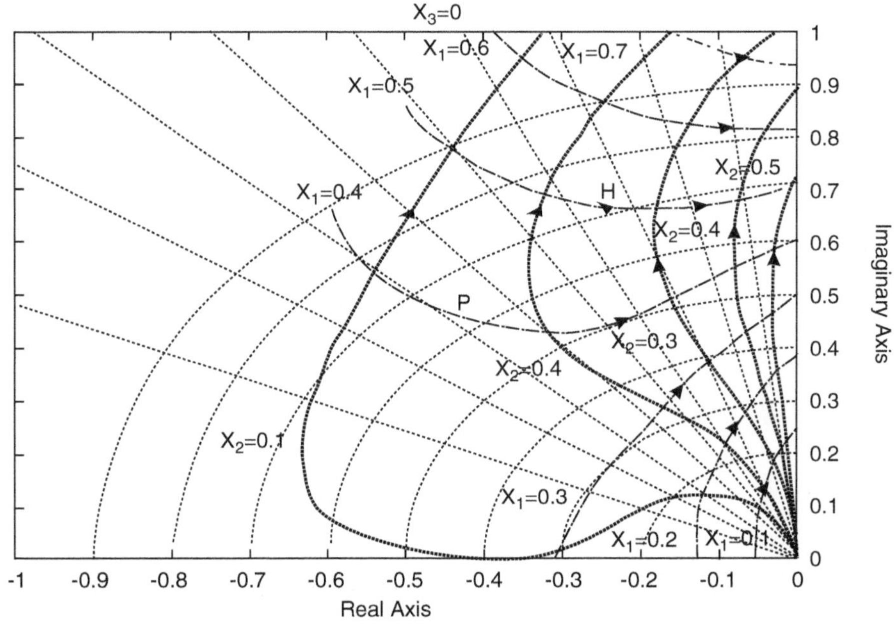

Fig. 9.3 Root loci of the system for $X_3 = 0$

Hagihara [32] predicted a change in the pattern of the root loci that limits the derivative gain to $K_d \leqslant K_p T_w / 3$. However, the root loci pattern changes occur at a higher value $X_3 = 0.45$ and yields an expansion in the K_d stability limits. The location of the open-loop poles of the system represented by Eq. 9.10 will be used in evaluating the system stability. Setting the system regulation $X_4 = 0$ (i.e. assuming frequency insensitive load), the locus of the poles for varying X_1 and X_2 for typical values of X_3 are plotted. Figure 9.3 shows the case of $X_3 = 0$, the solid lines illustrate root loci for a fixed value of X_2; increasing the value of X_1 will lead the complex root to move towards the imaginary axis and eventually enter the unstable region.

The root loci for fixed values of X_1 is illustrated by the dashed lines, which similarly move towards the imaginary axis and enter the unstable plane with increasing X_2. The points P and H mark the locations of the poles for optimum values of X_1 and X_2 as recommended by Paynter [69] and by Hovey [118], respectively. It can be seen that Paynter's operating point will result in a damping ratio $\zeta \approx 0.73$, which produces an overshoot of 5%, which represents good system tuning. Hovey's operating point has a damping ratio $\zeta \approx 0.38$, which results in a bigger overshoot.

The effect of the derivative gain on the root locus is investigated for the cases $X_3 = 1/3$ and $1/2$ are illustrated in Figs. 9.4 and 9.5, respectively. Adding X_3 expands the stability zone and for $X_3 = 1/3$ the optimum setting is marked as M, giving a

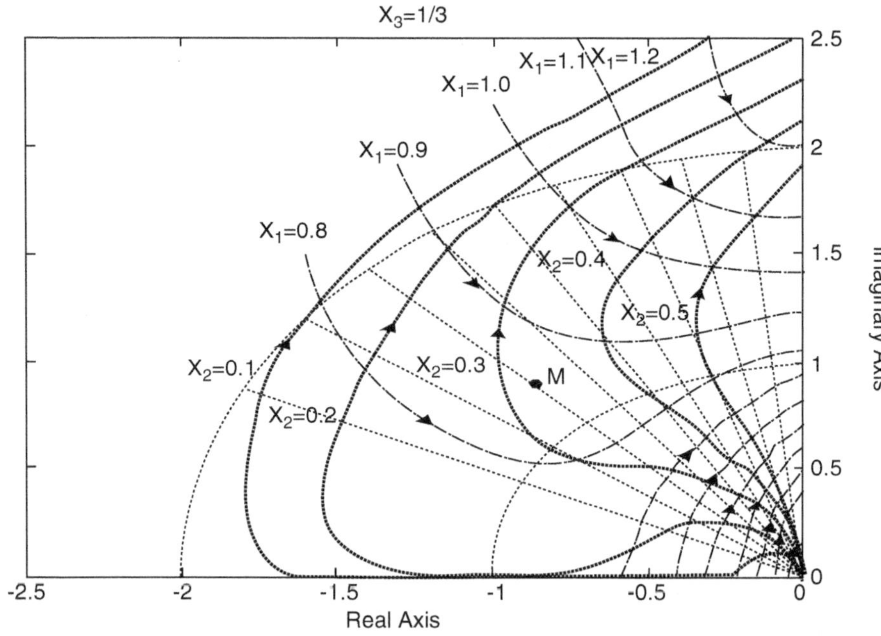

Fig. 9.4 Root loci of the system for $X_3 = 0.3$

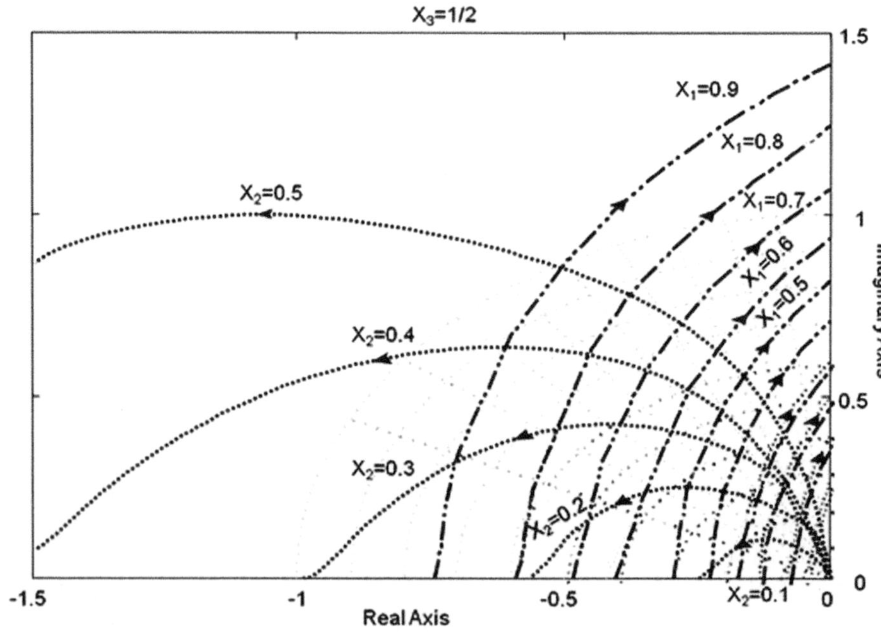

Fig. 9.5 Root loci of the system for $X_3 = 0.5$

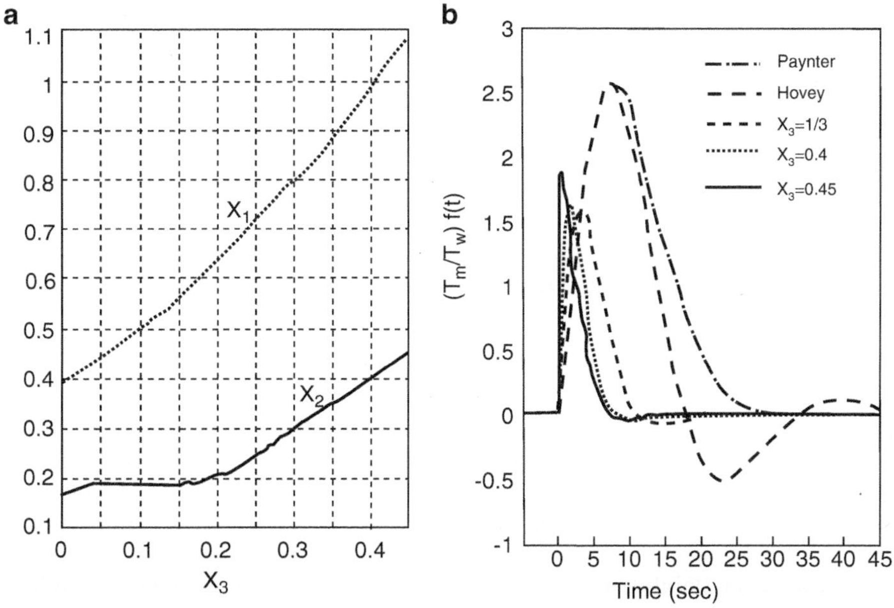

Fig. 9.6 (**a**) Optimum values of X_1 and X_2 for different values of X_3. (**b**) Optimum transient responses for $X_4 = 0$

damping ratio $\zeta \approx 0.707$. Operating with $X_3 = 0.5$ results in a different pattern of root locus that moves the pole towards the real axis while increasing X_1. The change in pattern of the root locus occurs when the value of $X_3 = 0.45$.

The optimum system settings change for each value of X_3; consequently, X_1 and X_2 should be adjusted accordingly as shown in Fig. 9.6a, where higher X_3 values result in increased values of X_1 and X_2. The effect of X_3 on the transient response of the system for a step load change has been obtained using a SIMULINK® model. Figure 9.6b illustrates the effect of adding the derivative term as it confirms that the system will stabilise rapidly with less overshoot compared to the cases where it is omitted. Meanwhile, operating with $X_3 = 0.45$ the system will be very active and may cause an extra overshoot due to the fast response caused by the high integral gain setting. The results imply that increasing the derivative gain will cause the enlargement of the stability boundaries to a predetermined value of $X_3 = 0.5$ where the final limits are reached according to Routh test of Sect. 9.2.2.1. The generating unit is required to deliver a fast response to load changes and selecting $X_3 = 0.4$ will achieve that aim. In this case, using Fig. 9.6a the optimum PID gains are adjusted to:

$$K_\mathrm{p} = 0.97\frac{T_\mathrm{m}}{T_\mathrm{w}} \quad K_\mathrm{i} = 0.39\frac{T_\mathrm{m}}{T_\mathrm{w}^2} \quad K_\mathrm{d} = 0.4T_\mathrm{m}.$$

The optimum response is based on a prototype ITAE response. Graham and Lathrop [120] worked out a prototype set of transient responses to minimise the

Integral of the Time multiplied by Absolute value of Error. Prototype models were developed based on pole location and the model chosen here to represent an optimum response has poles at $(s + 0.707 \pm 0.707j)$.

This analysis has extended the stability boundaries and provides tuning guidelines, which yield a better response than popular methods currently used during plant commissioning.

9.3 Stability of Plant Connected to a Power System

The response of a unit when used for system frequency control is determined by the behaviour of its governor, the stiffness of the connected power system and the interaction between them. An investigation into the governor settings at Dinorwig was carried out in order to establish the optimum governor settings for different power system (grid) loading conditions, providing as close to an ideal response as possible without endangering the stability of system frequency control. The gain and phase margins of the open-loop transfer function of the system were examined to determine the stability of the plant, for different droop settings and power system sizes. The stability conditions are defined as having 10 dB gain and 30° phase margins, as recommended by the British National Grid Company.

9.3.1 Plant Configuration

The study assumes that Dinorwig is the only frequency regulator on the power system and is operating at a full load (six units online) and that they can all be represented as one unit. The hydraulic coupling between the units caused by the common tunnel is ignored. The system under investigation is configured as shown in Fig. 9.8 and is divided into the following sections.

9.3.1.1 Speed Governor

The governor used is based on a typical PID controller consisting of two feedbacks. The droop considered here is the normal station setting of either 1% or 4%. The governor has an averaging filter for the power system frequency whose effect was included as an equivalent continuous transfer function. MATLAB® was used to find a best-fit transfer function. Figure 9.7 shows the result of the curve fitting obtained by using a continuous representation of Eq. 9.11:

$$G_{\text{av}}(s) = \left(\frac{1}{0.165s + 1}\right)\left(\frac{1}{0.165s + 1}\right). \tag{9.11}$$

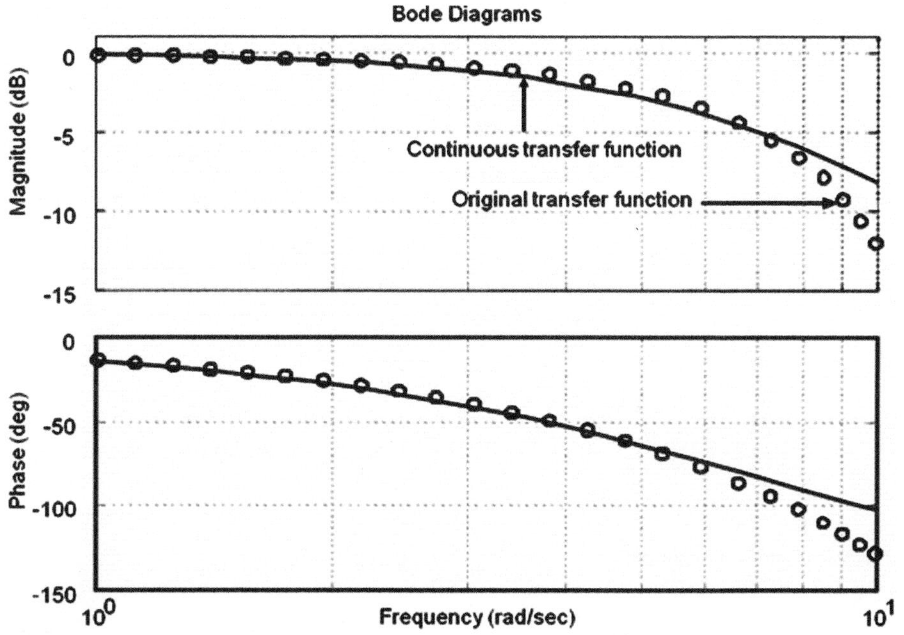

Fig. 9.7 Averaging filter frequency response

9.3.1.2 Hydraulic System

The linearised model for the turbine with elastic water column represents the hydraulic system because Dinorwig has a long penstock with high head that leads to travelling wave effects in the system. The transfer function of Eq. 9.12 relating the power output to guide vane position is described in detail in Chap. 4.

$$\frac{P_{\mathrm{m}}(s)}{G(s)} = \frac{1 - T_{\mathrm{w}}s + \frac{4}{\pi^2}T_{\mathrm{e}}^2 s^2 - \frac{T_{\mathrm{w}}T_{\mathrm{e}}^2}{\pi^2}s^3}{1 + 0.5T_{\mathrm{w}}s + \frac{4}{\pi^2}T_{\mathrm{e}}^2 s^2 + 0.5\frac{T_{\mathrm{w}}T_{\mathrm{e}}^2}{\pi^2}s^3}. \tag{9.12}$$

The guide vane dynamics are represented by a two-stage actuator whose transfer functions are obtained from step response tests.

9.3.1.3 Power System

The power system is assumed to be a single area with the tie line power set to zero so there is no power transfer to or from a second system. The linearised first-order system described in equation (5.20) was used to represent the total rotating inertia of the system (M) and the load/frequency damping (D).

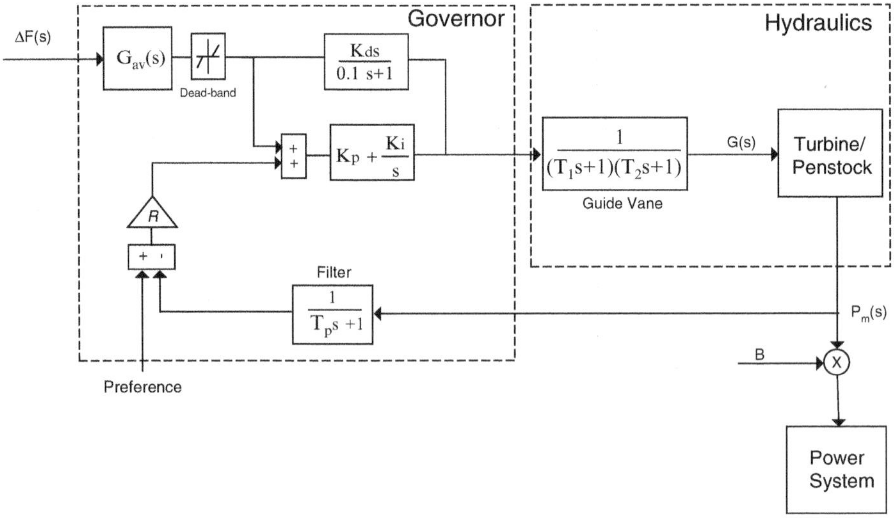

Fig. 9.8 System configuration

The plant block diagram of Fig. 9.8 was manipulated to obtain the transfer function between power and frequency. An alternative notation to the block diagram is given by the signal-flow graph introduced by Mason [121]. The method involves characterising the system by a network of directed branches and associated transfer functions connected at nodes. Using these rules, the complete plant transfer function can be written as

$$\frac{\Delta P_{\mathrm{m}}(s)}{\Delta F(s)} = \frac{(T_{\mathrm{p}}s+1)G_{\mathrm{av}}\left[K_{\mathrm{p}}+\frac{K_{\mathrm{i}}}{s}+\frac{K_{\mathrm{d}}s}{0.3s+1}\right]A(s)}{(T_{\mathrm{p}}s+1)+R\left(K_{\mathrm{p}}+\frac{K_{\mathrm{i}}}{s}\right)A(s)}, \tag{9.13}$$

where $A(s)$ is the transfer function representing the dynamics of the guide vane and the turbine-penstock:

$$A(s) = \frac{1}{(T_1s+1)(T_2s+1)} \times \frac{1-T_{\mathrm{w}}s+\frac{4}{\pi^2}T_{\mathrm{e}}^2s^2-\frac{T_{\mathrm{w}}T_{\mathrm{e}}^2}{\pi^2}s^3}{1+0.5T_{\mathrm{w}}s+\frac{4}{\pi^2}T_{\mathrm{e}}^2s^2+0.5\frac{T_{\mathrm{w}}T_{\mathrm{e}}^2}{\pi^2}s^3}.$$

9.3.2 *Stability Margins*

A MATLAB® programme was written to evaluate the performance of the control system at Dinorwig power station by determining the stability margins of the system for various operating conditions. Table 9.1 shows the system parameters used in the

Table 9.1 System parameters

D:	Load damping factor	1 p.u. MW/1 p.u. Hz
M:	Inertia of the power system	10
P_{m}:	Power output of the plant	6×300 MW
T_1:	First-stage actuator	0.19
T_2:	Second-stage actuator	0.4
T_{p}:	Power measurement filter	2

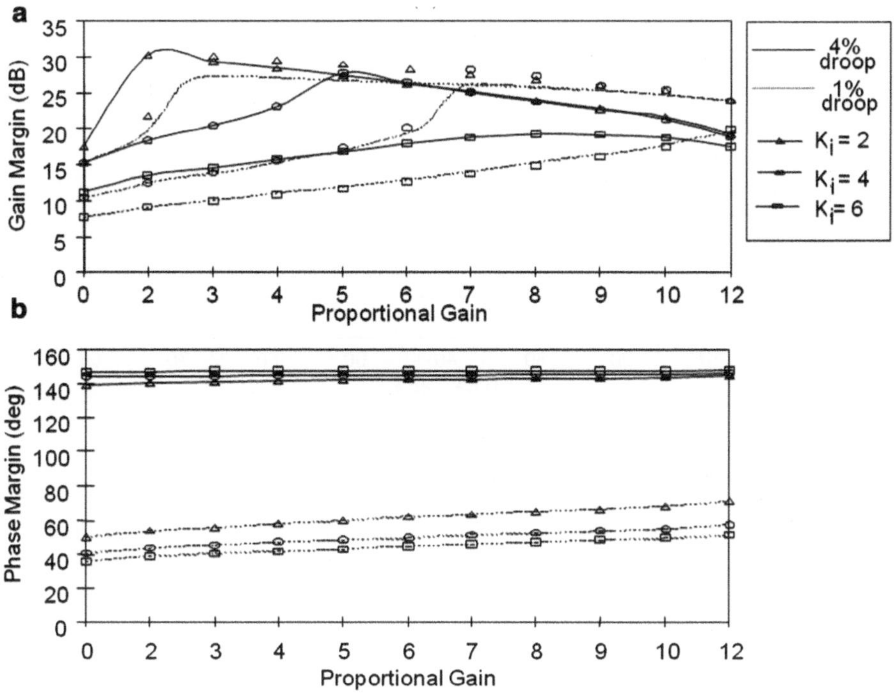

Fig. 9.9 Stability margins of the plant operating with 40 GW power system

investigation, Dinorwig operates with two droop settings of 4% and 1%; therefore, it is necessary to assess the effect of droop on the stability of the system. The open-loop frequency response of the system was plotted for different K_{p} and K_{i} gains with $K_{\mathrm{d}} = 10$ and for both droop settings; in each case the gain and phase margins were determined.

Figure 9.9a shows the gain margin for a 40 GW system in which the fraction of the total generation capacity supplied by Dinorwig corresponds to a base conversion of $B = 0.045$. The gain margin is initially improved by increasing the proportional gain. However, once the peak value is reached, indicating the ideal setting for the controller, the gain margin starts to decline. Figure 9.9b reveals that increasing the proportional gain will improve the phase margin and that unsatisfactory stability margins can result from reducing the droop value to below 1% at low values of

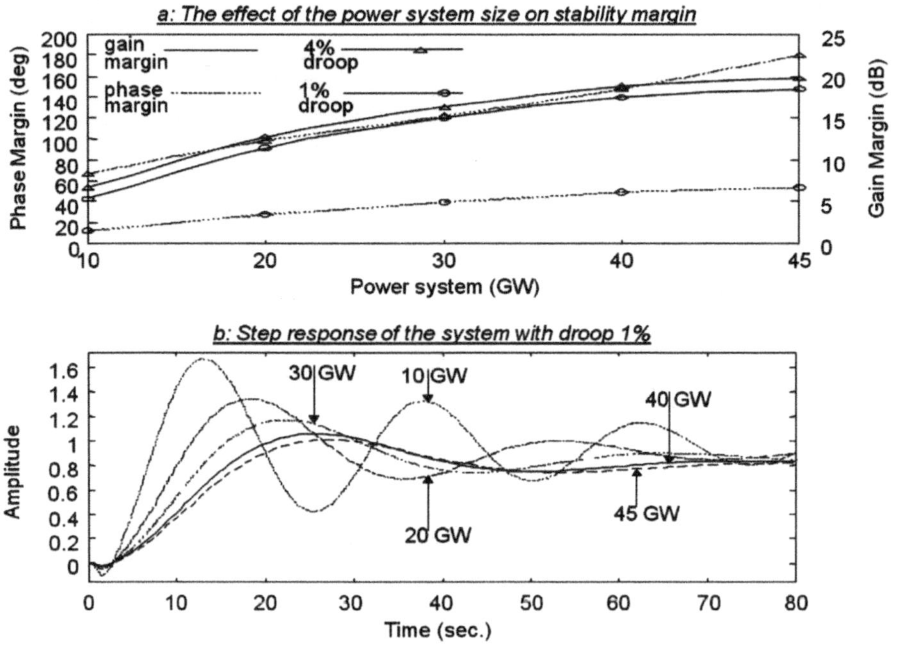

Fig. 9.10 The effect of power system size on the stability

proportional gain. Hence, increasing the integral gain to speed up the response will require an increase in the proportional gain to maintain the stability margins.

The effect of the power system size on the stability margins is shown in Fig. 9.10a, representing specific controller settings; $K_p=10$, $K_i=6$, and $K_d=10$. It is apparent that the system loading has a significant effect on the response of the plant, since the margins expand with increased system loading. Therefore, when the power system is heavily loaded, it is possible to increase the governor gain settings to speed up the unit response. The correlation between the power system size and the governor settings is evident when applying a step input into the system as shown in Fig. 9.10b. High governor settings will cause the response to oscillate when operating with low power system loading, while the response will stabilise with larger power system size.

9.4 Stability of Plant Operating with a Deadband

For commercial reasons, Dinorwig is often used as a fast base load unit to increase the system generation (raise the frequency) temporarily until slower response units are brought online. This is achieved by the introduction of the deadband of ±0.5 Hz on the frequency control loop and setting the unit to supply a fixed amount of power (adjusted by the operator) usually at 288 MW. The deadband associated with a speed governor is defined as 'the total magnitude of the change in steady state speed (frequency) within which there is no resulting measurable change in the position of the guide vane' [122].

When operating in this mode the power system will have no effect on the unit response as long as the frequency deviation is within the deadband limits. Consequently, the plant stability boundaries will be different from the frequency regulation mode. Dinorwig is configured to utilise the same PI governor gain settings for both operating conditions; therefore, it was necessary to investigate the plant phase and gain margins for the alternative mode of operation.

When operating in this mode the governor is configured as a PI type controller with the derivative action omitted from the control loop. The block diagram of Fig. 9.8 was used to obtain a new transfer function of the plant operating with deadband on. The transfer function relating the plant power output to power deviation (compared with the power set point) can be written as

$$\frac{\Delta P_{\mathrm{m}}(s)}{\Delta P(s)} = \left[R \left(K_{\mathrm{p}} + \frac{K_{\mathrm{i}}}{s} \right) \right] A(s). \tag{9.14}$$

A similar approach was used to assess the performance of the control system by identifying the gain and phase margins for Eq. 9.14. The open-loop frequency response of the system was plotted for different K_{p} and K_{i} gains for droop settings of 1% and 4%; in each case the gain and phase margins were determined. The results shown in Fig. 9.11 represent the plant margins when all the six units are operating with their deadband on. It is evident that operating with low droop gain (1%) increases the margins of the system and hence its stability; however, with high droop (4%), there is a dramatic decrease in the margins. The reason is that the droop represents a loop gain in the system as expressed by Eq. 9.14, thus using 4% droop means that the power error signal into the governor will be four times bigger than when operating with 1% droop yield. Meanwhile, increasing the integral gain also results in a reduction in the margins and the correlation between the integral and the proportional gains is clear in the gain margin plot as there is a maximum gain for each value of K_{i} which occurs at a particular proportional gain.

The effect of plant loading (number of units online) on the stability was also analysed, by identifying the margins for a single unit having deadband on and its droop set to 4%. As before the margins are obtained for different governor gain settings and the results compared with the margins for six units online. The results shown in Fig. 9.12 exhibit an increase in the margins due to the reduction in the number of units. Meanwhile, increasing the proportional gain will lead to a fall in the gain margins but an increase in the phase margins. The stability of the plant decreases when more units are operating with a high droop setting.

To the best of our knowledge, this is the first time that such a sensitivity study has been performed and it has considerably improved the understanding of how the station will react under different circumstances (both internal and external) and what could be done to compensate for varying conditions. The result has been published in [104].

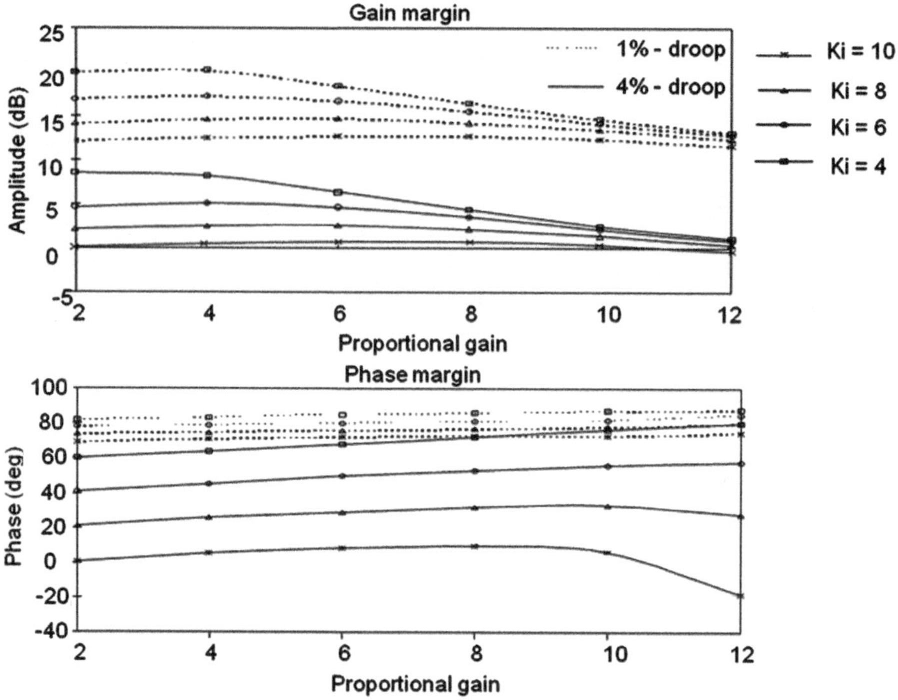

Fig. 9.11 Stability margins of the plant when operating with deadband

9.5 Tuning the Controllers

As has been studied, classical controllers can satisfy a wide range of specifications, but their performance depends strongly on the value of their internal parameters. A good balance between sensitivity, control effort and speed of response is generally the main objective of tuning [123]. The three mode controllers are robust and simple and these features make them suitable for manual adjustment. As was stated before, the current speed governor at Dinorwig uses a PI control in the power control loop with derivative feed-forward from the frequency control loop. In this section, the processes to tune PI and PI with anti-windup controllers are evaluated.

9.5.1 Proportional and Integral

The proportional (P), integral (I) and derivative (D) actions are standard modes of control for industrial applications. The controllers that contain these actions are robust and simple and these features make them suitable for manual adjustment.

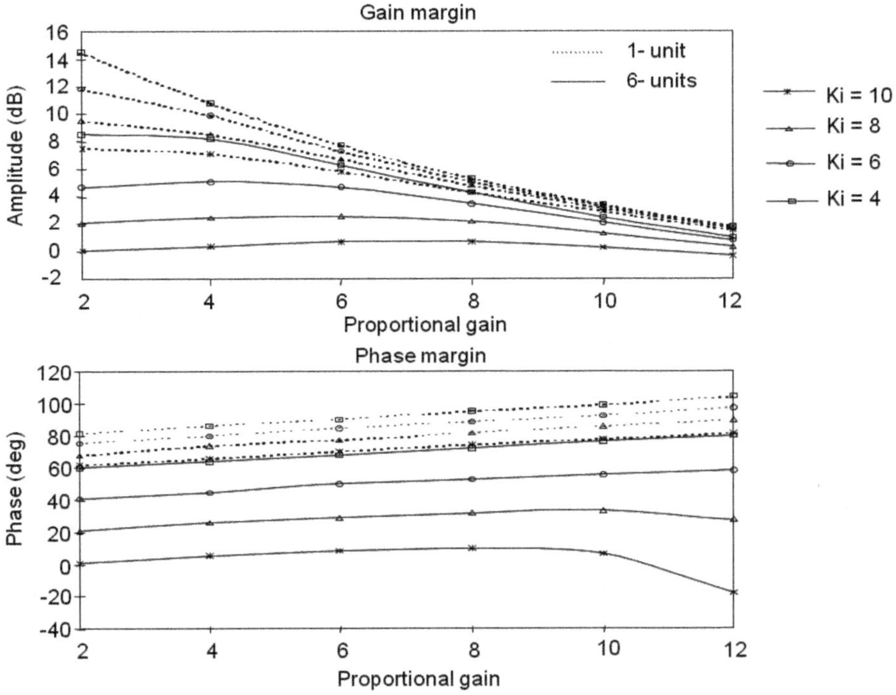

Fig. 9.12 The effect of the plant loading on the margins

The current speed governor at Dinorwig uses a PI control in the power control loop with derivative feed-forward from the frequency control loop. In this section, the process followed to tune a PI controller for the hydraulic station is discussed.

Tuning a proportional and integral controller (PI) requires selection of the correct values of K and K_i that allow the control to achieve a desired plant performance. Mansoor has studied the selection of the control parameters for the speed governor at Dinorwig [33]. The values that are currently implemented in the hydroelectric plant are $K = 0.1$ and $K_i = 0.12$ and will be used as a basis of comparison. The PI controller with these parameters has a performance which is a compromise between one and six unit operation. Therefore, to optimise the performance of the plant, different sets of parameters were selected for the extreme cases, one and six units operational. The evaluation accomplished in former sections shows that suitable ranges for the parameters of the PI are $0.23 \geq K \geq 0.1$ and $0.5 > K_i > 0.1$.

As discussed in previous sections, using classical Bode and Root Locus techniques plus a final manual adjustment, different sets of parameters were selected for the PI controller. The goal is to optimise the response of the plant in the one- and six-unit operational modes, reducing the primary response of the system without producing large overshoot. Table 9.2 shows the set of values selected for each case.

Table 9.2 Parameters of the PI controller

	Compromise	Optimised one unit	Optimised six units
K	0.10	0.175	0.165
K_i	0.12	0.210	0.110

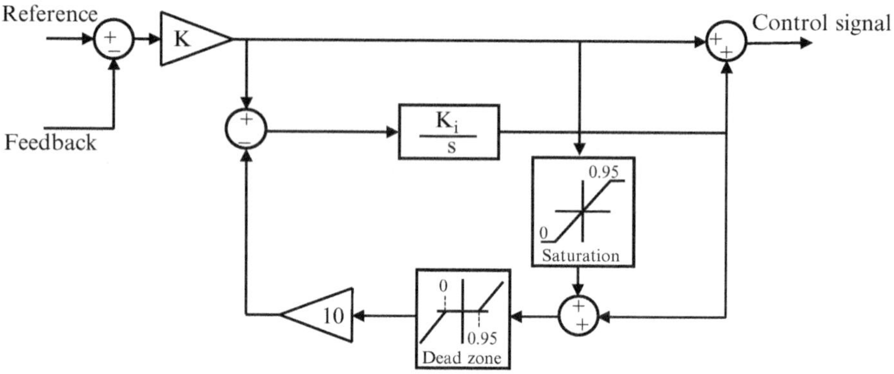

Fig. 9.13 General scheme of PI anti-windup

9.5.2 *PI Anti-windup*

A PI carefully tuned can offer good performance but, as was stated before, all processes are subject to constraints. The plant alters its behaviour when constraints are activated, and the performance of a linear controller, such as PI, can deteriorate significantly [124]. When the plant has actuator saturation, the integrator value becomes excessively large compared to a linear response (an actuator without saturation), it 'winds up'. In addition, a higher integrator output and a longer settling time are caused by the saturation effect [124–126].

The windup is produced when the control signal saturates the actuator, because an additional increase of the control signal will not accelerate the response of the plant. If this behaviour persists, the integrator value becomes very large, without affecting the output of the plant. To bring back the system to its steady state value, the control error has to be of the opposite sign for a long time, resulting, as was pointed out, in a large overshoot and a longer settling time [125].

Figure 9.13 shows a general scheme of a PI with tracking anti-windup structure [125]. This controller has an internal feedback path, which drives the integrator to a negative value and forces the output of the system to be in the linear range. The internal saturation is used to reduce the integrator input. As can be seen from Fig. 9.13, the signal to be integrated is modified by the proportional gain (K); therefore, the values of the integral gain (K_i) are adjusted in order to maintain equivalence with the PI. Table 9.3 shows these values. The saturation limit and the dead zone depend on the constraints fixed by the operator; a value of 0.95 p.u. is commonly used.

Table 9.3 Parameters of the PI controller with anti-windup

	Compromise	Optimised one unit	Optimised six units
K	0.10	0.175	0.165
K_i	1.2	1.2	0.66

9.6 Conclusions

Tuning the speed governor is a very important issue for control engineers in the power industry. A general guide for optimum adjustment of the gains of the governor for a unit supplying an isolated load (or a small power system) has been provided, where the performance of the governor and hence the plant response can be improved by appropriate derivative gain. In the case of a plant regulating the frequency of a large power system, it has been demonstrated that the stability boundaries of the system expand as the power system size increases.

In the case of Dinorwig, to fulfil its commercial obligation it is essential to respond quickly to any changes in the power system loading. This could be achieved by operating close to the stability boundaries. Since Dinorwig's control strategy is based on using one set of gain settings for all operating conditions, when the power system size reduces this may cause operation to be near or even outside these boundaries. Ensuring stability under all circumstances requires conservative gain settings which are well away from optimum for normal operating conditions. Avoiding instability while maximising the commercial potential of the plant suggests that it is necessary to schedule the gains according to operating conditions.

The original governor gains at Dinorwig were very conservative being based during commissioning on a relatively small power system. The results of this work have been implemented on the plant where the higher gains have resulted in faster and more accurate responses being achieved. However, gain scheduling for different power system sizes has not been implemented on the plant to date because the power system properties are not available – consultation with the British National Grid Company would be necessary to make such data continuously available to the control system at Dinorwig.

Another important factor for gain scheduling at Dinorwig is operation with deadband on, where the speed of the response is determined by the power setting ramping time as the unit is not required to be very fast in this mode of operation. In this case, the power system size has no influence on the plant response and the main cause for instability is the droop. Operating with a high droop setting of 4% reduces the stability boundaries, especially when several units are online and operating in this mode. Therefore, to avoid the risk of instability, it is recommended that the droop be set to 1% on these units in order to expand the boundaries. The nonlinear simulation of oscillatory behaviour in Chaps. 11 and 13 reinforces this conclusion. This recommendation is currently implemented as part of the control strategy at Dinorwig.

Chapter 10
Feed-Forward Characteristic

10.1 Introduction

Due to their rapid response capability, hydroelectric stations and pumped storage stations, in particular, are eminently suited to regulating grid frequency. In a hydroelectric station that has more than one generator, it is common practice to use one of them in closed loop for automatic frequency control while the other units are despatched under manual control, for load-shifting or peak-lopping purposes. Most often the frequency control algorithm will be a form of PID control as discussed in previous chapters.

A primary goal of any power network (grid) coordinator is to maintain the system frequency at all times within a narrow band around its nominal value, usually 50/60 Hz $\pm$ 0.5 Hz. At a given moment there will be several frequency regulators, of varying capacity and speed of response, connected to the grid. That means that no single station will be called upon to supply all the power necessary to regulate frequency. Instead, stable sharing of the power load between multiple generators is achieved by including in the governors a characteristic that causes generator speed to drop as the load is increased. This is known as the speed regulation or 'droop' characteristic, which was discussed in Chap. 5. The goal of a power station providing a frequency regulation (ancillary) service is to supply its targeted power contribution in an accurate and timely fashion, rather than the exclusive control of grid frequency. In this chapter, this will be treated as a problem of tracking a power target that continually varies with grid frequency error. The method, as described by Jones and Mansoor [127], uses:

- A linearised model for the hydroelectric power station
- A grid model identified from power and frequency records
- An algorithm for predicting frequency deviations a few seconds in the future
- A feed-forward loop that provides an additional signal which is added to the usual output of the PI feedback controller

G.A. Munoz-Hernandez et al., *Modelling and Controlling Hydropower Plants*, Advances in Industrial Control, DOI 10.1007/978-1-4471-2291-3_10,

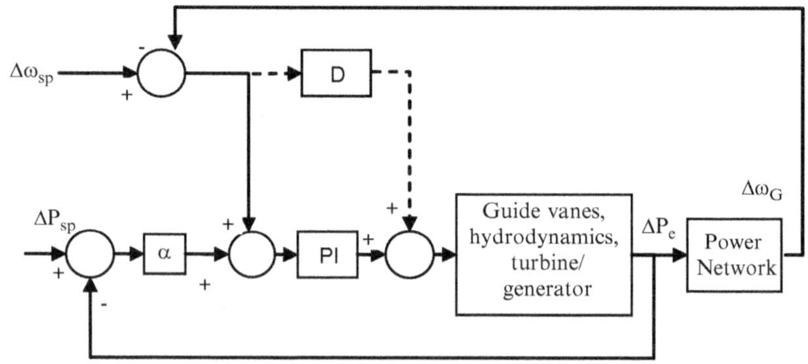

Fig. 10.1 Conventional implementation of PID controller

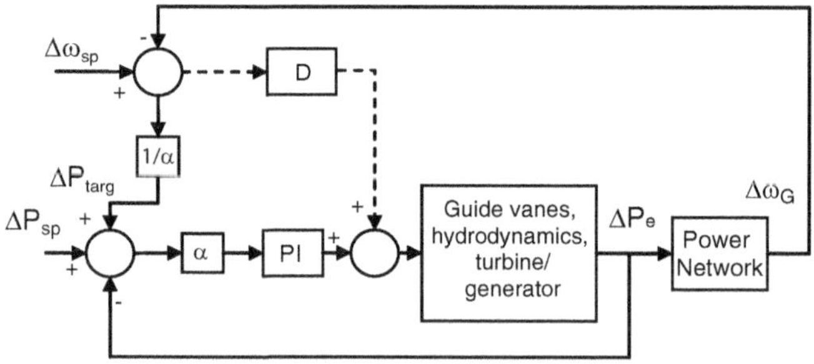

Fig. 10.2 Block diagram with droop gain moved into the forward loop

The concept of a power target is useful when assessing the response of the system in frequency-control mode and has been used in a proposal to extend the performance specifications for hydroelectric plant [97]. The usual implementation of a PID frequency control loop is shown in Fig. 10.1, where the variables are expressed as per-unit incremental deviations from a steady state operating point.

In Fig. 10.1:

$\Delta\omega_{\mathrm{G}}$ = change in power network frequency
$\Delta\omega_{\mathrm{sp}}$ = change in frequency set-point
ΔP_{e} = change in electrical power supplied by the generator
ΔP_{sp} = change in power set-point
α = droop gain

The rules of block diagram manipulation allow the droop gain block in Fig. 10.1 to be moved into the forward branch.

Now define the power target, ΔP_{targ}, in Fig. 10.2, which is seen to be an additional demand for power due to the frequency error:

$$\Delta P_{\mathrm{targ}} = \frac{1}{\alpha}(\Delta\omega_{\mathrm{sp}} - \Delta\omega_{\mathrm{G}}). \tag{10.1}$$

The condition that must be satisfied in steady state at the input to the integral term in the PI controller is

$$\alpha(\Delta P_{\mathrm{sp}} - \Delta P_{\mathrm{e}}) + (\Delta\omega_{\mathrm{sp}} - \Delta\omega_{\mathrm{G}}) = 0. \tag{10.2}$$

Substituting Eq. 10.1 into Eq. 10.2:

$$\begin{aligned}
&\Delta P_{\mathrm{targ}} + (\Delta P_{\mathrm{sp}} - \Delta P_{\mathrm{e}}) = 0 \\
&\therefore \Delta P_{\mathrm{targ}} = -(\Delta P_{\mathrm{sp}} - \Delta P_{\mathrm{e}}) = \frac{1}{\alpha}(\Delta\omega_{\mathrm{sp}} - \Delta\omega_{\mathrm{G}}) \\
&\therefore -\frac{\Delta P}{\Delta\omega} = \frac{1}{\alpha},
\end{aligned} \tag{10.3}$$

where $\Delta P = (\Delta P_{\mathrm{sp}} - \Delta P_{\mathrm{e}})$ and $\Delta\omega = (\Delta\omega_{\mathrm{sp}} - \Delta\omega_{\mathrm{G}})$.

Note that the final expression in (10.3) is the definition of droop gain. Note also that if ω_{G} falls below ω_{sp} then the RHS of (10.3) becomes positive, i.e. a low grid frequency demands increased power. Under the normal condition that $\Delta P_{\mathrm{sp}} = \Delta\omega_{\mathrm{sp}} = 0$, Eq. 10.3 reduces to:

$$\begin{aligned}
&\Delta P_{\mathrm{targ}} = \Delta P_{\mathrm{e}} = -\frac{1}{\alpha}\Delta\omega_{\mathrm{G}} \\
&\frac{\Delta P_{\mathrm{targ}}}{\Delta\omega_{\mathrm{G}}} = -\frac{1}{\alpha}.
\end{aligned} \tag{10.4}$$

Integral action ensures that $\Delta P_{\mathrm{targ}} = \Delta P_{\mathrm{e}}$ in steady state. Under transient conditions, ΔP_{targ} acts as a reference input to the control loop whose function is to produce changes in electrical power ΔP_{e} that track the input as closely as possible. The tracking error $(\Delta P_{\mathrm{targ}} - \Delta P_{\mathrm{e}})$ is a measure of how well this is done.

10.2 Linearised Model for the Hydroelectric Plant

A hydroelectric power station is a complex, nonlinear and time-varying multivariable system but a relatively simple linearised model that represents the salient dynamics is sufficient for this study. Figure 10.3 shows a block diagram of the system for a single conduit and turbine/generator. It is derived from Fig. 10.1 with the new feed-forward loop included. The transfer functions are:

$C(s)$ for the feedback (PI) controller
$G_{\mathrm{V}}(s)$ for the guide vane servo
$G_{\mathrm{H}}(s)$ for the hydraulics
$G_{\mathrm{T}}(s)$ for the turbine/generator
$G_{\mathrm{G}}(s)$ for the grid

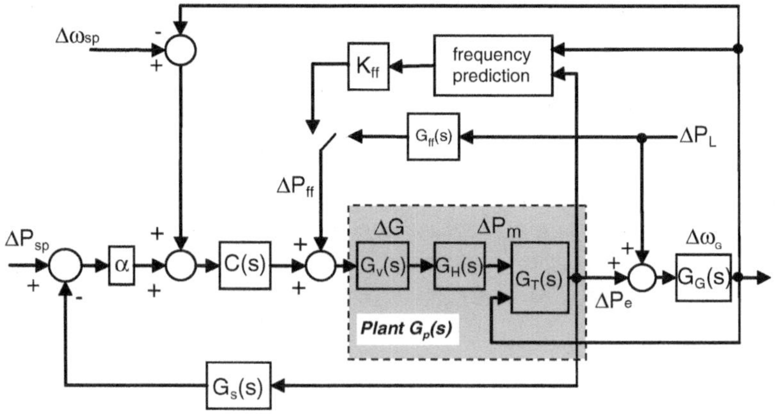

Fig. 10.3 Block diagram of the linearised system model

$G_s(s)$ for the power transducer
$G_{ff}(s)$ for the ideal feed-forward loop

and

ΔP_m is the change in mechanical power
ΔG is the change in guide vane opening
ΔP_L is the (unknown) change in load power

The principle of the method is first developed in terms of an ideal feed-forward loop (transfer function $G_{ff}(s)$); this will be replaced by the practical version at a later stage. Note that the derivative (D) loop in Fig. 10.1 has been omitted from Fig. 10.3. The derivative term is used to anticipate future values of the frequency error by examining its rate of change. The control signal is then enhanced by adding the derivative component to the PI output, via a suitable value of gain. Essentially, the 'frequency prediction' block in Fig. 10.3 replaces the function of the derivative term by estimating a future value of the frequency error $\Delta\hat{\omega}_G$. The feed-forward gain K_{ff} is analogous to the derivative gain.

The PI controller transfer function (see Chap. 6) is

$$C(s) = \frac{K_p s + K_i}{s}. \quad (10.5)$$

The noise reduction filter on the transducer is

$$G_s(s) = \frac{1}{s+1}. \quad (10.6)$$

The transfer function for the guide vane servo, as described in Chap. 6, is

$$G_v(s) = \left(\frac{1}{T_{S1}s+1}\right)\left(\frac{1}{T_{S2}s+1}\right). \quad (10.7)$$

Assuming an inelastic water column and negligible hydraulic friction, it was shown in Chap. 4 that a linearised transfer function relating per-unit (p.u.) changes in the mechanical power produced by the turbine (ΔP_{m}) to per-unit changes in the guide vane position (ΔG) is

$$G_{\mathrm{H}}(s) = \frac{\Delta P_{\mathrm{m}}}{\Delta G} = \frac{-T_{\mathrm{w}}s + 1}{0.5T_{\mathrm{w}}s + 1}, \tag{10.8}$$

where the water time constant T_{w} is chosen for the nominal operating point.

The transfer function (10.8) is nonminimum phase and this causes a delay between the control (ΔG) and the output (ΔP_{m}) which is a serious impediment to good tracking performance.

Using the 'swing' equation derived in Chap. 5, the electrical power (ΔP_{e}) is related to the mechanical power ΔP_{m} and frequency deviation $\Delta\omega_{\mathrm{G}}$ by

$$\Delta P_{\mathrm{e}} = \frac{K_{\mathrm{s}}\omega_0}{2Hs^2 + K_{\mathrm{D}}s + K_s\omega_0}\Delta P_{\mathrm{m}} + \frac{K_{\mathrm{s}}\omega_0(2Hs + K_{\mathrm{D}})}{2Hs^2 + K_{\mathrm{D}}s + K_{\mathrm{s}}\omega_0}\Delta\omega_{\mathrm{G}}, \tag{10.9}$$

where H is the per-unit inertia constant of the turbine/generator, K_{D} the per-unit coefficient of damping torque, K_{s} the per-unit coefficient of synchronising torque and ω_0 the generator's rated value of angular velocity.

10.3 Model for the Power Network

The model for the power network is a critical component because it is central to predicting how the frequency will respond to a change in power. Clearly, the dynamic characteristics of a power network depend on its exact configuration at any moment. However, studies have shown that a sharp injection of power onto a network will often cause the frequency to increase, reaching a peak after some tens of seconds, after which it will fall back slightly. A second-order transfer function will capture this 'macroscopic scale' behaviour better than the first-order relationship derived in Chap. 6. This is the approach taken by Anderson and Mirheydar [128] whose model is composed of two time constants, one associated with the sum of all the inertias of the rotating machines and the other associated with all regulatory mechanisms connected to the grid. Their transfer function relates changes in grid frequency to load imbalance:

$$\Delta\omega_{\mathrm{G}} = K\omega_{\mathrm{n}}^2 T_{\mathrm{R}}\left(\frac{s + 1/T_{\mathrm{R}}}{s^2 + 2\zeta\omega_{\mathrm{n}}s + \omega_{\mathrm{n}}^2}\right)(\Delta P_{\mathrm{e}} - \Delta P_{\mathrm{L}}), \tag{10.10}$$

where K is the Grid 'stiffness' (i.e. the steady state sensitivity of grid frequency to changes in power), T_{R} is the time constant of the regulatory mechanisms, ω_{n} the grid natural frequency and ζ the grid damping factor.

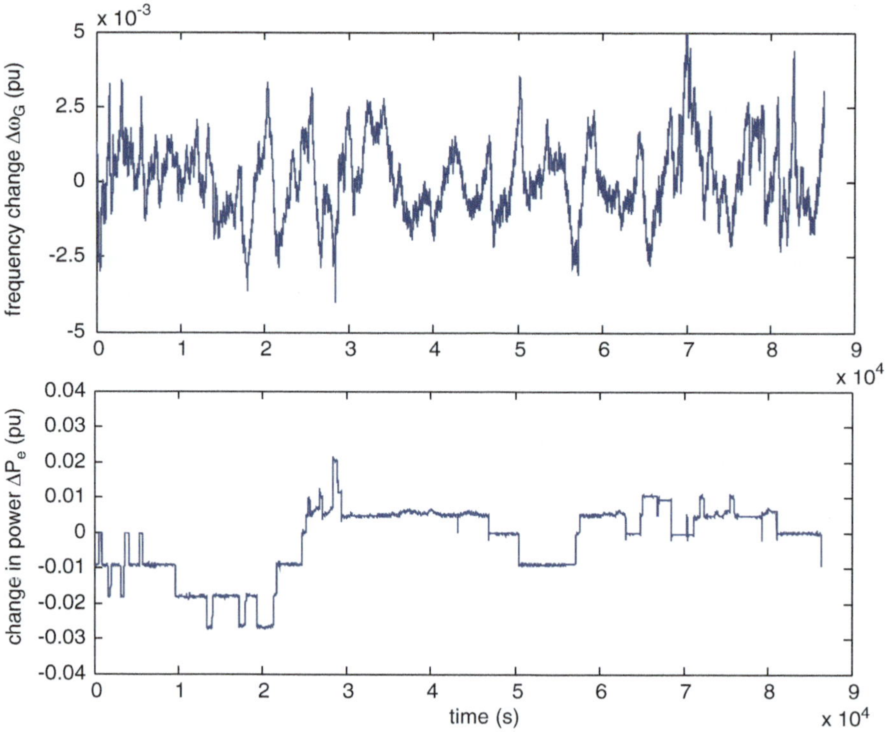

Fig. 10.4 Record of per-unit grid frequency and total power flow at Dinorwig (positive = generation, negative = pumping) for the 24-h period of March 4, 2001

The grid acts as an accumulator whose stored energy is proportional to the synchronous frequency of all the rotating machines connected [19]. The total 'size' of the grid load (which ranges from 30 to 55 GW in Great Britain) produces variations in the parameters K, T_R, ω_n and ζ. While ΔP_e is known, ΔP_L is the sum of a vast number of time-varying contributions, as discussed in Chap. 6, so ΔP_L is included in the model as an additive, random disturbance. The goal of system identification is to evaluate the parameters of (10.10) from recorded data.

Considerable insight into the dynamics of the power network can be obtained by nonparametric methods [129] which are implemented in the MATLAB® Systems Identification Toolbox [130]. Consider the graphs in Fig. 10.4 which show the per-unit grid frequency deviation and total power flow at Dinorwig over a 24-h period, sampled at 1 s intervals. Sharp changes in the power flow occur as units are brought online and offline to generate or pump. The frequency is a random variable which remains mostly within a band of about ±0.1 Hz (0.002 p.u.) of nominal.

The power spectral density function of the grid frequency record is shown in Fig. 10.5. Except for the small peak which appears in the range 0.3–0.4 r/s, it has

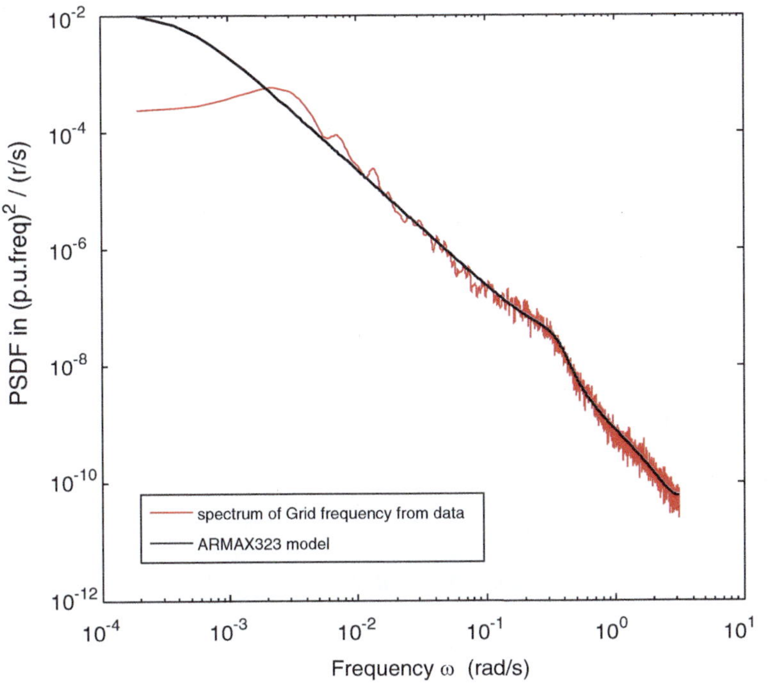

Fig. 10.5 Power spectral density function of the frequency record of Fig. 10.4 and the model identified from the data

a 20 dB/decade roll-off with frequency. This suggests that the basic grid model consists of:

- A low-order dynamic relationship between $\Delta\omega_{\mathrm{G}}$ and ΔP_{e}, which is consistent with the transfer function (10.10)
- An additive 'random walk' term to represent the effect on $\Delta\omega_{\mathrm{G}}$ of the unknown variation in ΔP_{L}

A parametric model can be found by applying the powerful prediction-error methods described by Ljung [129]. Transfer function (10.10) indicates that both the input ΔP_{e} and disturbance ΔP_{L} are affected by the grid dynamics, which is characteristic of the equation-error family of model structures.

A full account of the identification method is given by Jones [131] and only a summary will be given here. The method has been used [18] to study the seasonal variation of the parameters in the transfer function (10.10) for the National Grid in Great Britain.

The best parametric representation was found to be the Auto Regressive Moving Average with eXogenous input (ARMAX) model, which expresses ΔP_{L} as a moving average of white noise. The relationship between $\Delta\omega_{\mathrm{G}}$ and ΔP_{e} can be written as shown in the discrete-time equation (10.11), where $w(k)$ is a zero mean Gaussian white noise sequence:

$$\begin{aligned}\Delta\omega_G(k) + a_1\Delta\omega_G(k-1) + \cdots + a_{n_a}\Delta\omega_G(k-n_a)\\ = b_1\Delta P_e(k-1) + b_2\Delta P_e(k-2) + \cdots + b_{n_b}\Delta P_e(k-n_b)\\ + w(k) + c_1 w(k-1) + c_2 w(k-2) + \cdots c_{n_c} w(k-n_c)\end{aligned} \quad (10.11)$$

In (10.11), the order of the model is determined by n_a, n_b and n_c and these are determined in practice as a trade-off between model accuracy (which usually improves with model order) and computation time (which increases with model order).

Using q as the forward shift operator, Eq. 10.11 can be represented in polynomial form by letting:

$$\begin{aligned}A(q) &= 1 + a_1 q^{-1} + \cdots a_{n_a} q^{-n_a}\\ B(q) &= b_1 q^{-1} + \cdots b_{n_b} q^{-n_b}\\ C(q) &- 1 + c_1 q^{-1} + \cdots a_{n_c} q^{-n_c}.\end{aligned} \quad (10.12)$$

Then

$$A(q)\Delta\omega_G(k) = B(q)\Delta P_e(k) + C(q)w(k) \quad (10.13)$$

and

$$\underline{\theta} = [a_1 \cdots a_{n_a} \quad b_1 \cdots b_{n_b} \quad c_1 \cdots c_{n_c}], \quad (10.14)$$

where $\underline{\theta}$ is the vector of parameters to be determined.

A one-step-ahead predictor for the frequency deviation is [129]

$$C(q)\Delta\hat{\omega}_G(k|\underline{\theta}) = B(q)\Delta P_e(k) + [C(q) - A(q)]\Delta\omega_G(k), \quad (10.15)$$

where $\Delta\hat{\omega}_G(k|\underline{\theta})$ is the conditional expectation of the frequency deviation at time k given information up to time $(k-1)$. The prediction error is given by

$$\varepsilon(k,\underline{\theta}) = \Delta\omega_G(k) - \Delta\hat{\omega}_G(k|\underline{\theta}). \quad (10.16)$$

The value of $\underline{\theta}$ which minimises the quadratic criterion of Eq. 10.17 is found by iterative numerical search:

$$V_N(\underline{\theta}, Z^N) = \frac{1}{N}\sum_{k=1}^{N}\frac{1}{2}\varepsilon^2(k,\underline{\theta}), \quad (10.17)$$

where Z^N is the batch data $[\Delta\omega_G(1), \Delta P_e(1), \Delta\omega_G(2), \Delta P_e(2) \ldots. \Delta\omega_G(N), \Delta P_e(N)]$ taken from the record of Fig. 10.4.

Table 10.1 Coefficient values for the ARMAX323 model for the data in Fig. 10.4

	1	2	3
a	2.643	−2.403	0.759
b	0.00959	−0.00958	
c	−1.131	0.160	0.192

Experimentation with this method applied to several daily records showed that the lowest order ARMAX function which gives a good approximation to the spectrum of Fig. 10.5 has $n_a = 3$, $n_b = 2$ and $n_c = 3$, henceforth called the ARMAX323 model. For the data in Fig. 10.4, the model is Eq. 10.11 with coefficient values of Table 10.1.

The continuous-time transfer function for the ARMAX323 model can be calculated using the zero-order-hold equivalent method, as described by Franklin and Powell [96] and implemented in the MATLAB® Control System Toolbox [94]. This yields a transfer function of the same form as Anderson and Mirheydar's model:

$$\Delta\omega_{\mathrm{G}} = \frac{A(s)}{F(s)}\Delta P_{\mathrm{e}} + \frac{B(s)}{F(s)}w. \tag{10.18}$$

The first part of (10.18) is the 'system' transfer function and the second part is the 'noise' transfer function; note that the denominator *F*(*s*) is common, being determined by the grid dynamics. The 'system' transfer function can be reduced to second order using standard methods [96] with little loss of accuracy:

$$G_{\mathrm{G}}(s) = \frac{\Delta\omega_{\mathrm{G}}}{\Delta P_{\mathrm{e}}} = \frac{0.005376s + 0.01107}{s^2 + 0.2747s + 0.1334}. \tag{10.19}$$

Comparing the coefficients of (10.19) and (10.10) gives $\omega_n \approx 0.36$ r/s, $\zeta \approx 0.38$, $T_{\mathrm{R}} \approx 0.49$ s and $K = 0.083$ p.u. as the estimated parameters for the grid, averaged over the 24-h record. When de-normalised, the value of the grid 'stiffness' (*K*) is 723 MW/0.1 Hz. Later work using an ARIMAX model [18] has shown that this is on the high side for the National Grid in Great Britain, but not unreasonably so given the variance on estimating this parameter.

It is concluded that this method forms the basis for estimating the parameters of the grid in terms of a simple linear model. Moreover, as will be seen in Sect. 10.5, it implicitly provides the recursive predictions of grid frequency that are required in Fig. 10.3.

10.4 Predictive Feed-Forward

In this section, the idea of disturbance feed-forward is summarised. First, a hypothetical (but unrealisable) version of the feed-forward controller is derived based on the supposition that the external load disturbance ΔP_{L} can be measured. A practical version of the controller is then developed.

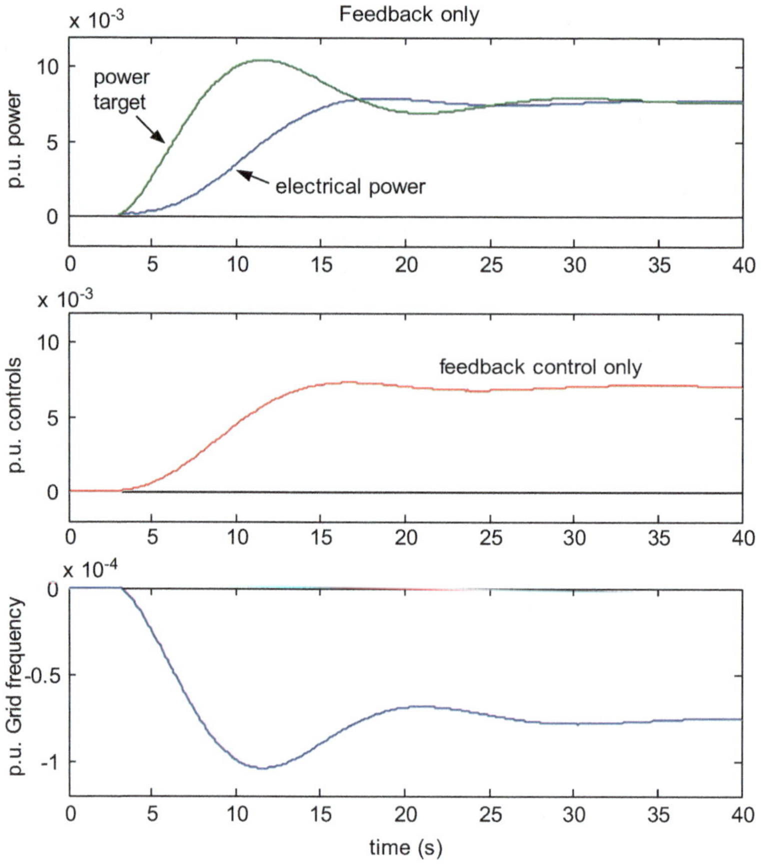

Fig. 10.6 Simulation results with only the feedback loop connected

Suppose that in the model of Fig. 10.3 a step increase in load ΔP_L occurs, causing the grid frequency to drop. This can be simulated by substituting the transfer functions derived in the previous two sections into the corresponding blocks in Fig. 10.3.

The simulation is run first with only feedback control and a droop gain $\alpha = 0.01$. Figure 10.6 shows the response to a step load applied at $t = 3$ s where the grid frequency is seen to fall according to the second-order response of (10.19). The corresponding power target is simply calculated from (10.4). With the chosen values of PI gains, the generated electrical power tracks the power target poorly and a substantial delay between the target and generated power is evident. It may be thought that the remedy is simply to increase the feedback gains but this would jeopardise system stability with longer water starting times (T_W), such as occur when multiple generators are online. A fixed gain governor is usually tuned conservatively so that it can be relied upon at all operating points.

Suppose now that it was somehow possible to forecast the incidence of a load disturbance before it was applied to the grid. Could the process of compensating

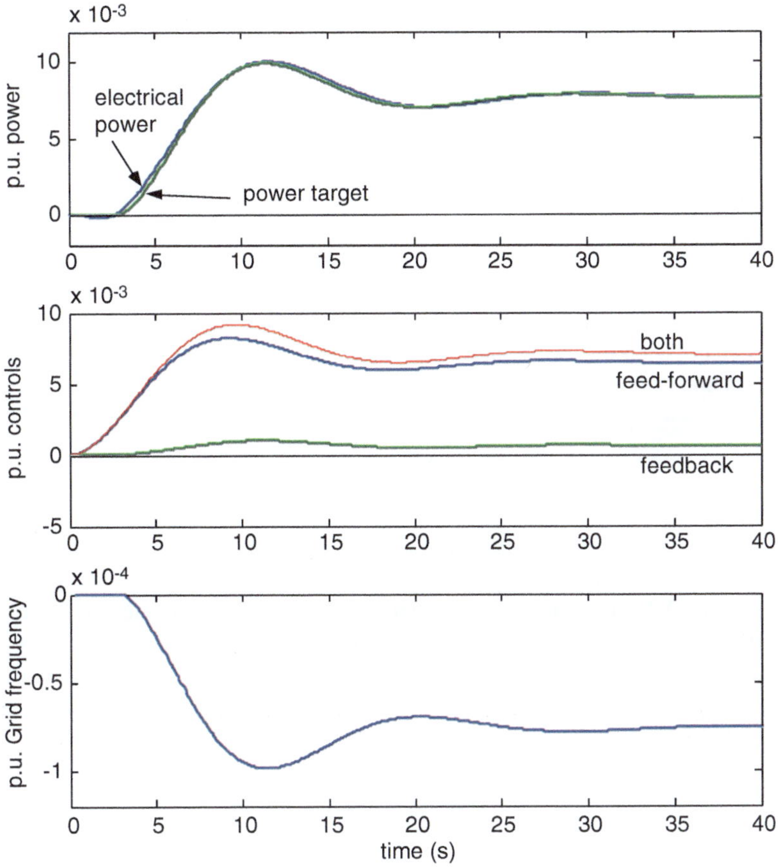

Fig. 10.7 Ideal disturbance feed-forward using a prescient controller

for the forthcoming disturbance be started early in order to circumvent the delay imposed by the hydraulics? The effectiveness of such a noncausal controller is demonstrated in Fig. 10.7, where the load step is applied to the grid model at $t = 3$ s as previously but as a signal by the feed-forward controller at $t = 0$. The feed-forward component generates ΔP_e that anticipates and cancels the disturbance. The feedback control is relegated to a minor corrective role, taking account of any differences between the real plant and its assumed model. Adjustment of the prediction time and feed-forward gain yields excellent tracking of the power target.

Clearly, this hypothetical controller is impossible to achieve so the prescience of the feed-forward controller must be approximated by a feasible controller. Of course, this will not be as effective as the ideal case but substantial gains may nevertheless be obtained.

The procedure for deriving a feed-forward disturbance controller is described by Goodwin et al. [126] and was used by Jones and Mansoor [127] but the same result is derived less formally here. Suppose that the hypothetical feed-forward loop in

Fig. 10.3 is connected so that the load disturbance ΔP_{L} is filtered by the transfer function $G_{\mathrm{ff}}(s)$ to give the feed-forward signal ΔP_{ff}:

$$\Delta P_{\mathrm{ff}} = G_{\mathrm{ff}} \Delta P_{\mathrm{L}}. \tag{10.20}$$

Let the plant transfer function be $G_{\mathrm{P}} = G_{\mathrm{v}} G_{\mathrm{H}} G_{\mathrm{T}}$. Assuming (as in Fig. 10.7) that the contribution made by the feedback control component is negligible compared to the feed-forward component:

$$\Delta P_{\mathrm{e}} = G_{\mathrm{P}} \Delta P_{\mathrm{ff}}. \tag{10.21}$$

For perfect tracking, we equate the generated power to its target, $\Delta P_{\mathrm{e}} = \Delta P_{\mathrm{targ}}$ in (10.21) and then use (10.4):

$$\Delta P_{\mathrm{e}} = G_{\mathrm{P}} \Delta P_{\mathrm{ff}} = -\frac{\Delta \omega_{\mathrm{G}}}{\alpha} = \Delta P_{\mathrm{targ}}$$

$$\therefore \Delta P_{\mathrm{ff}} = -\frac{1}{G_{\mathrm{P}}} \frac{1}{\alpha} \Delta \omega_{\mathrm{G}}. \tag{10.22}$$

Equation 10.22 indicates that the feed-forward signal is formed by filtering the measured frequency through the inverse of G_{P}. Now G_{P}, which consists of the hydrodynamics and gate vane hydraulics blocks, is essentially a delay. A reasonable approximation to $1/G_{\mathrm{P}}$ will be a negative delay or, in other words, a prediction. Equation 10.22 can be re-written as

$$\Delta P_{\mathrm{ff}} = -\frac{K_{\mathrm{ff}}}{\alpha} \Delta \hat{\omega}_{\mathrm{G}}, \tag{10.23}$$

where $\Delta \hat{\omega}_{\mathrm{G}}$ is a *k-step-ahead prediction* of the frequency error and the gain K_{ff} has been introduced as a tuning parameter.

10.5 Recursive Frequency Prediction

The prediction required in (10.23) is easily generated from the ARMAX model using (10.15). For the ARMAX323 model, the one-step-ahead predictor is given by

$$\begin{aligned}
\Delta \hat{\omega}_{\mathrm{G}}(k) = & -2.643 \Delta \omega_{\mathrm{G}}(k-1) + 2.403 \Delta \omega_{\mathrm{G}}(k-2) - 0.759 \Delta \omega_{\mathrm{G}}(k-3) \\
& + 0.00959 \Delta P_{\mathrm{e}}(k-1) \\
& - 0.00958 \Delta P_{\mathrm{e}}(k-2) - 1.131[\Delta \omega_{\mathrm{G}}(k-1) - \Delta \hat{\omega}_{\mathrm{G}}(k-1)] \\
& + 0.16[\Delta \omega_{\mathrm{G}}(k-2) - \Delta \hat{\omega}_{\mathrm{G}}(k-2)] + 0.192[\Delta \omega_{\mathrm{G}}(k-3) - \Delta \hat{\omega}_{\mathrm{G}}(k-3)].
\end{aligned} \tag{10.24}$$

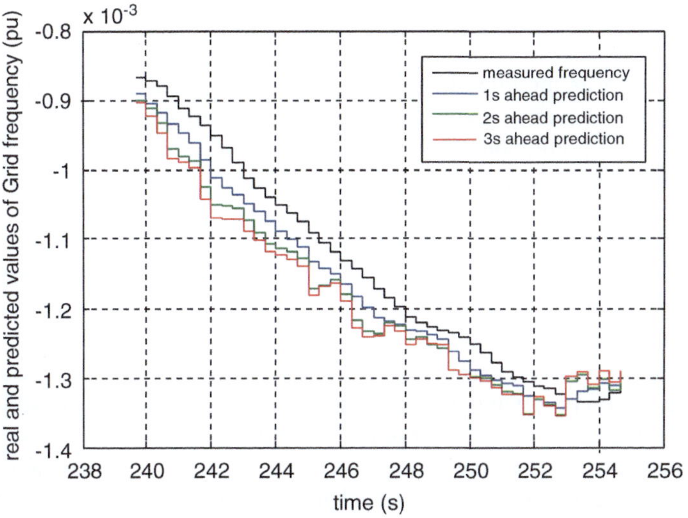

Fig. 10.8 Comparison of 1, 2 and 3 s-ahead predictors with the real measured frequency based on data re-dampled at 1/3 s intervals

It is straightforward to iterate Eq. 10.24 forward in time to obtain two-step-ahead predictors etc.

Applying (10.24) to the power and frequency data of Fig. 10.4 allows the predictions $\hat{\omega}_{\mathrm{G}}$ to be made and their accuracy to be determined by comparison with the actual values $\Delta\omega_{\mathrm{G}}$. At this stage, a detailed assessment [127] revealed that the prediction accuracy could be improved by re-sampling the data from 1 to 1/3 s (by interpolation with pre-filtering) and calculating a new ARMAX model. Following this procedure, an ARMAX969 model was obtained for the complete, re-sampled record of Fig. 10.4 – this order was chosen simply to match its maximum delays to the original ARMAX323 model.

The quality of the predictions may be gauged from Fig. 10.8 where the 1, 2 and 3 s-ahead predictions have been plotted at the time they were produced. The ideal result would therefore be the plot of true values reproduced 1, 2 and 3 s respectively to its left. Clearly this is not the case because the prediction quality deteriorates with length of look-ahead. Figure 10.8 shows that there is good agreement between the 1 s-ahead prediction and the value subsequently measured. The 2 s-ahead predictions are more noisy and also 'under-predict' by about 1/3 s. This becomes a second or more in the case of the 3 s-ahead predictor. This is likely to affect the feed-forward controller adversely when operating with a high value of T_{W}. It is concluded that the ARMAX969 model is capable of predicting what the grid frequency will be, a short time in the future, and can be used as a feed-forward signal.

10.6 Results

The simulation was modified to include the practical feed-forward loop shown in Fig. 10.3. The ‘frequency prediction’ block consists of the ARMAX969 predictor set for 2 s-ahead prediction. The water starting time was set at $T_W = 1$ s and the grid represented by its continuous-time model. For the purposes of comparison, the simulation was run with the feed-forward gain set at $K_{ff} = 0$ and $K_{ff} = 65$.

The extract in Fig. 10.9 shows that the frequency predictor anticipates the actual value quite well and is especially effective on trends lasting several seconds. However, it is quite a ‘noisy’ signal, as would be expected, and tends to overshoot when a sudden change of slope occurs in the actual frequency. Again, this would be expected because the only circumstance in which a rapid change can be anticipated is when the station producing the electrical power ΔP_e is itself responsible (e.g. due to another generator being brought online or offline).

This signal is used for feed-forward according to (10.23) via the gain K_{ff} which, together with the prediction look-ahead time, form a pair of tuning parameters for the feed-forward control. How these parameters affect the tracking error is considered in more detail by Jones and Mansoor [127].

Figure 10.10 is an extract from the power record over the same interval. When the feed-forward is absent, the electrical power tracks poorly – there is a significant phase lag and the peak amplitudes are much smaller than the target. The tracking is much improved by addition of the feed-forward signal, although the electrical power now exhibits more short-term variation.

A quantitative assessment of the effect of feed-forward is obtained by computing the integral square error between the delivered power and the target [97] and then taking the RMS:

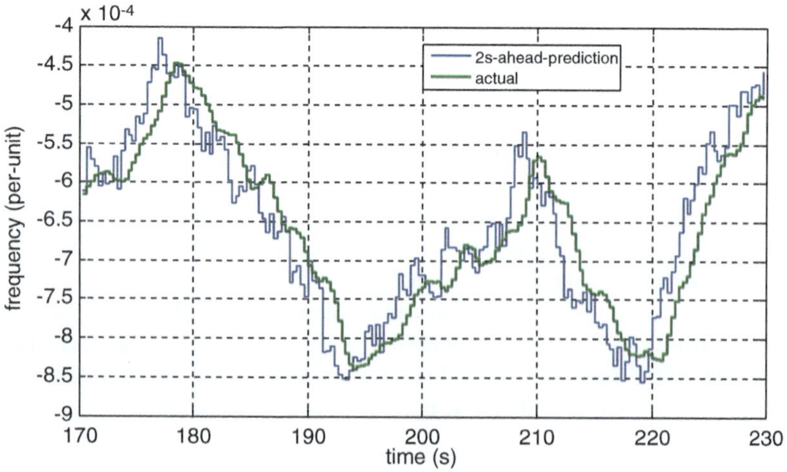

Fig. 10.9 Comparison of 2 s-ahead predicted frequency with the actual value

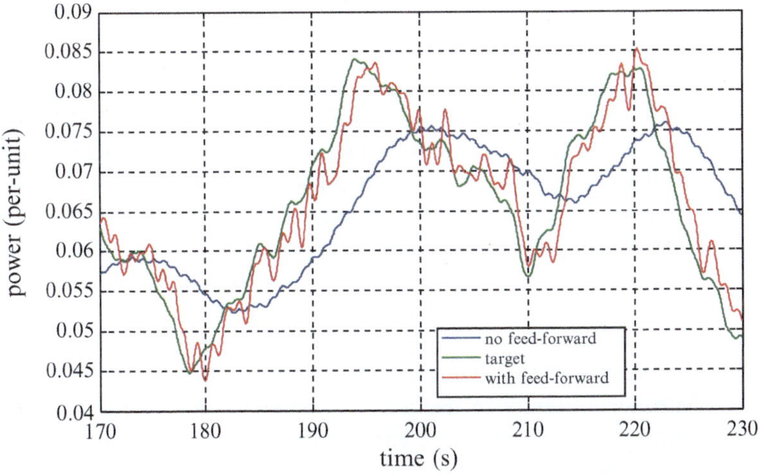

Fig. 10.10 Comparison of the generated electrical power with the target for feed-forward gains of 0 and 65

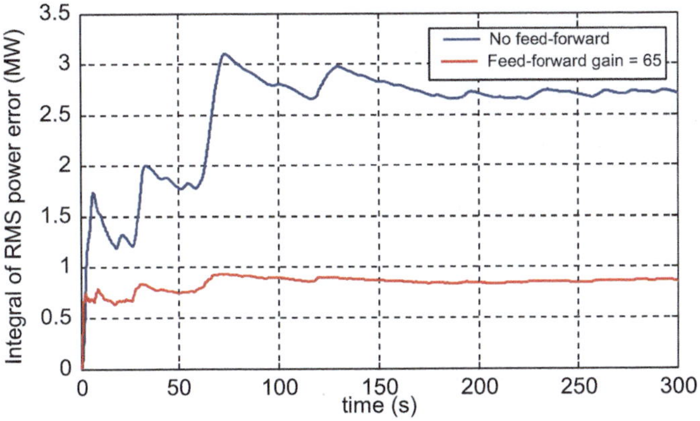

Fig. 10.11 RMS value error between the electrical power delivered and the power target

$$E_{\rm rms}(T) = \frac{1}{T}\sqrt{\int_0^T \left(\Delta P_{\rm targ}(t) - \Delta P_{\rm e}(t)\right)^2 {\rm d}t}. \tag{10.25}$$

Figure 10.11 shows that the power error settles to 2.7 MW RMS (on a 300 MW base) when feedback alone is used but this reduces to 0.88 MW RMS when feed-forward is introduced, a 67% improvement in tracking error.

10.7 Conclusions

In this chapter, it has been shown that predictive feed-forward can achieve a substantial improvement in delivering target power when a hydroelectric station is operating in frequency-control mode. Further simulation shows that the feed-forward controller is most effective at reducing power tracking error for low values of water starting time (T_W), when the frequency predictions $\Delta\hat{\omega}_G$ are at their best. Fortunately, this coincides with the common case of two to three generators (of the six available) being in operation simultaneously. However, modest but worthwhile improvement is also obtained for higher values of T_W.

An advantage of the feed-forward controller is that it is relatively easy to implement and commission. It does not require any changes to the well-tested feedback controller currently in use and can be switched in or out as necessary during operation. This makes testing easy and relatively safe, which is important when there are severe financial consequences if the plant should fail during operation. A possible drawback is that the controller produces more movement of the guide vane and could contribute to mechanical wear.

Chapter 11
Model Predictive Controller

11.1 Introduction

Model predictive control (MPC) methods, with special emphasis on generalised predictive control (GPC), are discussed in this chapter. Various studies [99, 132–135] have shown that this controller is a good candidate to improve performance of power plants since it provides an integrated approach to station control and also because it has the following characteristics:

- Deals with multivariable processes
- Takes actuator limitations into account
- Allows the process to run near constraints
- Has a short updating time

Section 11.2 presents a descriptive introduction of MPC and also discusses some MPC approaches and their applications in electrical power generation. The derivation for the GPC control is shown in Sect. 11.3 and is followed by some tuning guidelines for SISO models in Sect. 11.4. The tuning guidelines for MIMO models are discussed in Sect. 11.5, also this section shows a comparison between MPC and PI controllers. Finally, some conclusions are drawn in Sect. 11.6.

11.2 Model Predictive Control in Electric Power Generation

11.2.1 Model Predictive Control Elements

MPC is a general name for those methods that find the future control signals by looking for the minimum of a cost function over a fixed horizon of prediction, Fig. 11.1. The receding horizon for predictive control is like the 'real' horizon for a person who is walking; the horizon moves as they walk. The principle of the method is to calculate, using a mathematical model of the plant, the predicted output signal

G.A. Munoz-Hernandez et al., *Modelling and Controlling Hydropower Plants*, Advances in Industrial Control, DOI 10.1007/978-1-4471-2291-3_11,

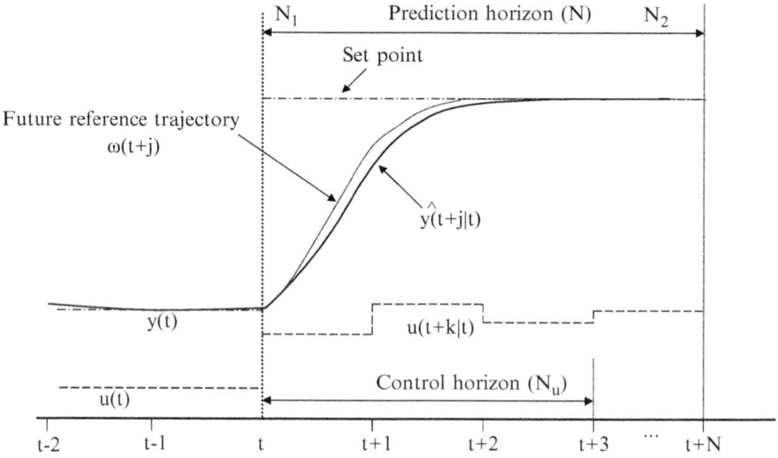

Fig. 11.1 MPC strategy

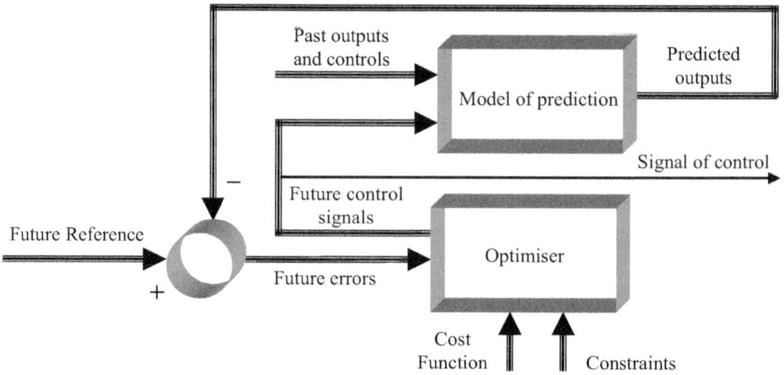

Fig. 11.2 General schema of a MPC

over some horizon into the future [123, 136, 137]. The control signal is calculated to minimise a defined performance index. Predictive control can also calculate the future reference trajectory that the output of the system must follow; this signal can be the operator setpoint or an approximation of it, for instance, in the case of a 'smooth' transition, Fig. 11.1.

The criterion for optimization is used to calculate the control signal. The criterion normally used is a quadratic function of the deviation of the predicted output signal from a future reference trajectory. The effort of control is typically included in the criterion for optimization. The control actions are produced in the optimizer, Fig. 11.2.

There are various factors that determine the complexity of the optimization problem, some of them are the number of variables and the size of the horizon and control predictions. The use of constraints also increases significantly the difficulty of finding the optimal solution. Only the first control signal is sent to the process while the remaining control signals that have been calculated are not used because, at the next sampling instant, a new output $y(t+1)$ is already known.

The prediction model is used to calculate the predicted outputs based on past inputs and outputs and future inputs (control signals), from the optimizer, Fig. 11.2. Different predictive control methods use different models of prediction; however, the process model must be able to represent the dynamics of the process in order to accurately predict the future outputs without consuming excessive computational time.

11.2.2 Brief Review of Some MPC Approaches

There are many variants of MPC, for example:

- Generalised predictive control (GPC)
- Dynamic matrix control (DMC)
- Extended prediction self-adaptive control (EPSAC)
- Model algorithmic control (MAC)
- Predictive functional control (PFC)
- Quadratic dynamic matrix control (QDMC)
- Sequential open loop optimisation (SOLO)

Several years before the first publications of predictive control appeared, industrial practitioners implemented the first applications of this controller [137]. Several people made the first advances in predictive control almost simultaneously.

Model predictive heuristic control was proposed by Richalet et al. in 1978 [138], and they called it a new method of digital process control. In this approach, the MIMO system is represented by its impulse responses; a long-range prediction by the digital control uses these responses online. The behaviour of the closed-loop system is set by reference trajectories that are initiated at each sample time. The control variables are computed using a heuristic method.

Predictor-based self-tuning control was introduced by Peterka in 1984 [139], which was defined as a time-domain method of quadratic-optimum control. In that approach the strategy of control was designed to minimize the expected error using a quadratic criterion over a control horizon. The system is modelled using two representations named positional and incremental. The positional model uses past output and input signals plus a noise signal. The incremental model uses incremental past and input signals plus an incremental noise signal. Peterka called it 'a new numerical method for LQ-optimum control synthesis', suggesting that the incremental model rather than the positional be used, because this has an integral action.

Dynamic matrix control (DMC) was proposed by Cutler and Ramaker, first named Cutler's method [136, 140]; this method uses the step response to model the process. It is assumed that the process is open-loop stable and has no integrators. In this approach, the system is modelled only taking into account the first N step response terms. The goal is to drive the output close to the setpoint by minimising a quadratic function that includes a penalty term on the input moves. As in other methods, only the first element of the computed control signal is sent to the plant. The calculation of the output signal is made every sample time.

The GPC method was proposed by Clarke et al. [141]. It finds the future control signals by looking for the minimum of a cost function over a horizon of prediction. The function to be optimised is a quadratic function of some sequence of reference and predicted outputs, both signals being calculated over a defined horizon of prediction.

11.2.3 Applications of MPC in Power Plants

Several applications of MPC to conventional power plant have been reported over the years. Prasad et al. [132] describe controlling a thermal (boiler-turbine) power plant using a hierarchical MPC approach. The strategy consists of two-level control structure: a lower level of conventional PI regulators and a higher level of a nonlinear physical model predictive controller (NPMPC). The stabilisation of the drum-boiler dynamics is performed by the lower level PI loops; it also allows faster governor valve action for power and grid-frequency regulation. An optimal load demand (or setpoint) transition is provided by the higher level NPMPC, with the effective handling of plant-wide interactions and system disturbances. The authors report excellent disturbance-rejection properties and alleviation of plant-wide interactions for simulations performed on a 200 MW power plant. Rossiter et al. [133] also consider the control of a fossil-fired power station where they are faced with a design constraint due to limitations of the hardware available for implementation. They proceed to derive a slightly suboptimal algorithm called efficient constrained GPC (ECGPC) with the interesting conclusion that a marginal improvement in the output of the plant is achieved, compared to using a PID, but with much reduced control activity. Ramond et al. [142] proposed MPC for controlling the level of the lake which supplies a hydroelectric plant. Sansevero and Botura [134] considered a small-perturbation SISO model relating turbine speed to input power. Their results indicate that, provided the prediction horizon has adequate length (compared with T_W), the MPC controller can be tuned less conservatively than PI while retaining closed loop stability. They concluded that even linear MPC is competitive with the standard PI controllers installed at two stations currently in operation.

11.3 Generalised Predictive Control

This section summarises the theory of one of the most popular predictive control methods, generalised predictive control (GPC). Clarke et al. proposed this method and it has been successfully implemented in many industrial applications [141, 143].

The objective of GPC is to drive future plant outputs close to a reference trajectory taking into account the control activity required to achieve this goal. As was stated before, this predictive control method finds the future control signals by computing the minimum of a cost function over a horizon of prediction. A quadratic function, of the difference (error) between some sequence of reference and the predicted output, is optimised; both signals are calculated over a prediction horizon. Normally the cost function includes a term to reduce or increase the control effort. As with other methods of predictive control, this method of control can be applied to linear system with the following characteristics [133, 136, 144–148]:

- A nonminimum-phase plant
- An open-loop unstable plant or plant with insufficiently damped poles
- A plant with variable or unknown dead time
- A plant with unknown order

11.3.1 Unconstrained GPC

The GPC calculates the future control signals by looking for the minimum of a quadratic cost function, Eq. 11.1, over one horizon of prediction:

$$J(N_1, N_2, N_u) = \sum_{j=N_1}^{N_2} [\hat{y}(t+j\,|t) - w(t+j)]^{\mathrm{T}} \bar{Q}\, [\hat{y}(t+j\,|t) - w(t+j)] + \sum_{j=1}^{N_\mathrm{u}} [\Delta u\,(t+j-1)]^{\mathrm{T}} \bar{R}\, [\Delta u\,(t+j-1)]\,, \tag{11.1}$$

where

$\hat{y}(t+j\,|t)$ – optimum system predicted output j steps ahead calculated at time t.
Δ – $(1 - q^{-1})$ operator.
N_1, N_2 – minimum and maximum of the prediction horizon.
N_u – control horizon.
$\bar{Q}$ and $\bar{R}$ – positive definite weighting matrices.
$w(t+j)$ – future reference trajectory, for constant references:
$w(t+k) = \alpha w(t+k-1) + (1-\alpha) r(t+k)$, with $1 > \alpha \geqslant 0$ and $r(t+k)$ is the real reference.

The future reference trajectory, $w(t+j)$, does not necessarily has to coincide with the real reference $r(t+k)$; it is a common practice that a smooth approximation from the current value of the output $y(t)$ towards the known reference $r(t)$ is performed by means of the first-order system, as in this case, where the parameter α constitutes an adjustable value that will influence the system's dynamic response, the closer the value of α to 1, the smoother the approximation.

A controlled auto-regressive and integrated moving average model (CARIMA), Eq. 11.2, is used to model the plant:

$$\boldsymbol{A}(q^{-1})y(t) = q^{-\mathrm{d}}\boldsymbol{B}(q^{-1})u(t-1) + \boldsymbol{C}(q^{-1})\frac{e(t)}{1-q^{-1}}, \tag{11.2}$$

where q^{-1} is the backward-shift operator, d is the dead time of the system, $y(t)$ is the $n \times 1$ output vector, $u(t)$ is the $m \times 1$ input vector, $e(t)$ is the $n \times 1$ noise vector at time t and $\boldsymbol{A}(q^{-1})$ and $\boldsymbol{C}(q^{-1})$ are monic polynomial matrices of size $n \times n$, and $\boldsymbol{B}(q^{-1})$ is a polynomial matrix of size $n \times m$, defined as

$$\begin{aligned}
\boldsymbol{A}\left(q^{-1}\right) &= I_{n\times n} + A_1 q^{-1} + A_2 q^{-2} + \cdots + A_{na} q^{-n_a} \\
\boldsymbol{B}\left(q^{-1}\right) &= B_0 + B_1 q^{-1} + B_2 q^{-2} + \cdots + B_{nb} q^{-n_b} \\
\boldsymbol{C}\left(q^{-1}\right) &= I_{n\times n} + C_1 q^{-1} + C_2 q^{-2} + \cdots + C_{nc} q^{-n_c}.
\end{aligned}$$

To derive a j-step ahead predictor of $\boldsymbol{y}(t+j)$ let us multiply (11.2) by $\Delta E_{\mathrm{j}}(q^{-1})q^{\mathrm{j}}$, where $E_{\mathrm{j}}(q^{-1})$ is a polynomial of degree j^{-1}, and consider $C(q^{-1}) = 1$, then

$$\begin{aligned}
\Delta E_j(q^{-1})q^j \boldsymbol{A}(q^{-1})y(t) &= \Delta E_j(q^{-1})q^j q^{-\mathrm{d}}\boldsymbol{B}(q^{-1})u(t-1) \\
&\quad + \Delta E_j(q^{-1})q^j \frac{e(t)}{1-q^{-1}}.
\end{aligned}$$

Defining $\tilde{A}(q^{-1}) = \Delta \boldsymbol{A}(q^{-1})$

$$\begin{aligned}
E_j(q^{-1})\tilde{A}(q^{-1})y(t+j) &= E_j(q^{-1})\boldsymbol{B}(q^{-1})\Delta u(t+j-d-1) \\
&\quad + E_j(q^{-1})e(t+j).
\end{aligned} \tag{11.3}$$

Considering the following Diophantine equation:

$$I_{n\times n} = E_j(q^{-1})\tilde{A}(q^{-1}) + q^{-j}F_j(q^{-1}) \tag{11.4}$$

Equation 11.3 is written as

$$\begin{aligned}
y(t+j) &= F_j(q^{-1})y(t) + E_j(q^{-1})\boldsymbol{B}(q^{-1})\Delta u(t+j-d-1) \\
&\quad + E_j(q^{-1})e(t+j).
\end{aligned} \tag{11.5}$$

The noise terms, $e(t+j)$, in Eq. 11.5 are all in the future because the degree of polynomial $E_j(q^{-1})$ is $j-1$. The best prediction of $\boldsymbol{y}(t+j)$ is therefore

$$\hat{y}(t+j\,|t) = G_j(q^{-1})\Delta u(t+j-d-1) + F_j(q^{-1})y(t).$$

where $G_j(q^{-1}) = E_j(q^{-1})B(q^{-1}) = \sum_{i=0}^{j-1} G_i q^{-i}$ then

$$\begin{bmatrix} \hat{y}(t+1|t) \\ \hat{y}(t+2|t) \\ \hat{y}(t+j|t) \\ \hat{y}(t+N|t) \end{bmatrix} = \begin{bmatrix} G_0 & 0 & 0 & 0 \\ G_1 & G_0 & 0 & 0 \\ G_{j-1} & G_{j-2} & G_0 & 0 \\ G_{N-1} & G_{N-2} & \cdots & G_0 \end{bmatrix} \begin{bmatrix} \Delta u(t) \\ \Delta u(t+1) \\ \Delta u(t+j-1) \\ \Delta u(t+N-1) \end{bmatrix} + \begin{bmatrix} f_1 \\ f_2 \\ f_j \\ f_N \end{bmatrix}$$

or:

$$y = Gu + f \tag{11.6}$$

f is the free response term and can be calculated recursively by

$$f_{j+1} = q\left(I - \tilde{A}\left(q^{-1}\right)\right) f_j + \boldsymbol{B}\left(q^{-1}\right) \Delta u\left(t + j\right) \tag{11.7}$$

with $f_0 = y(t)$ and $\Delta u\,(t + j) = 0$ for $j \geqslant 0$.

If after N_u steps of control, the control signal variation, between consecutive steps, is slight, then N_u could be considered as a constant, the set of predictions will be

$$y_{N_{12}} = \left[\hat{y}(t + N_1\,|t)^{\mathrm{T}}, \hat{y}(t + N_1 + 1\,|t)^{\mathrm{T}}, \ldots, \hat{y}(t + N_2\,|t)^{\mathrm{T}}\right]$$

using this equation affecting the cost function, $y_{N_{12}}$ can be expressed as

$$y_{N_{12}} = G_{N_{12}} u_{N_u} + f_{N_{12}}$$

$$u_{N_u} = \left[\Delta u(t)^{\mathrm{T}}, \ldots, \Delta u(t + N_u - 1)^{T}\right]^{\mathrm{T}}, \tag{11.8}$$

where

$$f_{N_{12}} = \left[f_{N_1}{}^{\mathrm{T}}, f_{N_1+1}{}^{\mathrm{T}}, \ldots, f_{N_2}{}^{\mathrm{T}}\right]^{\mathrm{T}}$$

and $G_{N_{12}}$ is the following submatrix of G, with $G_i = 0$ for $i < 0$.

$$G_{N_{12}} = \begin{bmatrix} G_{N_1-1} & G_{N_1-2} & \cdots & G_{N_1-N_u} \\ G_{N_1} & G_{N_1-1} & \cdots & G_{N_1+1-N_u} \\ \vdots & \vdots & \ddots & \vdots \\ G_{N_2-1} & G_{N2-2} & \ldots & G_{N_1-N_u} \end{bmatrix}$$

Equation 11.1 can be rewritten as

$$J = (G_{N_{12}} u_{N_u} + f_{N_{12}} - w)^{\mathrm{T}} \bar{Q}\,(G_{N_{12}} u_{N_u} + f_{N_{12}} - w) + u_{N_u}^{\mathrm{T}} \bar{R} u_{N_u}, \tag{11.9}$$

where $\bar{Q}$ and $\bar{R}$ (the positive definite weighting matrices) are diagonal matrices:

$$\bar{Q} = \mathrm{diag}\,(Q, \ldots, Q) \quad \text{and} \quad \bar{R} = \mathrm{diag}\,(\lambda, \ldots, \lambda)\,.$$

If constraints are not considered, the optimum control signal can be calculated by

$$u = \left(G_{N_{12}}^{\mathrm{T}} \bar{R} G_{N_{12}} + \bar{Q}\right)^{-1} G_{N_{12}}^{\mathrm{T}} \bar{R}\,(w - f_{N_{12}})\,. \tag{11.10}$$

11.3.2 Constrained GPC

When an equation of control is formulated, it is common to assume that all signals possess an unlimited range. Nevertheless, this is not realistic because, in the real world, all plants are subject to constraints. There are constraints for constructional and safety reasons. Levels in tanks and pressure in vessels are examples of process variables that are often constrained in value for safety reasons. Limited values for actuators input and outputs and limited slew rate are some examples of constructional constraints [149–151].

Normally three kinds of constraints are considered in predictive control: limits on control signals (saturation), rate limits on control signals and limits on output signals. These constraints can be represented by the following equations:

$$\begin{aligned} &\underline{u} \leqslant u(t) \leqslant \overline{u} \quad \forall t \\ &\underline{\Delta u} \leqslant u(t) - u\,(t-1) \leqslant \overline{\Delta u} \quad \forall t \\ &\underline{y} \leqslant y(t) \leqslant \overline{y} \quad \forall t. \end{aligned}$$

If Eq. 11.1 is expressed in the following form:

$$J(u) = \frac{1}{2} u^{\mathrm{T}} H u + b^{\mathrm{T}} u + f_0 \tag{11.11}$$

where

$$\left.\begin{aligned} &f_0 = (f_{N_{12}} - w)^{\mathrm{T}} (f_{N_{12}} - w) \qquad b^{\mathrm{T}} = 2(f_{N_{12}} - w)^{\mathrm{T}} G_{N_{123}} \\ &\text{and} \quad H = 2\left(G_{N_{123}}^{\mathrm{T}} G_{N_{123}} + \lambda I\right) \end{aligned}\right.,$$

the predictive control formulation can be expressed as a quadratic programming problem, with

$$J(u) = \frac{1}{2} u^{\mathrm{T}} H u + b^{\mathrm{T}} u \tag{11.12}$$

as the function to optimize under the following constraints: $A_{\mathrm{qp}} u \leqslant b_{\mathrm{c}}$.

In Eq. 11.12 f_0 is not used because, at every stage of the optimization, it is a constant whose value does not depend on u. A_{qp} and b_{c} are the following matrix and vector of size $6N_{\mathrm{u}} \times m$ and $6N_{\mathrm{u}}$ respectively:

$$A_{\mathrm{qp}} = \begin{bmatrix} I \\ -I \\ I \\ -I \\ G_{N_{123}} \\ -G_{N_{123}} \end{bmatrix} \quad b_{\mathrm{c}} = \begin{bmatrix} \mathbf{1}\,(\bar{u} - u) \\ -\mathbf{1}\,(\underline{u} + u) \\ \mathbf{1}\Delta u\,\mathrm{max} \\ -\mathbf{1}\Delta u\,\mathrm{max} \\ \mathbf{1}\bar{y} - f \\ -\mathbf{1}\underline{y} + f \end{bmatrix} \quad \text{where } \mathbf{1} = \begin{bmatrix} 1 \\ 1 \\ \vdots \\ 1 \end{bmatrix}$$

11.4 Tuning Guidelines: SISO GPC

Controllers are designed to satisfy an ample range of specifications and their performances has a strong dependence on their parameters. The purpose is generally to achieve a balance between good sensitivity, input activity and speed of response [123]. The performance of the system is determined by the cost function to be optimised; in GPC a quadratic function (11.1) is utilised. The parameters to be tuned in the GPC controller are:

N_1,N_2 – minimum and maximum of the prediction horizon
N_{u} – control horizon
α – future reference trajectory weight factor
λ – cost function control signal weight factor

N_1 is fixed as 1 plus the dead time of the system, similar to the guidelines provided by Clarke et al. [141]. N_2 is set close the rise-time of the plant. N_{u} can be set to 1 to obtain a reasonable performance; however, increasing its value makes the control signal and the response of the system more active. Nonetheless, there is a point after which any further increase in N_{u} makes little difference to the response of the system [141, 152].

Many equations can be used to calculate the future reference trajectory [136, 141]; here the equation stated in Sect. 11.3 will be employed. Values of α near zero produce fast responses and values of α near to 1 produce slow responses. Here a value of $\alpha = 0$ is normally selected.

λ has a strong effect on the effort of control, low values of this parameter producing an 'active' control while high values produce a 'passive' control [123, 141, 152]. There is a relationship between the lowest feasible values of λ and the accuracy of the prediction model; only accurate prediction models support low values of λ and therefore an active control.

The prediction model plays a central role in all MPC approaches. Linear stable systems could be modelled with a first order plus dead time transfer function. On the other hand, nonlinear systems with complex dynamics must have an accurate prediction model in order to achieve a precise prediction [123, 136, 141].

11.4.1 Prediction Model

As mentioned in early sections, the performance of the GPC depends on the predictive model and control parameters selected. The process of selecting the parameters of the GPC controller utilised in this study begins with the selection of an approximated prediction model, for instance from the reaction curve. This technique is probably the most popular method used in industry for setting controllers and is used in the pretune stage of some commercial adaptive and auto-tuning regulators [87, 136]. Then with a high value of λ the horizon of prediction is fixed. It is desirable to maintain the latter parameter as low as possible in order to reduce computational time. The horizon of control is normally fixed to the same value as the horizon of prediction, but it can be fixed to a lower value or even $N_u = 1$ if the effort of control is not under consideration. Finally, λ is reduced to the lowest value possible; its optimum value depends on the characteristics of the system and the desired response [136].

As was stated before, the predictive model used by GPC in this chapter is obtained by the reaction curve technique; as a result, the prediction model is a first order plus dead time (FOPDT) transfer function. This simple model is computationally economical but cannot reproduce accurately some system's characteristic such as: NMP behaviour. However, it will be shown that its short-term predictive capability is adequate. Representing the plant's nonminimum phase response by a transport delay, the model obtained for Dinorwig on the basis of a 10–50% (0.1–0.5 p.u.) increase in power output is given by (11.13) with one unit operational, see Fig. 11.3 where an operating point of 0.1 p.u. was first established.

$$G(s) = \frac{1.12e^{-T_d s}}{1.2954s + 1}. \tag{11.13}$$

Converting to discrete time with a sample period $T_s = 0.25$ s and a transport delay of $T_d = 2.07$ s gives (11.14)

$$G(z) = \frac{0.197}{z^8(z - 0.8245)}. \tag{11.14}$$

Figure 11.4 shows the open loop step response of the linear SISO model and the prediction model for GPC (11.14). As the reaction curve model does not represent the NMP response, the first 3 s of simulation have a large error. On the other hand, the reaction curve model is close to the linear model after this delay.

11.4.2 Controller Parameters

Simulations were carried out to find the allowable range of parameters for the GPC algorithm when using (11.14) as the prediction model and $T_s = 0.25$. The

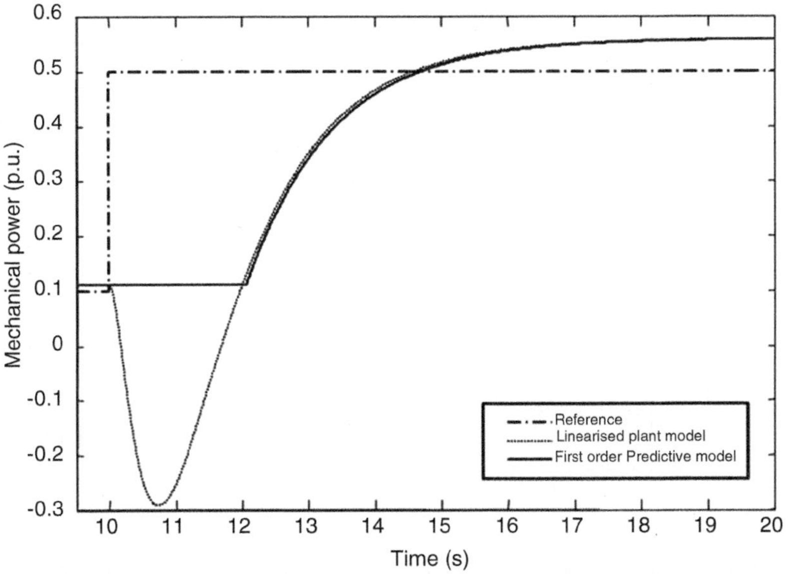

Fig. 11.3 Mechanical power output of the linearised plant model and prediction continuous time model to a 0.4 p.u. step in control u

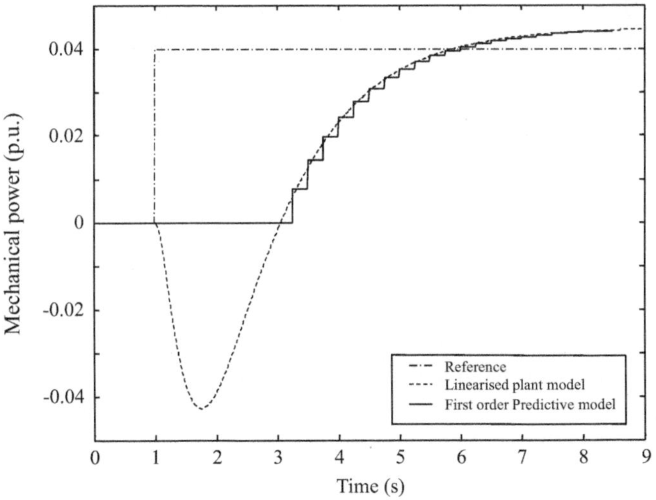

Fig. 11.4 Mechanical power output of the linearised plant model and prediction discrete time model to a 0.04 p.u. step in control u

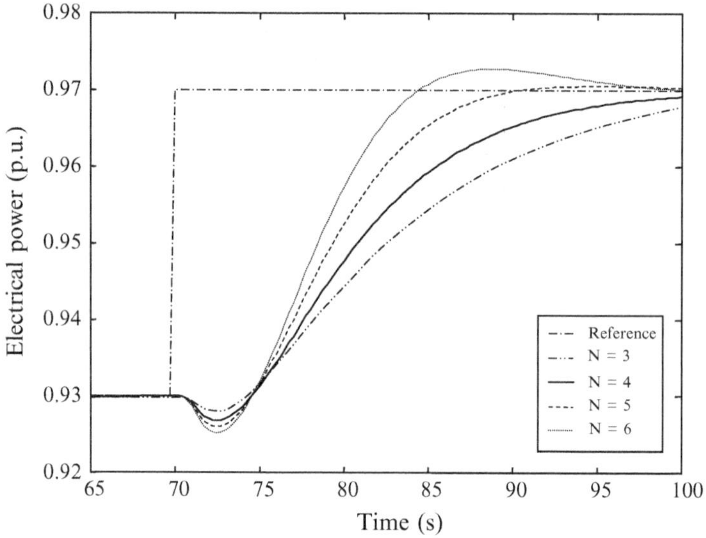

Fig. 11.5 Response of electrical power to a small step under GPC control with different values horizon of prediction N

hydroelectric plant (Dinorwig) was modelled as a linear system with six units operational. The effect of varying four parameters is studied below:

- N – horizon of prediction.
- λ – effort of control weighting factor.
- N_u – horizon of control.
- α – future reference trajectory-tracking factor.

11.4.2.1 Horizon of Prediction

Varying the horizon of prediction, N, leads to a variety of responses as shown in Fig. 11.5. In this simulation, the other parameters were fixed as follows: $N_u = 5$, $\lambda = 200$ and $\alpha = 0$. α is fixed to this value in order to eliminate the tracking process that produces a slower response. Short horizons of prediction generate inaccurate estimation of the future output signals. It is desirable that the selected value of N allows the future output to settle or at least to leave the NMP region, as in this case. A value of $N = 5$ was selected because it produces a fast response with modest overshoot.

11.4.2.2 Effort of Control Weighting Factor

With the horizons of prediction and control fixed to $N = N_u = 5$ and $\alpha = 0$, simulations were conducted in order to determine the value of λ. Low values of λ

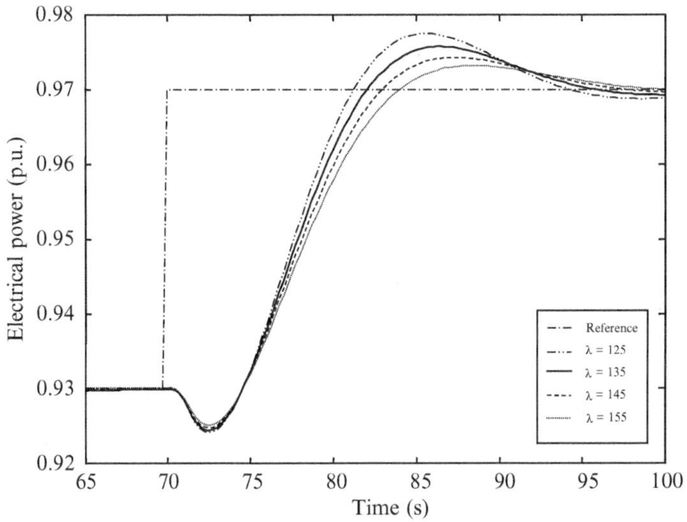

Fig. 11.6 Response of electrical power to a small step under GPC control with different values of effort of control weighting factor λ

make the responses faster but both under- and overshoot increase, Fig. 11.6. In order to obtain a response with an overshoot about 25% of the change in the input signal, the evaluate range of λ was $125 \leq \lambda \leq 155$. To some extent it acts as the inverse of the loop gain in a conventional compensator because, as discussed in Sect. 11.2, this parameter directly affects the value of the control signal in the quadratic cost function of GPC.

11.4.2.3 Horizon of Control

The control horizon is in general fixed to the same value as the prediction horizon. As shown in Fig. 11.7, when $N = 5$, $\lambda = 135$ and $\alpha = 0$, higher values of N_u yield a reduction of the overshoot without increasing the undershoot. It was found that values higher than 10 have no visible effect on the response; therefore, $N_u = 10$ was selected.

11.4.2.4 Future Reference Trajectory-Tracking Factor

Figure 11.8 shows that the system allows values of $0 \leq \alpha < 1$, which is the normal range of values for this parameter. A value of $\alpha = 0$ was selected because the tracking process, which reduces the speed of the future reference trajectory and the NMP response, was unnecessary due to the relatively small undershoot. Table 11.1 summarises the values of the parameters selected for the GPC.

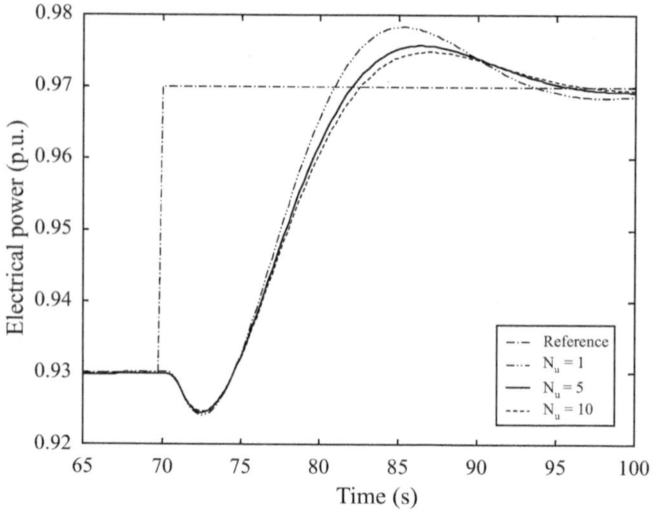

Fig. 11.7 Response of electrical power to a small step under GPC control with different values of horizon of control N_u

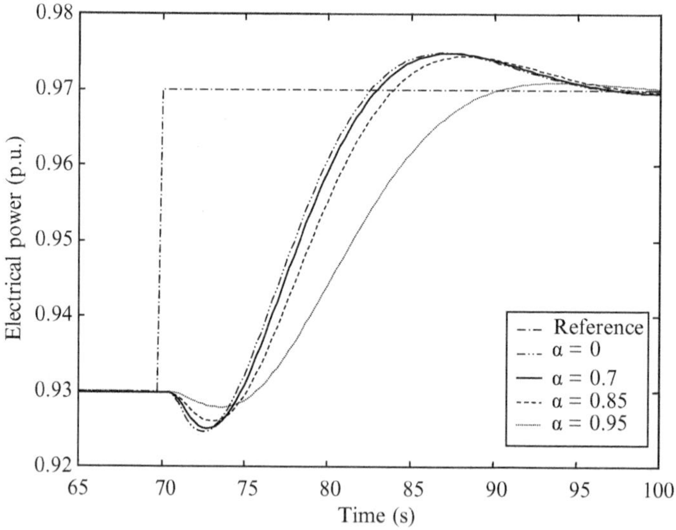

Fig. 11.8 Response of electrical power to a small step under GPC control with different values of future reference trajectory-tracking factor α

Table 11.1 Parameters of the GPC controller

	Compromise	Optimised One unit	Optimised Six units	Constrained
N	5	20	20	20
N_u	10	20	20	20
λ	135	150	600	20
α	0	0	0	0

11.5 Tuning Guidelines: MIMO GPC

In this section, the performance of GPC is evaluated when used with two MIMO models of a hydropower plant described in Sects. 4.3, 4.5 and 4.6:

- The MIMO linear model of section
- The nonlinear MIMO model of section, which includes the effect of an elastic water column

The MIMO linear model of a hydropower plant can represent the multivariable hydraulic interactions that take place between the penstocks and main tunnel, as well as the effects of the synchronisation of the generators. In hydraulic linear models, the water column is considered to be an inelastic fluid. This causes the effective water starting time to vary, depending on the number of units active. Consequently, the effects of controllers in the reduction of cross-coupling interaction and their performance with different number of units active can be evaluated.

The MIMO nonlinear elastic model of hydropower plants can be considered to be a 'full nonlinear' model because it integrates all the nonlinearities and multivariable effects discussed in Chap. 4. Again, the behaviour of the system with different operational conditions can be evaluated, where the 'operating point' now changes as the simulation proceeds. Also, both large and small signals can be used as inputs to the system. In summary, the scope of the simulation is greatly expanded and can be considered comparable to the 'real' plant. Nevertheless, it must be noted that even the MIMO nonlinear elastic model remains only a mathematical representation of the hydroelectric power plant. The analysis will be focused on Dinorwig Power Plant; the specific values of the parameters are taken from the work of Mansoor [33].

11.5.1 MIMO GPC

In Sect. 11.4, a model derived from the step response of the system, for a single active unit, was used as the prediction model. In this chapter, two types of predictive model are investigated. The first is a discrete-time analytical model, obtained using zero-order hold on the inputs, based solely on the hydraulic and guide-vane subsystems models of Chap. 4. For instance, in the case of six units operating at

$G_o = 0.95$ and a sample period $T_s = 0.25$ s the direct and cross-coupling transfer functions of the analytical model are:

$$G_6(z) = \frac{-0.2312z^3 + 0.5145z^2 - 0.1481z - 0.07865}{z^4 - 1.802z^3 + 1.093z^2 - 0.2616z + 0.02113} \tag{11.15}$$

$$X_6(z) = \frac{-0.04757z^3 - 0.02751z^2 + 0.06817z + 0.006913}{z^4 - 1.802z^3 + 1.093z^2 - 0.2616z + 0.02113}. \tag{11.16}$$

The second predictive model is obtained empirically from the reaction curve of the nonlinear nonelastic model using a small step (0.04 p.u.) when six units are active. In this model, the direct transfer function (11.15) is replaced by the simpler transfer function calculated from the reaction curve with six units operational but the cross-coupling transfer function (11.16) remains unchanged.

$$G_{6c}(z) = \frac{0.29044}{z^4 - 0.7422z^3} \tag{11.17}$$

The corresponding polynomials of a CARIMA model using (11.16) and (11.17) are:

$$A_6(z^{-1}) = 1 - 2.5442z^{-1} + 2.4304z^{-2} - 1.0728z^{-3} - 0.2153z^{-4} - 0.0157z^{-5} \tag{11.18}$$

$$B_6(z^{-1}) = 0.2904z^{-4} - 0.5234z^{-5} + 0.3175z^{-6} - 0.076z^{-7} + 0.0061z^{-8} \tag{11.19}$$

$$b_6(z^{-1}) = -0.0476z^{-1} + 0.0078z^{-2} + 0.0886z^{-3} - 0.0437z^{-4} - 0.0051z^{-5} \tag{11.20}$$

where A_6, B_6 are diagonal elements and b_6 are the off-diagonal elements.

Actually, the predictions produced by both models are similar and agree well with the MIMO linear model output as shown in Figs. 11.9 and 11.10, for the direct and cross-coupling responses, respectively. Both predictive models are derived for the extreme operational case, six units in operation at 95% output. The predictive model is fixed, regardless of how many units are actually in operation or their loading; this restriction will be relaxed in Chap. 12.

GPC controllers were designed using both Eqs. 11.15 and 11.17 as the direct transfer function in the predictive model. The results were similar, so the more parsimonious transfer function (11.17) was used for the remainder of the study. Reducing the complexity of the predictive model is very important in the constrained GPC (CGPC) case, which involves quadratic programming iterations. In order to minimise the computational time, this method is usually tuned using low values of control and prediction horizons. Table 11.2 summarises the values of the parameters for the GPC and CGPC controllers used in this section.

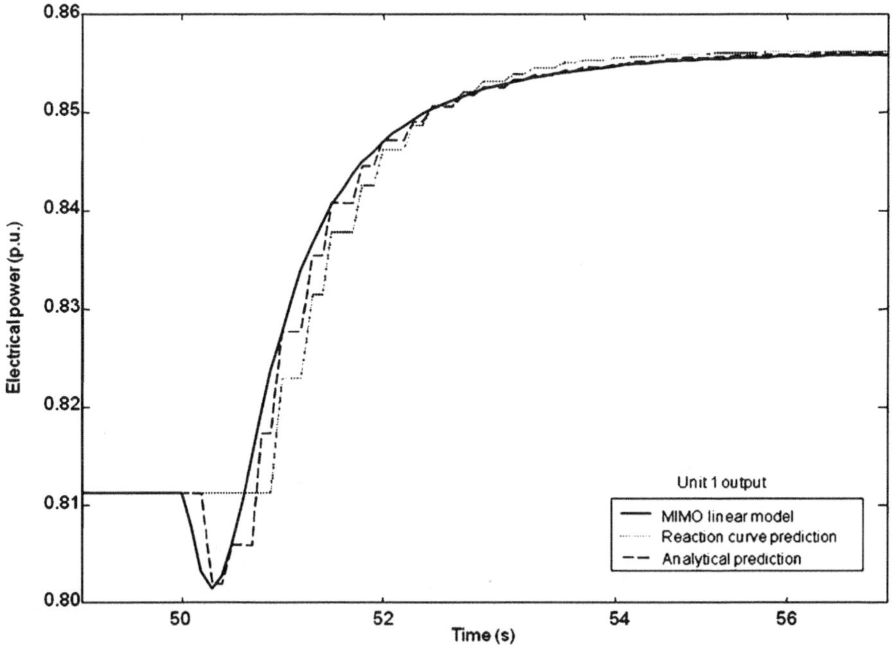

Fig. 11.9 Response of the plant model and the predictive models to a step in control, applied to unit 1, with the controls for units 2–6 fixed at their initial values

11.5.2 MIMO Linear Model

In order to evaluate the performance of the GPC, several simulations were carried out. Step power demand signals are applied to the MIMO linear model of Dinorwig with both the PI and GPC controllers.

11.5.2.1 Discrete-Time Transfer Function Matrix Dinorwig

The hydrodynamics block of the extended MIMO model has the ability to change the matrix transfer function, which relate mechanical power with gate opening, depending on the number of units active. The elements of the matrix transfer function ($G(s)$) are the direct transfer function (diagonal) $G_i(s)$ and cross-coupling transfer function (symmetric) $X_i(s)$. In Fig. 11.11, ΔG is the guide vane opening, ΔP_m is the turbine's mechanical power and ΔP_e is the electrical power. The current governor comprises an individual classical PI controller on each turbine. The rules to determine whether a unit is online or offline are:

- Unit n comes online when $P_{di} > 0$
- Unit n goes offline when $[(\Delta P_{di} = 0)$ and $(\Delta P_{mi}) = 0]$

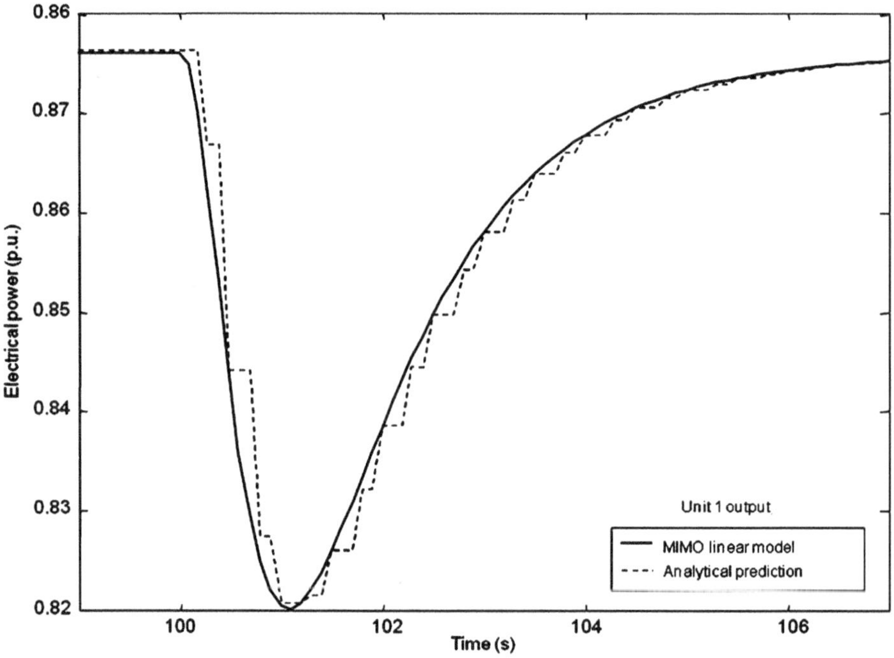

Fig. 11.10 Response of the plant model and the predictive models to a step in control, applied simultaneously to units 2–6, with the control for unit 1 fixed at its initial value

Table 11.2 Parameters of the GPC controller

	MIMO linear		MIMO nonlinear	
	Unconstrained (GPC)	Constrained (CGPC)	Unconstrained (GPC)	Constrained (CGPC)
N	40	40	40	40
N_u	20	10	40	10
λ	425	350	300	250
α	0	0	0	0

Figure 11.11 describes the MIMO linearised model when two units are active. If this diagram is extended to the case when n units are active, then the system can be described by Eq. 11.21:

$$\Delta P_m(s) = [G(s)]\,\Delta G(s), \tag{11.21}$$

where

$$G(s) = C\,[sI - A]^{-1}B + D \tag{11.22}$$

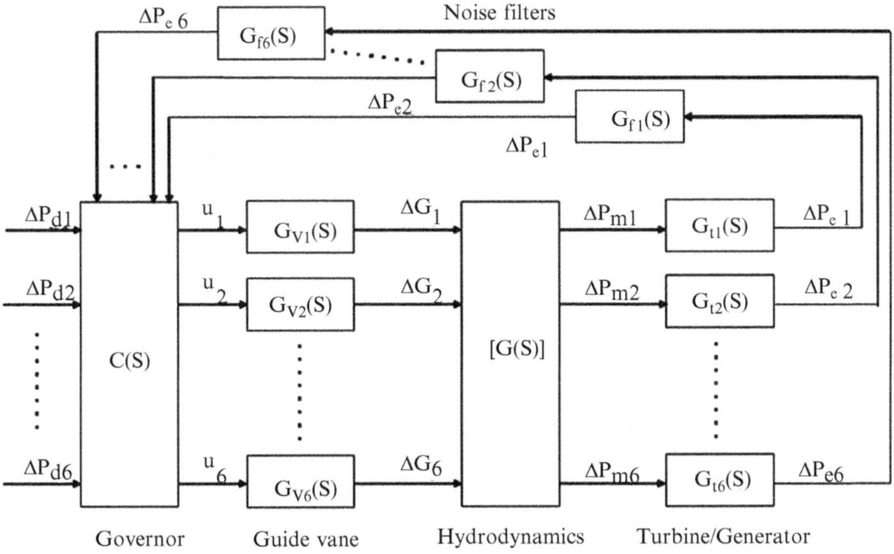

Fig. 11.11 MIMO model of the hydroelectric plant

and

$$C = I c_1 \tag{11.23}$$

$$c_1 = A_t \left(1 + \frac{2q_0}{G_0}\right)$$

$$D = I d_1 \tag{11.24}$$

$$d_1 = -\frac{2A_t q_0}{G_0}$$

$$A = -\begin{bmatrix} a_1 & a_2 & \cdots & a_2 \\ a_2 & a_1 & \cdots & a_2 \\ \vdots & \vdots & \ddots & \vdots \\ a_2 & a_2 & \cdots & a_1 \end{bmatrix} \tag{11.25}$$

$$B = \begin{bmatrix} a_1 & a_2 & \cdots & a_2 \\ a_2 & a_1 & \cdots & a_2 \\ \vdots & \vdots & \ddots & \vdots \\ a_2 & a_2 & \cdots & a_1 \end{bmatrix} \tag{11.26}$$

Table 11.3 Elements of the matrix transfer function

Units active	a_1	a_2
1	$\frac{2}{G_0 T_{\mathrm{wt}}}$	0
2	$\frac{2T_{\mathrm{wt}}}{G_0\left(T_{\mathrm{wt}}{}^2 - T_{\mathrm{mt}}{}^2\right)}$	$\frac{-2T_{\mathrm{mt}}}{G_0\left(T_{\mathrm{wt}}{}^2 - T_{\mathrm{mt}}{}^2\right)}$
3	$\frac{2\left(T_{\mathrm{wt}} + T_{\mathrm{mt}}\right)}{G_0\left(T_{\mathrm{wt}}{}^2 + T_{\mathrm{mt}}T_{\mathrm{wt}} - 2T_{\mathrm{mt}}{}^2\right)}$	$\frac{-2T_{\mathrm{mt}}}{G_0\left(T_{\mathrm{wt}}{}^2 + T_{\mathrm{mt}}T_{\mathrm{wt}} - 2T_{\mathrm{mt}}{}^2\right)}$
4	$\frac{2\left(T_{\mathrm{wt}} + 2T_{\mathrm{mt}}\right)}{G_0\left(T_{\mathrm{wt}}{}^2 + 2T_{\mathrm{mt}}T_{\mathrm{wt}} - 3T_{\mathrm{mt}}{}^2\right)}$	$\frac{-2T_{\mathrm{mt}}}{G_0\left(T_{\mathrm{wt}}{}^2 + 2T_{\mathrm{mt}}T_{\mathrm{wt}} - 3T_{\mathrm{mt}}{}^2\right)}$
5	$\frac{2\left(T_{\mathrm{wt}} + 3T_{\mathrm{mt}}\right)}{G_0\left(T_{\mathrm{wt}}{}^2 + 3T_{\mathrm{mt}}T_{\mathrm{wt}} - 4T_{\mathrm{mt}}{}^2\right)}$	$\frac{-2T_{\mathrm{mt}}}{G_0\left(T_{\mathrm{wt}}{}^2 + 3T_{\mathrm{mt}}T_{\mathrm{wt}} - 4T_{\mathrm{mt}}{}^2\right)}$
6	$\frac{2\left(T_{\mathrm{wt}} + 4T_{\mathrm{mt}}\right)}{G_0\left(T_{\mathrm{wt}}{}^2 + 4T_{\mathrm{mt}}T_{\mathrm{wt}} - 5T_{\mathrm{mt}}{}^2\right)}$	$\frac{-2T_{\mathrm{mt}}}{G_0\left(T_{\mathrm{wt}}{}^2 + 4T_{\mathrm{mt}}T_{\mathrm{wt}} - 5T_{\mathrm{mt}}{}^2\right)}$

The values of a_1 and a_2 depend on the number of units active. These values are based on the matrix of relations described for Eq. 4.21, duplicated here as (11.27):

$$\begin{bmatrix} \dot{q}_1 \\ \dot{q}_2 \\ \vdots \\ \dot{q}_n \end{bmatrix} = \begin{bmatrix} T_{\mathrm{wt}n} & T_{\mathrm{mt}} & \cdots & T_{\mathrm{mt}} \\ T_{\mathrm{mt}} & T_{\mathrm{wt}n} & \cdots & T_{\mathrm{mt}} \\ \vdots & \vdots & \ddots & \vdots \\ T_{\mathrm{mt}} & T_{\mathrm{mt}} & \cdots & T_{\mathrm{wt}n} \end{bmatrix}^{-1} \begin{bmatrix} h'_1 \\ h'_2 \\ \vdots \\ h'_n \end{bmatrix}. \tag{11.27}$$

For instance, if only one unit is active, then Eq. 11.27 can be written as

$$\dot{q}_1 = T_{\mathrm{wt}n} h'_1 \tag{11.28}$$

and if four units are active

$$\begin{bmatrix} \dot{q}_1 \\ \dot{q}_2 \\ \dot{q}_3 \\ \dot{q}_4 \end{bmatrix} = \left(\frac{1}{G_0\left(T_{\mathrm{wt}}{}^2 + 2T_{\mathrm{mt}}T_{\mathrm{wt}} - 3T_{\mathrm{mt}}{}^2\right)} \right)$$

$$\times \begin{bmatrix} 2\left(T_{\mathrm{wt}n} + 2T_{\mathrm{mt}}\right) & -2T_{\mathrm{mt}} & -2T_{\mathrm{mt}} & -2T_{\mathrm{mt}} \\ -2T_{\mathrm{mt}} & 2\left(T_{\mathrm{wt}n} + 2T_{\mathrm{mt}}\right) & -2T_{\mathrm{mt}} & -2T_{\mathrm{mt}} \\ -2T_{\mathrm{mt}} & -2T_{\mathrm{mt}} & 2\left(T_{\mathrm{wt}n} + 2T_{\mathrm{mt}}\right) & -2T_{\mathrm{mt}} \\ -2T_{\mathrm{mt}} & -2T_{\mathrm{mt}} & -2T_{\mathrm{mt}} & 2\left(T_{\mathrm{wt}n} + 2T_{\mathrm{mt}}\right) \end{bmatrix}^{-1} \begin{bmatrix} h'_1 \\ h'_2 \\ h'_3 \\ h'_4 \end{bmatrix}. \tag{11.29}$$

Table 11.3 shows the values of the coefficients a_1 and a_2 for the matrices A and B.

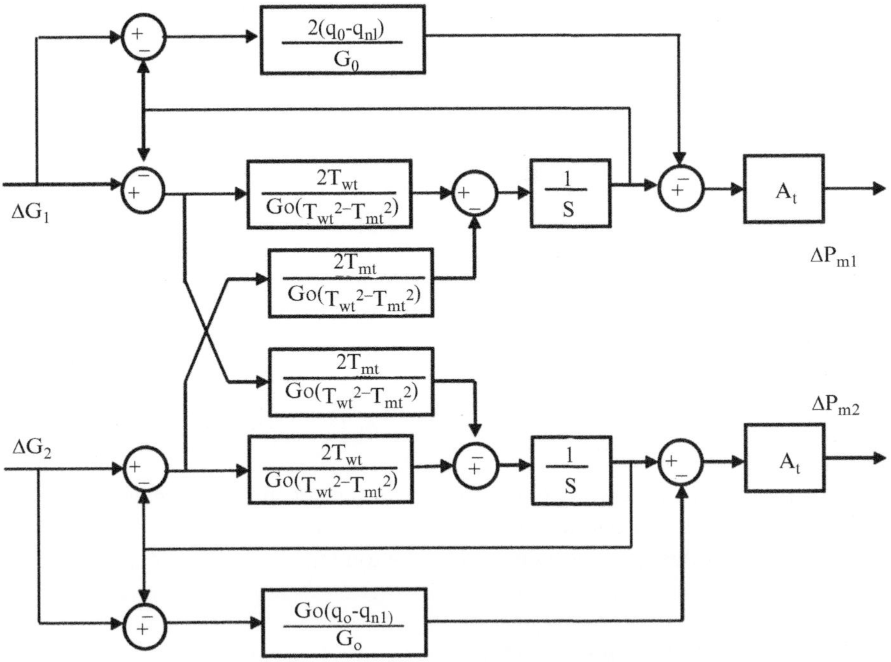

Fig. 11.12 MIMO linear model of the hydraulic subsystem with two units

Solving Eq. 11.21 with appropriate coefficients and multiplying by the transfer function of the guide vanes (6.16) gives the transfer function matrix of the MIMO linearised model:

$$G(s) = \begin{bmatrix} c_1 & 0 \\ 0 & c_1 \end{bmatrix} \left[\begin{bmatrix} s & 0 \\ 0 & s \end{bmatrix} - \begin{bmatrix} -a_1 & -a_2 \\ -a_2 & -a_1 \end{bmatrix} \right]^{-1} \begin{bmatrix} a_1 & a_2 \\ a_2 & a_1 \end{bmatrix} + \begin{bmatrix} d_1 & 0 \\ 0 & d_1 \end{bmatrix}. \tag{11.30}$$

For example, considering the case of two units active:

$$\begin{bmatrix} T_{\mathrm{wt}} & T_{\mathrm{mt}} \\ T_{\mathrm{mt}} & T_{\mathrm{wt}} \end{bmatrix}^{-1} = \begin{bmatrix} \dfrac{2T_{\mathrm{wt}}}{G_{\mathrm{o}}\left(T_{\mathrm{wt}}^{2} - T_{\mathrm{mt}}^{2}\right)} & \dfrac{-2T_{\mathrm{mt}}}{G_{\mathrm{o}}\left(T_{\mathrm{wt}}^{2} - T_{\mathrm{mt}}^{2}\right)} \\ \dfrac{-2T_{\mathrm{mt}}}{G_{\mathrm{o}}\left(T_{\mathrm{wt}}^{2} - T_{\mathrm{mt}}^{2}\right)} & \dfrac{2T_{\mathrm{wt}}}{G_{\mathrm{o}}\left(T_{\mathrm{wt}}^{2} - T_{\mathrm{mt}}^{2}\right)} \end{bmatrix}. \tag{11.31}$$

In Eq. 11.31, the two penstocks are considered equal, that is $T_{\mathrm{wt}} = T_{\mathrm{wt1}} = T_{\mathrm{wt2}}$. Figure 11.12 shows this case.

If $G_0 = 0.95$, $T_{\mathrm{wt}} = 0.6946$, $T_{\mathrm{wt}} = 0.388$, $A_{\mathrm{t}} = 1.12$ and $q_0 = 1$, then

$$G(s) = \begin{bmatrix} \dfrac{-2.36s^2 - 5.45s + 15}{s^2 + 8.8s + 13.4} & \dfrac{-8.56s}{s^2 + 8.8s + 13.4} \\ \dfrac{-8.56s}{s^2 + 8.8s + 13.4} & \dfrac{-2.36s^2 - 5.45s + 15}{s^2 + 8.8s + 13.4} \end{bmatrix}.$$

Multiplying $G(s)$ by the guide vane transfer function (6.16):

$$Gv(s)G(s) = \begin{bmatrix} \dfrac{-2.36s^2 - 5.45s + 15}{0.076s^4 + 1.26s^3 + 7.213s^2 + 16.69s + 13.35} & \dfrac{-8.56}{0.076s^4 + 1.26s^3 + 7.213s^2 + 16.69s + 13.35} \\ \dfrac{-8.56}{0.076s^4 + 1.26s^3 + 7.213s^2 + 16.69s + 13.35} & \dfrac{-2.36s^2 - 5.45s + 15}{0.076s^4 + 1.26s^3 + 7.213s^2 + 16.69s + 13.35} \end{bmatrix}. \tag{11.31}$$

This process is followed to produce the transfer function matrices for all operational cases, from one to six units active. A programme using the symbolic toolbox of MATLAB® was written to obtain automatically the solutions.

11.5.2.2 CARIMA Models for One to Six Units Operational

In this section, the process to obtain the CARIMA model, used by the GPC controller, will be described. The CARIMA model is based on the discrete-time direct ($G(z^{-1})$) and cross-coupling ($X(z^{-1})$) transfer functions, which form the discrete-time transfer function matrix of the plant (11.32):

$$Y\left(z^{-1}\right) = \begin{bmatrix} G\left(z^{-1}\right) & X\left(z^{-1}\right) & \cdots & X\left(z^{-1}\right) \\ X\left(z^{-1}\right) & G\left(z^{-1}\right) & \cdots & X\left(z^{-1}\right) \\ \vdots & \vdots & \ddots & \vdots \\ X\left(z^{-1}\right) & X\left(z^{-1}\right) & \cdots & G\left(z^{-1}\right) \end{bmatrix} U\left(z^{-1}\right). \tag{11.32}$$

If the elements of the matrix transfer function are considered as a polynomial product:

$$G\left(z^{-1}\right) = \frac{N_{\mathrm{G}}}{D_{\mathrm{G}}} \tag{11.33}$$

and

$$X\left(z^{-1}\right) = \frac{N_X}{D_X}. \tag{11.34}$$

Multiplying both sides of (11.32) by (11.35):

$$A_z = D_{\mathrm{G}} D_{\mathrm{X}} \tag{11.35}$$

gives

$$\begin{bmatrix} A_z & 0 & \cdots & 0 \\ 0 & A_z & \ddots & 0 \\ \vdots & \vdots & \ddots & \vdots \\ 0 & 0 & \cdots & A_z \end{bmatrix} Y\left(z^{-1}\right) = \begin{bmatrix} B_z & b_z & \cdots & b_z \\ b_z & B_z & \cdots & b_z \\ \vdots & \vdots & \ddots & \vdots \\ b_z & b_z & \cdots & B_z \end{bmatrix} U\left(z^{-1}\right), \tag{11.36}$$

where

$$B_z = D_{\mathrm{X}} N_{\mathrm{G}} \tag{11.37}$$

and

$$b_z = D_{\mathrm{G}} N_{\mathrm{X}}. \tag{11.38}$$

For example, if only one unit is active at an operating point of 0.95 p.u., with a sample time of 0.25 s, then

$$\begin{aligned} A_z &= 1 - 1.74z^{-1} + 1.12z^{-2} - 0.31z^{-3} + 0.03z^{-4} \\ B_z &= -0.34z^{-1} + 0.52z^{-2} + 0.008z^{-3} - 0.08z^{-4} \\ b_z &= 0. \end{aligned}$$

If two units are active with the same conditions:

$$\begin{aligned} A_z &= 1 - 3.20z^{-1} + 4.34z^{-2} - 3.26z^{-3} + 1.47z^{-4} - 0.41z^{-5} + 0.07z^{-6} \\ &\quad - 0.006z^{-7} + 0.0003z^{-8} \\ B_z &= -0.30z^{-1} + 0.92z^{-2} - 0.98z^{-3} + 0.40z^{-4} - 0.011z^{-5} - 0.04z^{-6} \\ &\quad + 0.01z^{-7} - 0.0008z^{-8} \\ b_z &= -0.11z^{-1} + 0.12z^{-2} + 0.12z^{-3} - 0.24z^{-4} + 0.11z^{-5} - 0.02z^{-6} \\ &\quad - 0.001z^{-7} + 0.0002z^{-8}. \end{aligned}$$

This process is followed to produce the CARIMA model for all operational cases, from one to six units active.

Table 11.4 Comparison of PI and GPC single unit responses

Test	Specification for single unit operation	PI	GPC	CGPC
P1	$P_1 \geq 90\%$ at $t_{p1} = 10$ s	80% at 10 s, 90% at 13.4 s	84% at 10 s, 90% at 11.6 s	91.5% at 10 s, 90% at 9.3 s
P2	$P_2 \leq 5\%$ and $t_{p2} \leq 20$ s	No overshoot	No overshoot	No overshoot
P3	$t_{p3} = 25$ s for $P_3 \leq 1\%$	25 s	21.3 s	16.4 s
P4	$t_{p4} = 60$ s for $P_4 \leq 0.5\%$	28.5 s	24.4 s	18.9 s
P5	$t_{p5} = 8$ s	11.82 s	9.35 s	7.15 s
P6	$P_6 = 2\%$	6.4%	2%	2.6%
P7	$t_{p7} = 1.5$ s	1.25 s	1.55 s	1.55 s

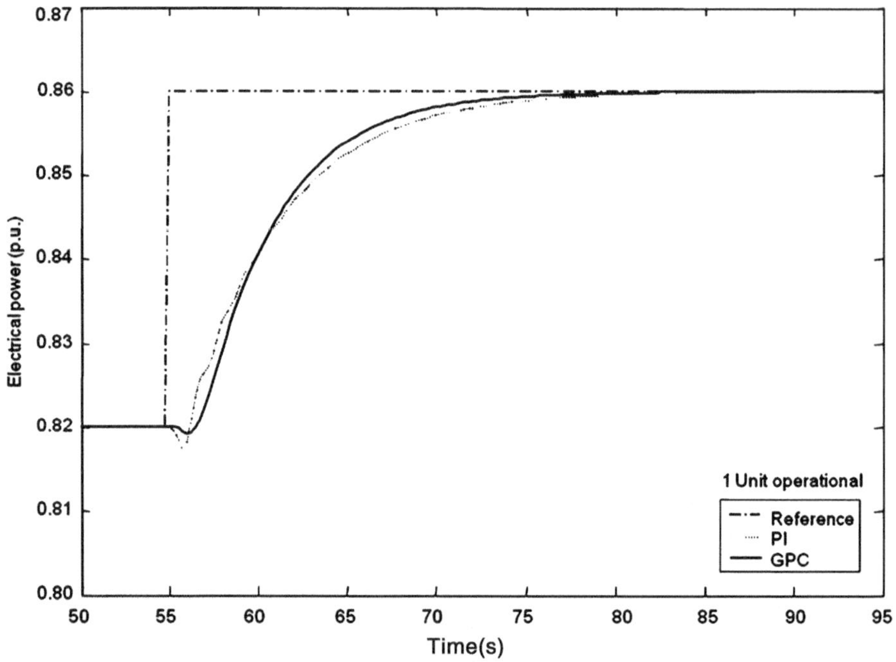

Fig. 11.13 Comparison of the step responses produced by GPC and PI controllers, one unit operational

11.5.2.3 Comparison of the Controllers, Unconstrained Case

A small step of 0.04 p.u. is applied to the units, after an operating point has been fixed, in order to evaluate the step response. At the one unit case with the other units offline, GPC produces a primary response (test P1), which is 13% faster, settles 14% sooner and has an NMP undershoot inside the specification limits (test P6), Table 11.4 and Fig. 11.13. Figure 11.14 shows the corresponding control signals for these responses. GPC also produces a smoother control that inhibits the rapid, poorly damped synchronous electrical mode which the relatively sharp PI control excites.

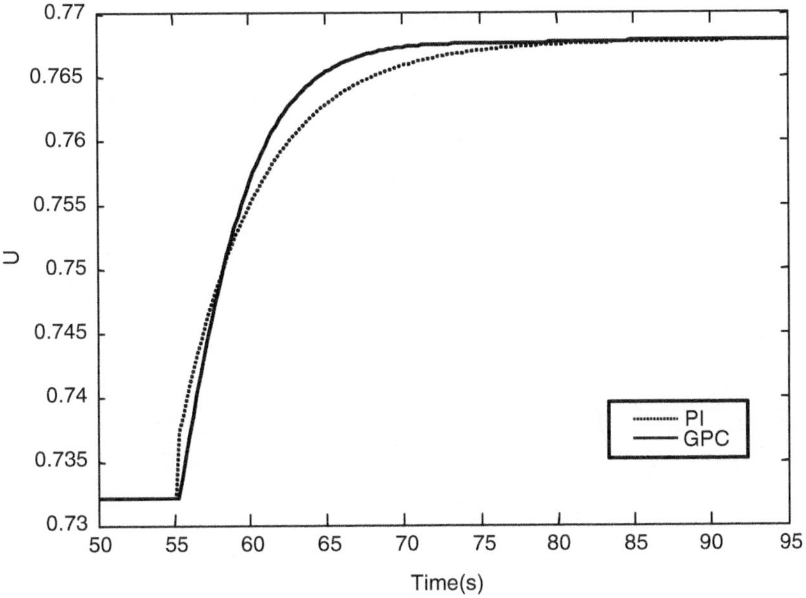

Fig. 11.14 Control signals of the step responses produced by GPC and PI controllers, one unit operational

In the six-unit case, Fig. 11.15, GPC almost eliminates the overshoot produced by the PI controller and also reduces the NMP undershoot, with no adverse effect on the primary response. Figure 11.16 shows the corresponding control signals for these responses.

It is worth stressing here that neither controller nor any other controller of practical interest can eliminate the NMP undershoot. Because when the guide vanes open, part of the mechanical power is diverted to accelerating the water column and is therefore denied to the output; this is a fundamental physical limitation. The controller may therefore distribute the power shortfall in time (by opening the guide vane slower) or by 'borrowing' from other turbines (decoupling control [28]), but it has no way to make up the power deficiency until the flow increases.

The cross-coupling responses of the two control systems are showed in Fig. 11.17 while Fig. 11.18 shows the corresponding control signals for these responses. At first, all six units are operating at 0.84 p.u. then at $t = 100$, a 0.04 p.u. step demand is applied to Unit 1. The GPC reduces the consequent perturbation on the outputs of the remaining five units. At $t = 140$, a simultaneous step demand is applied to Units 2–6. Here, GPC eliminates the overshoot, reduces the initial undershoot and offers a small improvement in the cross-coupling into Unit 1. While the PI controller can be retuned to reduce the overshoot, it was found that this always slowed the response and caused a larger initial undershoot. It is concluded that GPC is superior to PI across the operating range of the plant.

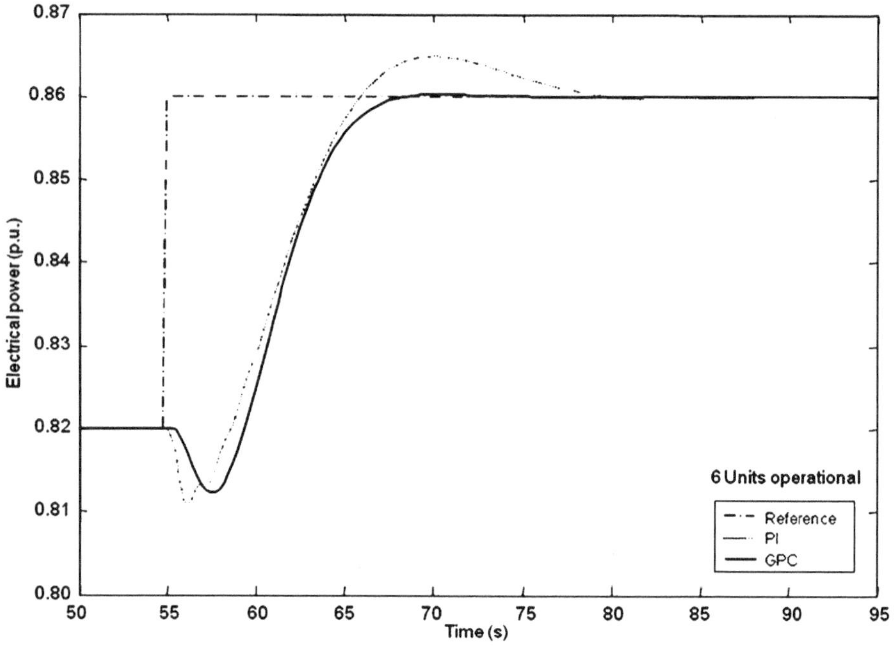

Fig. 11.15 Comparison of the step responses produced by GPC and PI controllers, six units operational

11.5.2.4 Comparison of the Controllers, Constrained Case

Although unconstrained GPC is somewhat superior to PI, the performance indicators in Table 11.4 still fall short of the specification, tests P1, P6 and P7. Reducing the value of λ yields a faster response but extensive simulations showed that it is always at the expense of other criteria. In this section, constraints on the guide vane rate and amplitude are incorporated within the theoretical framework of GPC. This allows the one-unit specification to be satisfied while preserving stability during multiple unit operation. The control rate-limit is fixed at $-0.2 \leq \Delta u \leq 0.2$ p.u. and the strategy for saturation involving P_d/A_t is followed, as explained in Sect. 11.3. Retuning the CGPC (constrained generalised predictive controller) yielded the new parameters $N_u = 10$, $N = 40$, $\alpha = 0$ and $\lambda = 350$ (compared to $\lambda = 425$ for the unconstrained case, thus effectively increasing the 'loop gain'). To perform a fair comparison, the PI controller was also modified to include control constraints. The same rate limit of 0.2 p.u. was used and the saturation limit fixed to $u = 1$ p.u. in the anti-windup configuration, see Chap. 9. The PI parameter values were chosen as $K = 0.165$ and $K_i = 0.66$ to give the best response for the six-unit case. The results are shown in Figs. 11.19 and 11.21, while Figs. 11.20 and 11.22 show the

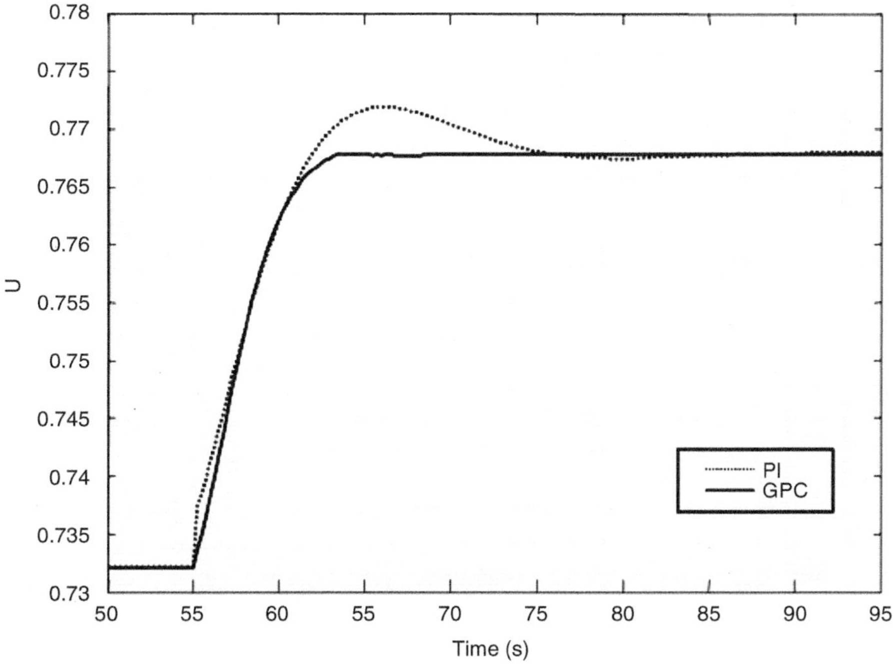

Fig. 11.16 Control signals of the step responses produced by GPC and PI controllers, six units operational

corresponding control signals for these responses. Here, the CGPC response for the one-unit case is faster than the unconstrained case but remains well damped and has a very small NMP undershoot. Its performance indicators (last column of Table 11.4) comply with the specification except for test P6. In contrast, the response produced by the modified PI controller is barely faster than the unconstrained case and the electrical oscillation is prominent; in effect, this prevents any further increase in the loop gain.

In the six-unit case, the advantage of the CGPC controller is retained, Fig. 11.21. CGPC produces a smooth and fast response with no overshoot despite the change in operating conditions. Embedding knowledge of the constraints in the predictive model allows CGPC to make informed decisions about their effect and produce better control (Fig. 11.22).

The cross-coupling responses for CGPC and the modified PI are shown in Fig. 11.23, where it is seen that the overshoot is improved in both the direct and coupled transients, with little effect on the speed response. However, as discussed previously, the NMP undershoot remains. Figure 11.24 shows the corresponding control signals for these responses.

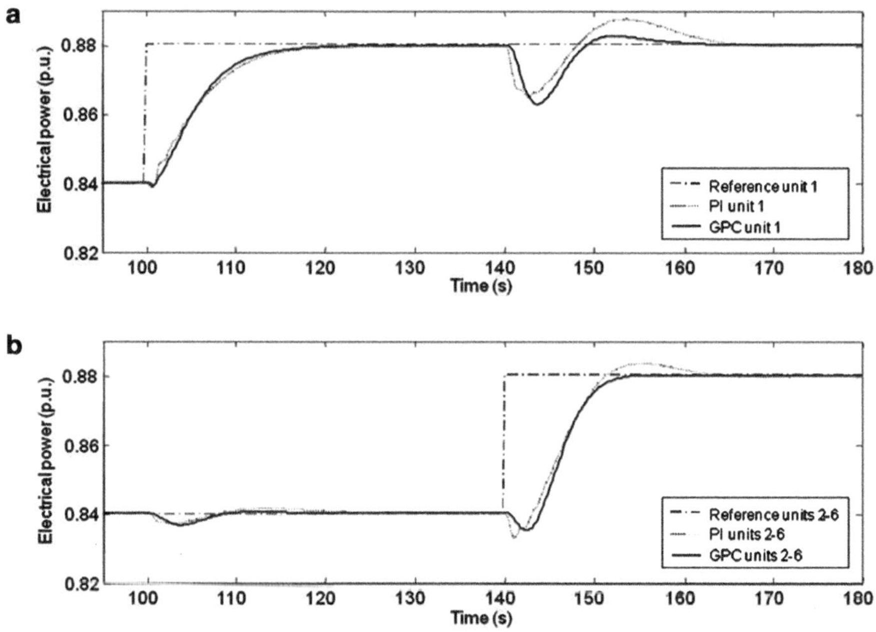

Fig. 11.17 Responses of unconstrained GPC and PI controllers to a step power demand, unit 1 (**a**) and units 2–6 (**b**)

11.5.3 MIMO Nonlinear Elastic Model

A MIMO nonlinear elastic model is used in this section to represent the Dinorwig power plant, in order to obtain a better assessment of GPC controllers. In these simulations, small step and large ramp input signals are applied to the model. The response of the plant to variations in frequency is also assessed. Finally, the robustness of the controllers with different operational conditions is investigated.

11.5.3.1 Small Step Responses, One Unit Operational

Here, the step responses produced by the GPC and CGPC controllers designed before in this chapter are compared with those of the PI controller with the full MIMO nonlinear elastic model of Dinorwig. The GPC and CGPC controllers were tuned with the values given in Table 11.2 (Sect. 11.4). The PI was tuned with its standard settings of $K = 0.1$ and $K_i = 0.12$.

Figure 11.25 shows the results for a 0.04 p.u. power demand, and Fig. 11.26 shows the corresponding control signals for these responses. The test criteria are compared in Table 11.5 and show that both GPC controllers produce primary

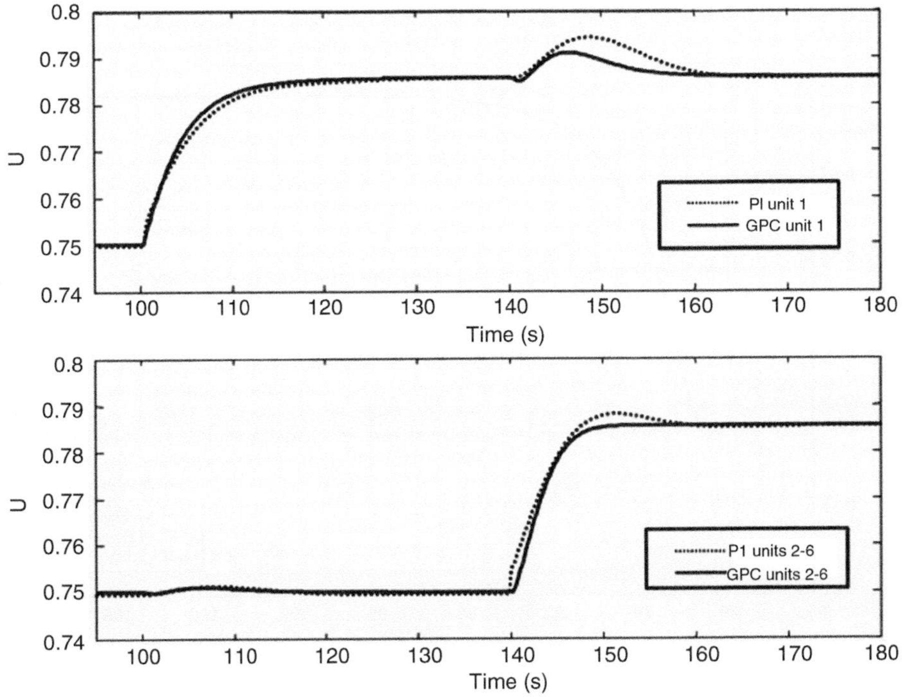

Fig. 11.18 Control signals of unconstrained GPC and PI controllers to a step power demand, unit 1 (**a**) and units 2–6 (**b**)

responses that are 23% and 38% faster, settle 27% and 44% sooner and have 71% and 57% lower NMP responses, respectively, than the response produced by PI control. Again, GPC produces control signals that reduce the rapid, poorly damped synchronous electrical mode which the relatively sharp PI control excites.

11.5.3.2 Small Step Responses, Six Units Operational

As can be seen from Fig. 11.27, all controllers have a different response when the number of active units varies. The figure shows the response of the system to a small step with the six units operating synchronously. Under this operating condition, the three controllers have a faster reaction than with one unit operational. The responses reach 90% of the new reference 14% (CGPC), 10% (GPC) and 23% (PI) sooner than for one unit operational. The NMP responses are more than ten times higher than in the one-unit operational case. The PI controller produces the lowest undershoot but has significant overshoot. In general, the GPC controllers seem to be less sensitive to the variation in the number of units active.

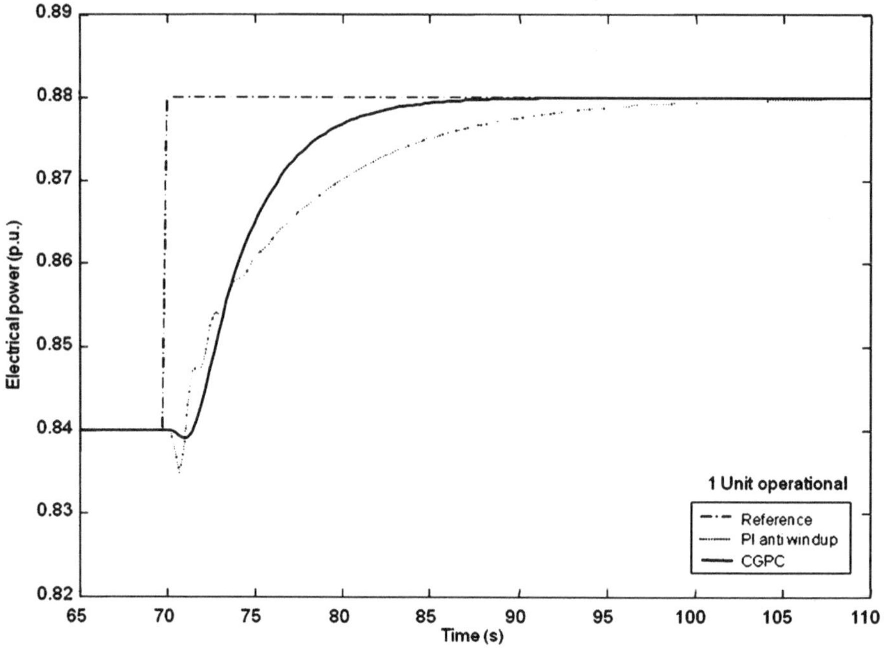

Fig. 11.19 Comparison of the step responses produced by CGPC and the modified PI controllers, one unit operational

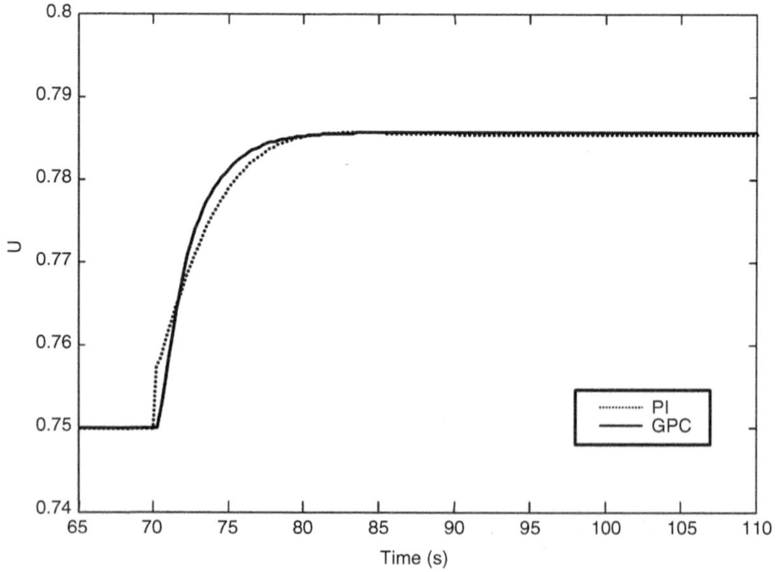

Fig. 11.20 Control signal of the comparison of the step responses produced by CGPC and the modified PI controllers, six units operational

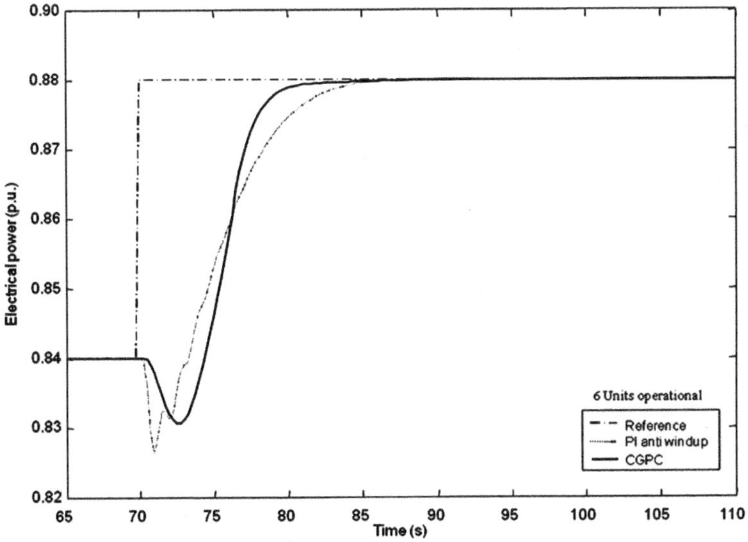

Fig. 11.21 Comparison of the step responses produced by CGPC and the modified PI controllers, six units operational

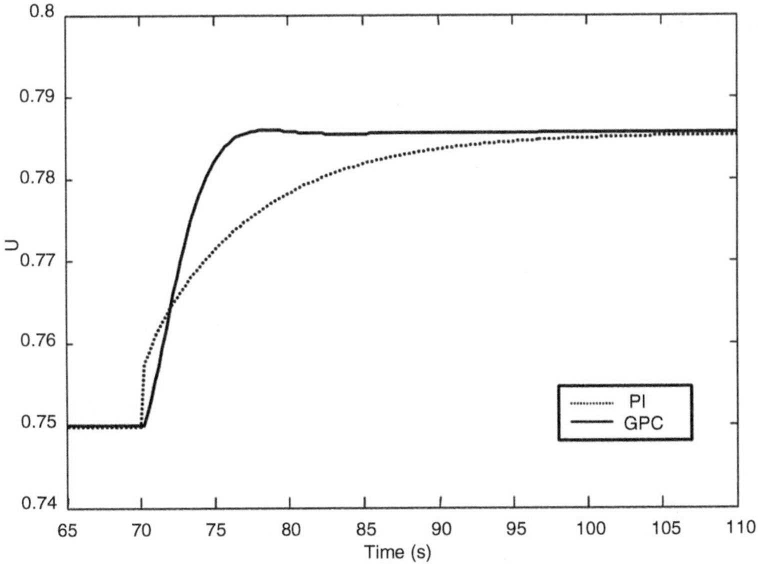

Fig. 11.22 Control signals of the comparison of the step responses produced by CGPC and the modified PI controllers, six units operational

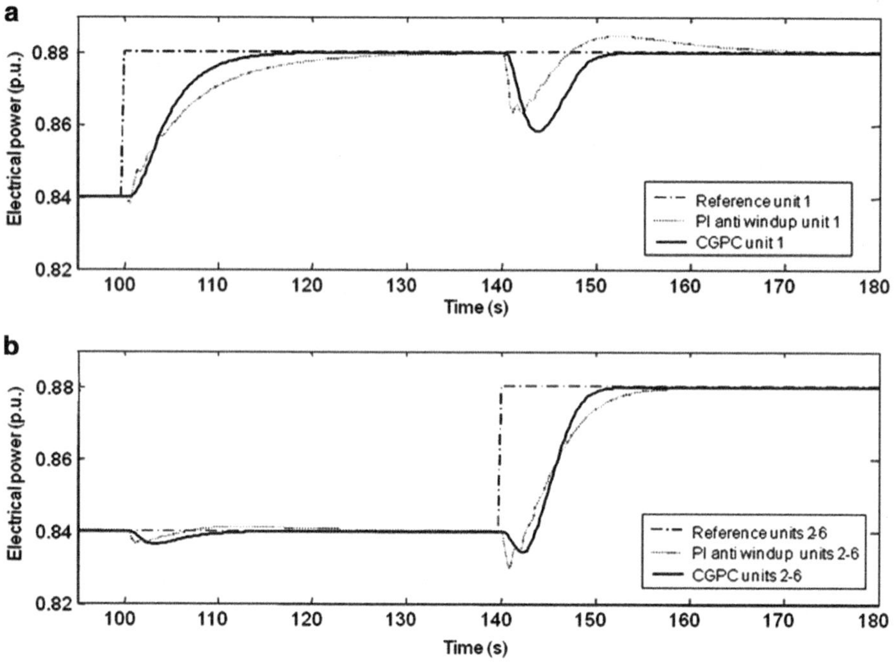

Fig. 11.23 Responses of CGPC and the modified PI controllers to a step power demand, unit 1 (**a**) and units 2–6 (**b**)

The controllers now give a fast response. Actually, the primary response is better than with only one unit operational, because the system's stability margin is decreased by cross-coupling (i.e. a longer T_w). However, the penalty is a very deep undershoot.

11.5.3.3 Large Ramp Responses

It is common in hydropower stations, and it is the case at Dinorwig, to use a large ramp, instead of step, when large operating point changes are required; the main idea is to reduce the NMP response. To explore the behaviour of the plant under these operational conditions, large ramp responses produced by GPC controllers are compared with those of the PI controller in Fig. 11.28. The GPC and PI controllers were tuned with the same values as the previous section. Figure 11.28 shows the response of the MIMO nonlinear elastic model to a large ramp. CGPC produces the faster response in the cases evaluated, one and six units operational, though the NMP response is bigger when the system is controlled by the GPCs.

A method that has been employed at Dinorwig to accelerate the response, when large changes in reference are applied, is the introduction of a component

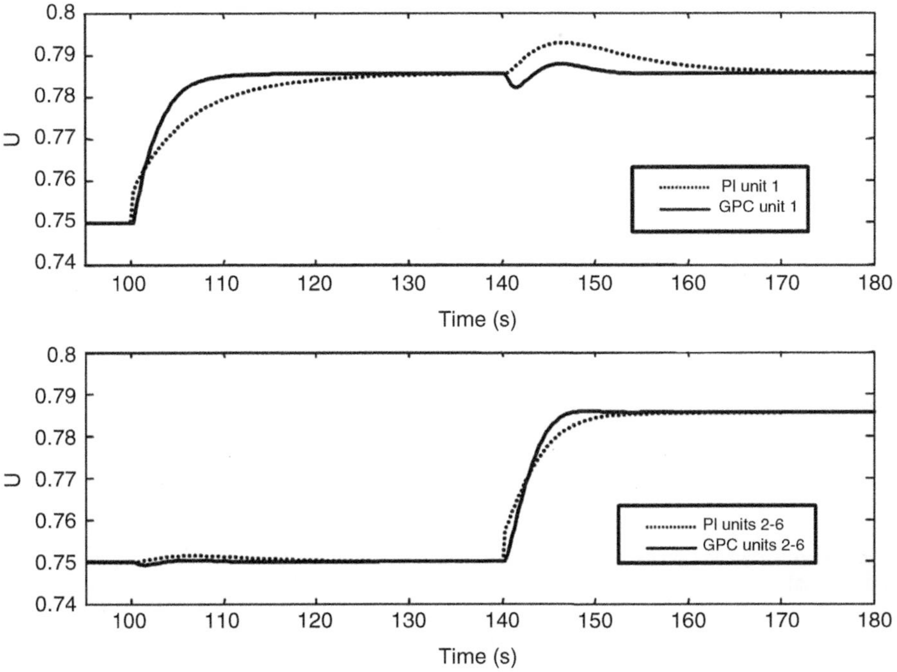

Fig. 11.24 Control signals of CGPC and PI controllers to a step power demand, unit 1 (**a**) and units 2–6 (**b**)

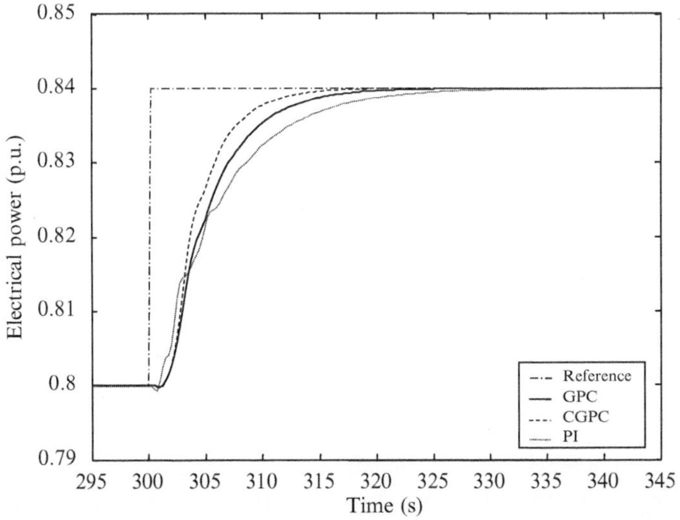

Fig. 11.25 Comparison of the step responses produced by GPC and PI controls, one unit operational

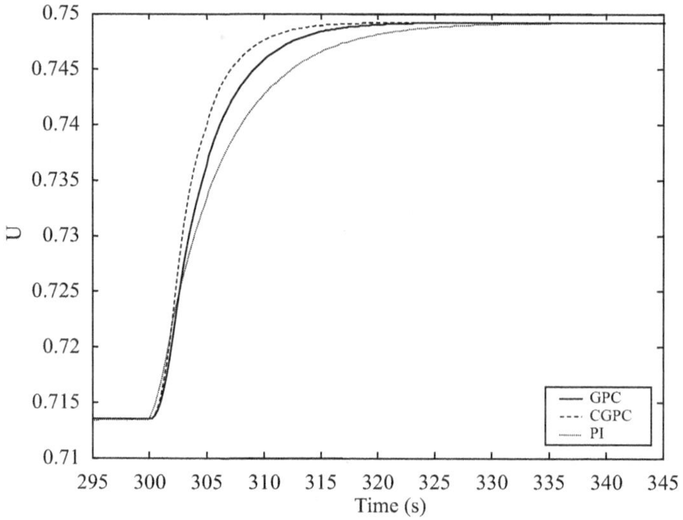

Fig. 11.26 Control signals of the step responses produced by GPC and PI controls, one unit operational

Table 11.5 Comparison of PI and GPC single unit step responses

Test	Specification for single unit operation	PI	GPC	CGPC
P1	$P_1 \geq 90\%$ at $t_{p1} = 10$ s	81% at 10 s, 90% at 13.68 s	88% at 10 s, 90% at 10.58 s	90% at 8.45 s
P2	$P_2 \leq 5\%$ and $t_{p2} \leq 20$ s	No overshoot	No overshoot	No overshoot
P3	$t_{p3} = 25$ s for $P_3 \leq 1\%$	25.93 s	18.87 s	14.65 s
P4	$t_{p4} = 60$ s for $P_4 \leq 0.5\%$	29.20 s	21.06 s	16.30 s
P5	$t_{p5} = 8$ s	12.05 s	8.33 s	6.26 s
P6	$P_6 = 2\%$	1.75%	0.50%	0.75%
P7	$t_{p7} = 1.5$ s	0.88 s	1.15 s	1.28 s

of the reference signal directly into the control signal [33]. Figure 11.29 shows the response of the MIMO nonlinear elastic model to a large ramp, when the controllers have a component of feed-forward in their control signal. All controllers produce a faster response as compared with Fig. 11.28, where feed-forward is not used. The PI controller has the fastest response in the one unit operational, but it produces a big overshoot in the six-unit operational case and also a slightly larger undershoot. CGPC produces the faster response in the six-unit operational case without producing large overshoot.

To evaluate the cross-coupling interaction, a 0.84 p.u. step was applied simultaneously at $t = 350$ s to units 2–6 and the perturbation of unit 1 was observed. As shown in Fig. 11.30 and Table 11.6, the PI control has a very high overshoot, 72% and 120% more than CGPC and GPC, respectively. Even though the PI has a lower undershoot, 34% and 44%, than the GPC controllers, the PI response has a longer

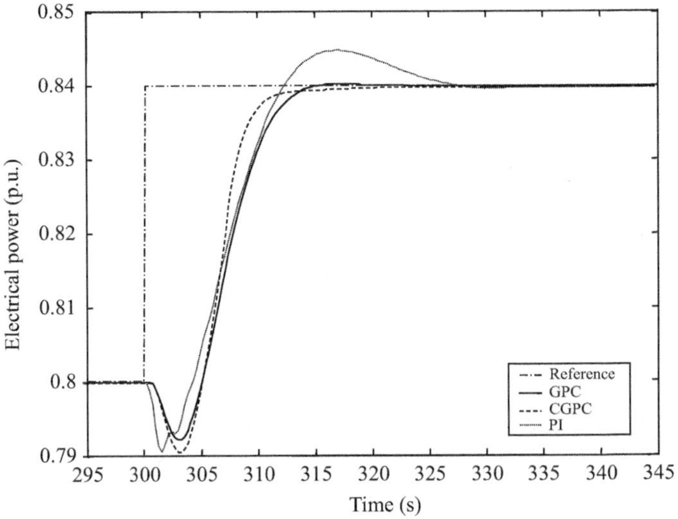

Fig. 11.27 Comparison of the step responses produced by the GPC six units operational

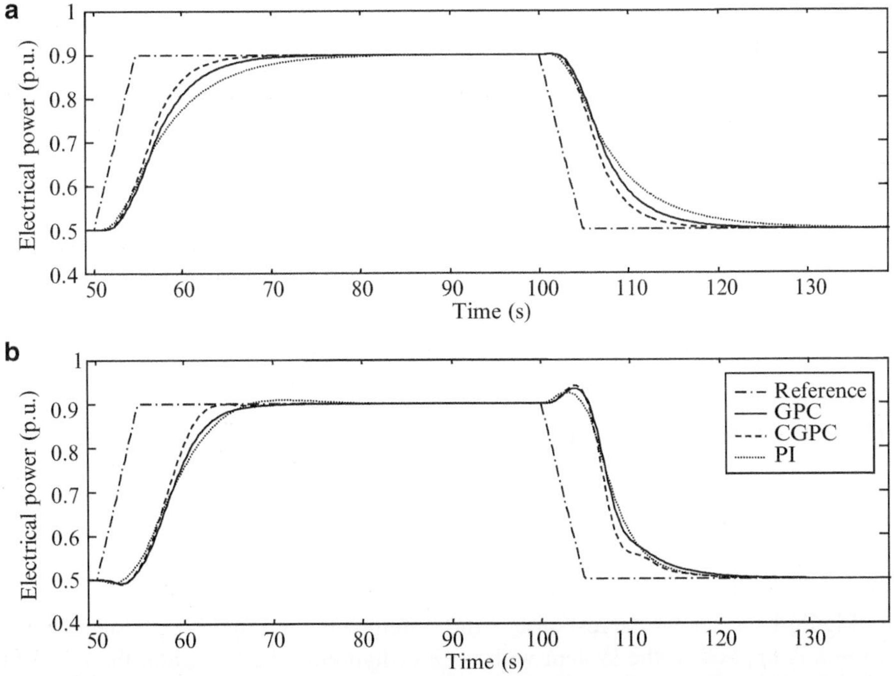

Fig. 11.28 Comparison of the large ramp responses produced by GPC and PI controls, (**a**) one unit operational and (**b**) six units operational

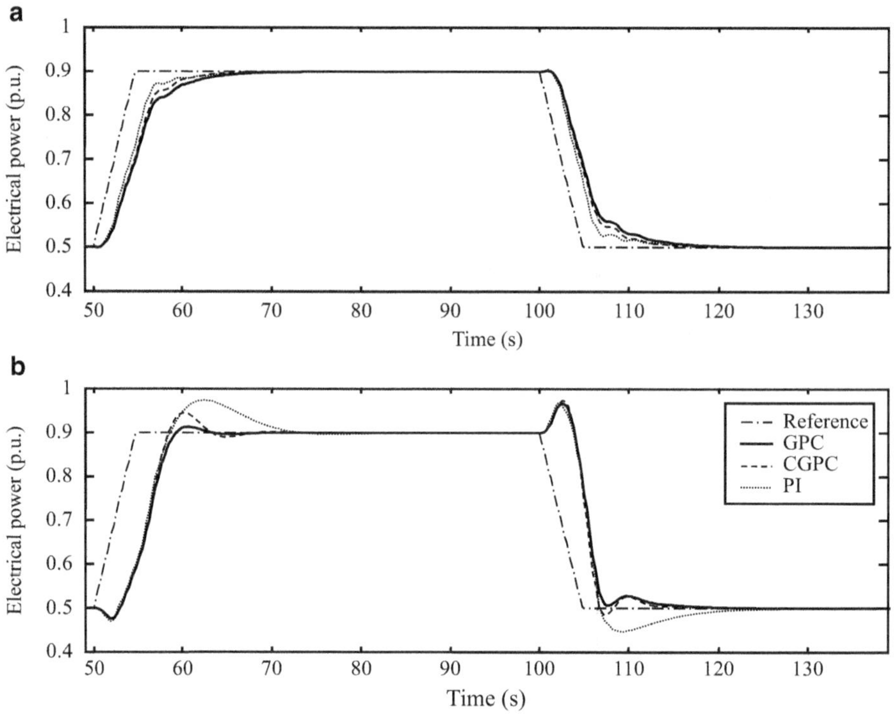

Fig. 11.29 Comparison of the large ramp responses produced by GPC and PI controls with feedforward, (**a**) one unit operational and (**b**) six units operational

settling time that is reflected in high values of the integral square error (ISE) and the integral absolute error (IAE) measures; these measures were used because they are very precise and give exact comparisons between different control schemes and, in this case, they can also be used to evaluate not only the maximum (overshoot) and the minimum (undershoot) but also the full cross-coupling effect.

11.5.3.4 Different Hydraulic Head

As was discussed in Chap. 4, the behaviour of the hydroelectric plant depends on the value of the hydraulic head. In order to determine the performance of the controllers when the plant is working under different conditions, simulations with different hydraulic heads were carried out.

Figure 11.31 shows the response of the system with GPC and PI controllers when a ramp is applied to the system with 90% of hydraulic head. Again, the one-unit operational response region with 100% of hydraulic head is shown for comparison purposes. All the responses are slower with reduced head, the primary response, t_{q1}, being increased by 18% in the PI control case but by only 6% and 3% respectively

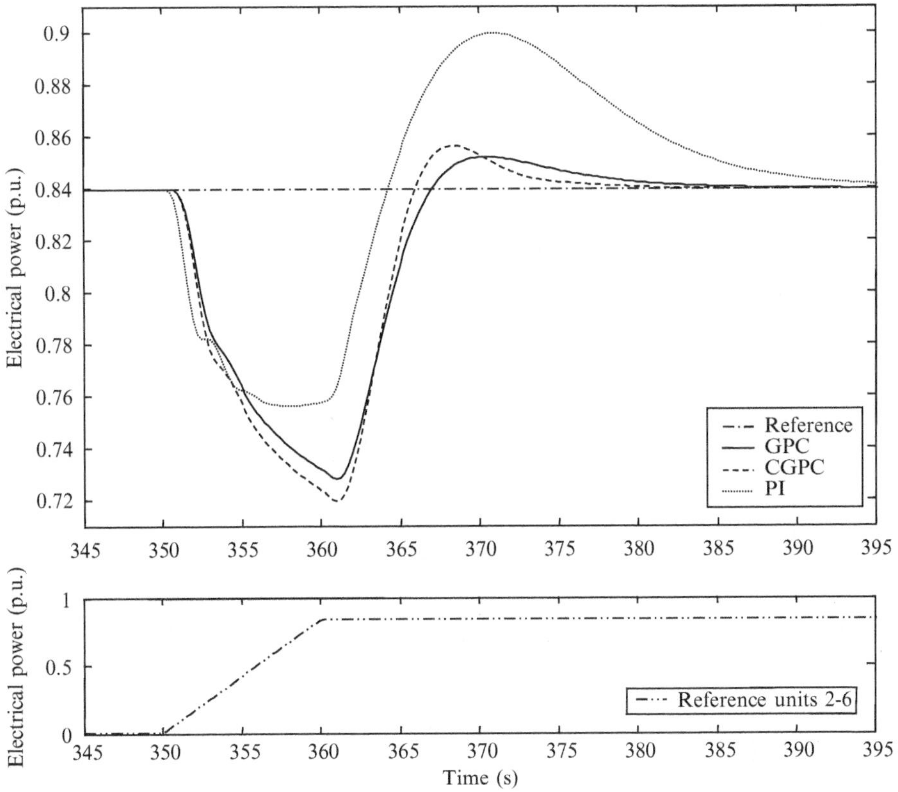

Fig. 11.30 Cross-coupling response of the plant using GPC and PI controls, six units operational

Table 11.6 Comparison of PI and GPC cross-coupling responses

	PI	GPC	CGPC
Undershoot	0.76	0.73	0.72
Overshoot	0.90	0.85	0.86
ISE	0.23	0.19	0.19
IAE	4.06	3.31	3.18

for GPC and CGPC. The settling time, t_{q3}, is increased by 19% for the PI control, 10% for GPC and by only 7% for CGPC. Therefore, GPC controllers seem to be less sensitive to variation of hydraulic head.

11.5.3.5 Effect of Different Rate Limits

In order to determine if reducing the rate limit can reduce the NMP response when a CGPC controller is used, simulations with different rate limits were carried out.

Figure 11.32 shows the large ramp response of the plant with different rate limits. It can be seen that the lower the rate limit, the lower the NMP response.

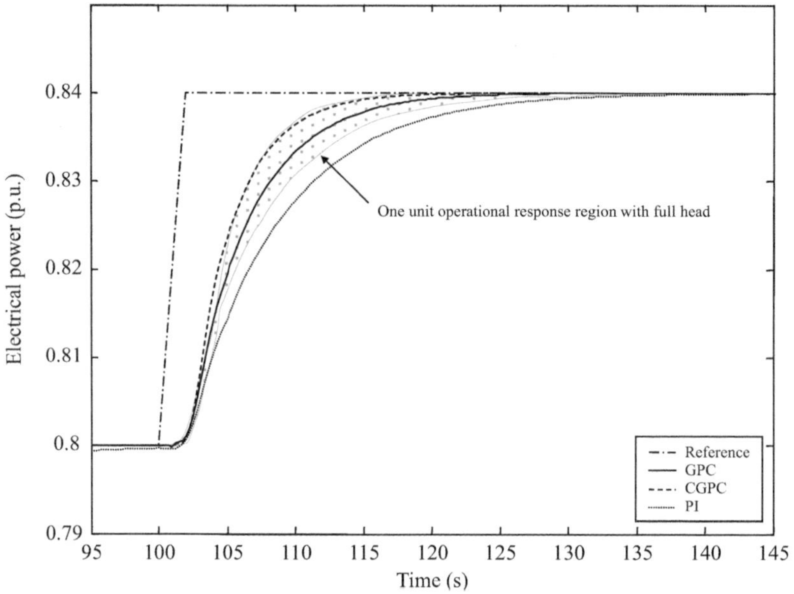

Fig. 11.31 Comparison of the ramp responses produced by the GPC (one unit operational) with 90% of hydraulic head

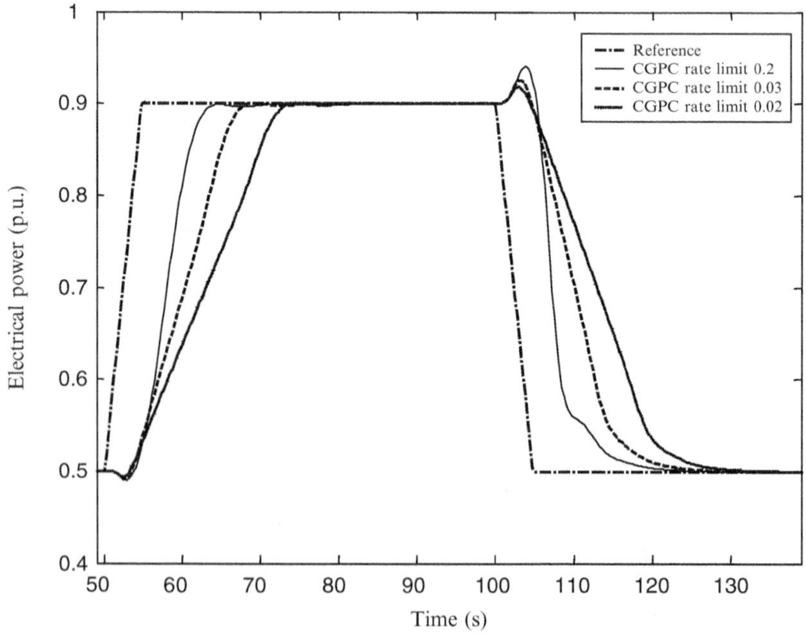

Fig. 11.32 Effects of different rate limits on large ramp response, six units operational

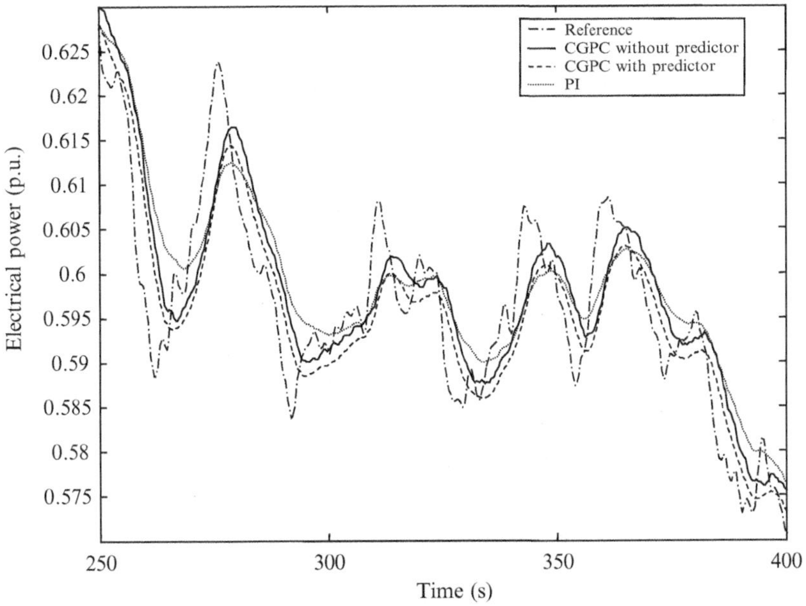

Fig. 11.33 Comparison of the power output variable from GPC and PI controls, six units operational, used in frequency control mode

This reduction on the NMP response is more evident when the set point is being decreased (negative ramp). However, to significantly reduce the NMP response a very low rate limit is necessary. This low rate limit produces a very slow reaction increasing the primary response of the system. Therefore, a value of rate limit that produces a balance between low NMP and short primary response is normally selected.

11.5.3.6 Response in Automatic Frequency Control Mode

Figure 11.33 shows the response of hydroelectric plant under PI and GPC controllers with six units operational and a power grid model based on the work of Jones and Mansoor [127]. As discussed in Chap. 4, the grid model has a band-limited white noise input to simulate changes in the power reference target. As the power grid model has a sample period of 1/3 s, the GPC controllers were retuned to fit this sample time. The parameters are chosen to match the step and ramp responses discussed in previous sections. For this simulation an ARMAX predictor [62], with prediction for four sample times ahead, is used to produce a feed-forward control, Eq. 11.39:

Table 11.7 Comparison of PI and GPC frequency responses

	PI	CGPC	CGPC predictor
ISE	0.0066	0.0057	0.0045
IAE	0.7858	0.7480	0.6639

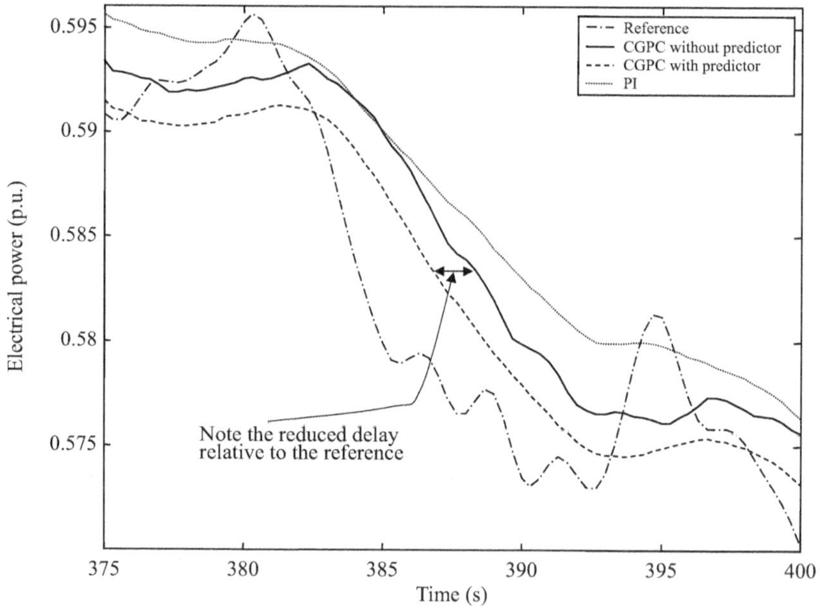

Fig. 11.34 Detail of the comparison of the power output variable from GPC and PI controls, six units operational, used in frequency control mode

$$\begin{aligned}\Delta \hat{f}(k) =& 2.19\Delta f\,(k-1) - 1.688\Delta f\,(k-2) + 0.5572\Delta f\,(k-3)\\ &- 0.07965\Delta f\,(k-4) + 1.451e^{-7}\Delta P_e(k) + 4.362e^{-7}\Delta P_e\,(k-1)\\ &+ 7.398e^{-7}\Delta P_e\,(k-2) + 9.697e^{-7}\Delta P_e\,(k-3) + e\,(k-1)\,,\end{aligned} \tag{11.39}$$

where $\Delta \hat{f}(k)$ is the predictive output, $\Delta f(k)$ is the output of the system, $\Delta P_e(k)$ is the input to the system (control signal) and $e(k)$ is the error between the predictive and output signals, all signals at the time k.

The hydroelectric plant has a better performance under GPC control; the responses follow the 'noisy' reference closer than under PI control. As shown in Table 11.7, the system under CGPC has lower values for ISE and IAE. The best performance is obtained when the predictor is used, 32% and 16% lower values in ISE and IAE, respectively, than PI (Fig. 11.34).

11.6 Conclusions

In the first part of the chapter, the general features of MPC were presented. Then the basic theory of GPC was described. The method has been easily implemented on a computer. The GPC algorithms have been programmed using MATLAB®. Simulations of simple systems have confirmed that the results are consistent with others presented in papers from the literature [136, 141]. The unconstrained and constrained cases of GPC were reviewed.

This chapter also has shown a method for optimum adjustment of the parameters of the GPC controller, where the selection of the appropriate values of these parameters yields an improved response of the linear SISO model of the hydroelectric plant.

The results from the unconstrained GPC controller were compared with the results from the PI governor, currently implemented in Dinorwig. It has been shown that applying the GPC method to the hydroelectric pumped-storage station can improve its dynamic response. The primary advantage of GPC, relative to conventional PI control, is its smooth treatment of the nonminimum phase response, resulting in minimisation of the undershoot. In order to perform a comparable evaluation, the PI controller was retuned, reducing the primary response of the system without allowing large overshoot.

When constraints on the control are introduced, their explicit inclusion within the GPC algorithm yields a response which is closer to the 'hydraulic limit' (i.e. using all the power available in the water column) than achieved by PI anti-windup control, even when the latter has been designed carefully for maximum performance.

In summary, analysis assuming a SISO linearised plant model and unconstrained controllers shows that the PI and GPC controllers produce different response characteristics but neither has a clear superiority. On the other hand, when constraints are introduced, the improved performance of the GPC controller is evident and it is to be expected that the gap between the two will widen as the analysis is extended to MIMO models that include plant nonlinearities.

Simulation has also shown that improved power delivery is obtained when the plant is operated in frequency control mode. The addition of reference feed-forward to the controller brings the response closer to the demanded target.

The sensitivity analysis has shown that both GPC and CGPC have better robustness to changes in the number of units active and variations in the hydraulic head than the PI controller. This robustness feature is extremely important because of the varying conditions which occur during day-to-day operation. The hydraulic head, for instance, normally decreases during the day. Therefore, controllers that can cope with this diversity of operational conditions without losing their performance, such as GPC and CGPC, are very valuable. However, the fixed parameters GPC and CGPC have essentially the same limitations as the PI plus anti-windup. It can be said that a fixed parameter controller does not give optimum response across the operating envelope. In the following chapter, a variation of GPC that can use different prediction models and different sets of tuning parameters, depending on the operation conditions, will be evaluated.

Chapter 12
Predictive Controller of Mixed Logical Dynamical Systems

12.1 Introduction

In earlier chapters, different control approaches for hydropower plants have been described; in this chapter, predictive control with a hybrid model will be introduced. The principal characteristic of this controller is its ability, a form of open-loop adaptive controller, to change with the operational conditions and subsequently to integrate logic rules.

Fixed linearised models have limited validity but are useful for control studies, designing and tuning. Nonlinear models have the accuracy required for plant simulation but are computationally demanding, which poses a difficulty for real-time implementation if they are used as prediction models. MLD provides an intermediate option that has good accuracy but low computational demand and it is a suitable choice as the prediction models for predictive controllers [153].

This chapter begins with a brief introduction to mixed logical dynamical (MLD) systems in Sect. 12.2. It is followed by comparison of the MLD model with the nonlinear nonelastic model of the hydroelectric plant, Sect. 12.3. After that the behaviour of the plant under generalised predictive control with constraints (CGPC) and MLD-GPC (GPC with a MLD system as the prediction model) are analysed in Sect. 12.4. These results show the improved response provided by MLD-GPC. Section 12.5 is a brief description of how MLD can represent high-level rules in the optimisation of power plant, and Sect. 12.6 is a short revision of MPC real-time features. Finally some conclusions are drawn.

12.2 MLD Theory

Linear dynamic equations, subject to linear mixed-integer inequalities, can describe MLD systems [153]. These inequalities might involve binary (logical) and continuous variables. MLD models are capable of representing a great variety

G.A. Munoz-Hernandez et al., *Modelling and Controlling Hydropower Plants*, Advances in Industrial Control, DOI 10.1007/978-1-4471-2291-3_12,

of systems, such as hybrid systems, finite state machines, constrained linear systems and nonlinear systems using piecewise linear functions to represent the nonlinearities [154].

12.2.1 Hybrid Systems

Hybrid systems are composed of dynamical components at the lower level which are governed by logical/discrete components at the upper level; hence, hybrid systems can be classed as hierarchical systems [153, 155–157]. Hybrid systems can be used in a large number of industrial and small-scale applications. There are two paradigms to deal with hybrid systems: aggregation and continuation. However, some unified frameworks appeared in the last years of the twentieth century [153, 156]. The continuous paradigm defines the whole system as a differential equation. That is done by two techniques: simulating the discrete actions by means of nonlinear ordinary differential equations or treating the discrete actions as disturbances of the differential equations that represent the system. The aggregation paradigm describes the entire system as a finite automaton or discrete-event dynamic system. It is common in this theory to partition the continuous state space and consider only the aggregate dynamics from cell to cell in the partition. The unified frameworks, such as MLD, aim to capture both discrete and continuous features to allow the designer free movement between the two domains, analogue and discrete [153, 156].

12.2.2 Inequalities and Integer Programming

To translate logical problems to linear mixed-integer inequalities, integer programming has been proved as an efficient method [158–160]. These inequalities are useful to model logical parts of processes and heuristic knowledge about plant operation; they are an important feature of MLD. Propositional logic problems are represented by inequalities of logical integer variables (δ_i) as in (12.1):

$$\begin{aligned} X_1 \vee X_2 &\Rightarrow \delta_1 + \delta_2 \geq 1 \\ X_1 \wedge X_2 &\Rightarrow \delta_1 = 1, \delta_2 = 1 \\ X_1 \oplus X_2 &\Rightarrow \delta_1 + \delta_2 = 1, \end{aligned} \tag{12.1}$$

where X_1 and X_2 are logical variables (true or false), δ_1 and δ_2 are integer variables (0 or 1), $\wedge$ is the logical operation AND, $\vee$ is the logical operation OR and $\oplus$ is the logical operation exclusive OR (XOR). Cavalier et al. have stated a complete set of these rules [158].

12.2.3 *Illustration of a MLD System*

To exemplify the characteristics of MLD, allow us to consider a linearised model of the hydraulic subsystem of a power plant consisting of only two units, with identical first-order transfer function $G_{\rm j}(s)$ relating changes in output ($\Delta P_{\rm m}$) to guide vane opening (ΔG). Note that here the change in the position of the guide vane is used as the input signal ($U(s)$). Although the dynamic cross-coupling is not modelled, the hydraulic coupling does appear as the change in operating condition and hence $T_{{\rm w}ti}$ (see Chap. 4):

$$G_j(s) = \frac{\Delta P_{\rm m}(s)}{\Delta G(s)} = \frac{A_{\rm t}(1 - G_0 T_{\rm wt1} s)}{\left(1 + \frac{G_0 T_{\rm wt1}}{2} s\right)}. \tag{12.2}$$

In (12.2) where $T_{\rm wti} = T_{\rm wt1}$, the turbine gain, $A_{\rm t}$, has a nonlinear dependence on h (hydraulic head), (12.3) shows how $A_{\rm t}$ is defined by the values of h:

$$A_{\rm t} = \begin{cases} 1.18 & \text{if} \quad 1.00 \geq h > 0.95 \\ 1.12 & \text{if} \quad 0.95 \geq h \geq 0.93 \end{cases}. \tag{12.3}$$

The water starting time of the main tunnel and a single penstock ($T_{{\rm w}ti}$) has a dependence on the number of units active ($U_{\rm o}$) as is shown in (12.4):

$$T_{{\rm w}ti} = \begin{cases} 0.70 & \text{if} \quad U_{\rm o} = 1 \\ 1.10 & \text{if} \quad U_{\rm o} = 2 \end{cases}. \tag{12.4}$$

The operating point ($G_{\rm o}$) typically varies from 0.5 to 0.95 for an active unit, on the per-unit system. In order to keep this example simple, a fixed value of 0.9 will be considered. Now defining the following integer variables:

$$p_1 = \begin{cases} 0 & \text{if} \quad U_{\rm o} = 1 \\ 1 & \text{if} \quad U_{\rm o} = 2 \end{cases} \tag{12.5a}$$

$$p_2 = \begin{cases} 0 & \text{if} \quad 0.95 \geq h \geq 0.93 \\ 1 & \text{if} \quad 1.00 \geq h > 0.95. \end{cases} \tag{12.5b}$$

If the following integer variables are defined:

$$\begin{aligned} \delta_1 &= 1 \quad \text{if} \quad p_1 = 0 \wedge p_2 = 0 \\ \delta_2 &= 1 \quad \text{if} \quad p_1 = 0 \wedge p_2 = 1 \\ \delta_3 &= 1 \quad \text{if} \quad p_1 = 1 \wedge p_2 = 0 \\ \delta_4 &= 1 \quad \text{if} \quad p_1 = 1 \wedge p_2 = 1 \end{aligned} \tag{12.5c}$$

there follows the condition:

$$\bigoplus_{i=1}^{4} [\delta_i = 1].$$

By defining: $\Delta P_{\mathrm{m}}(s) = Y(s)$ and $\Delta G(s) = u(s)$ the system (12.2) can be described as a MLD system in discrete time with $T = 0.25$, see Eq. 12.6a,b:

$$y_j(k) = \sum_{i=1}^{4} z_i(k), \tag{12.6a}$$

where the auxiliary variable $z_i(k)$ is defined by

$$\begin{array}{lll} z_1(k) = 0.52y(k-1) - 2.10u(k) + 2.58u(k-1) & \text{if} & \delta_1 = 1 \\ z_2(k) = 0.60y(k-1) - 2.10u(k) + 2.52u(k-1) & \text{if} & \delta_2 = 1 \\ z_3(k) = 0.52y(k-1) - 2.36u(k) + 2.95u(k-1) & \text{if} & \delta_3 = 1 \\ z_4(k) = 0.60y(k-1) - 2.36u(k) + 2.83u(k-1) & \text{if} & \delta_4 = 1 \end{array} \tag{12.6b}$$

under the following constraints:

$$z_i(k) \le M\delta_i \qquad z_i(k) \ge m\delta, \tag{12.6c}$$

where $M = 1$ is the maximum value of $y(k)$ and $m = 0$ is the minimum value of $y(k)$.

For example, if $h = 0.97$ and $U_o = 2$ then $\delta_3 = 1$ and

$$\begin{aligned} z_1(k) &= 0 \\ z_2(k) &= 0 \\ z_3(k) &= 0.52y(k-1) - 2.36u(k) + 2.95u(k-1) \\ z_4(k) &= 0 \end{aligned}$$

then $\quad y_{1,2}(k) = 0.52y(k-1) - 2.36u(k) + 2.95u(k-1)$
where $\quad 1 \ge y_{1,2}(k) \ge 0.$

On the other hand, if $h = 0.93$ and $U_o = 1$, then: $\delta_1 = 1$ and

$$\begin{aligned} z_1(k) &= 0.52y(k-1) - 2.10u(k) + 2.58u(k-1) \\ z_2(k) &= 0 \\ z_3(k) &= 0 \\ z_4(k) &= 0 \end{aligned}$$

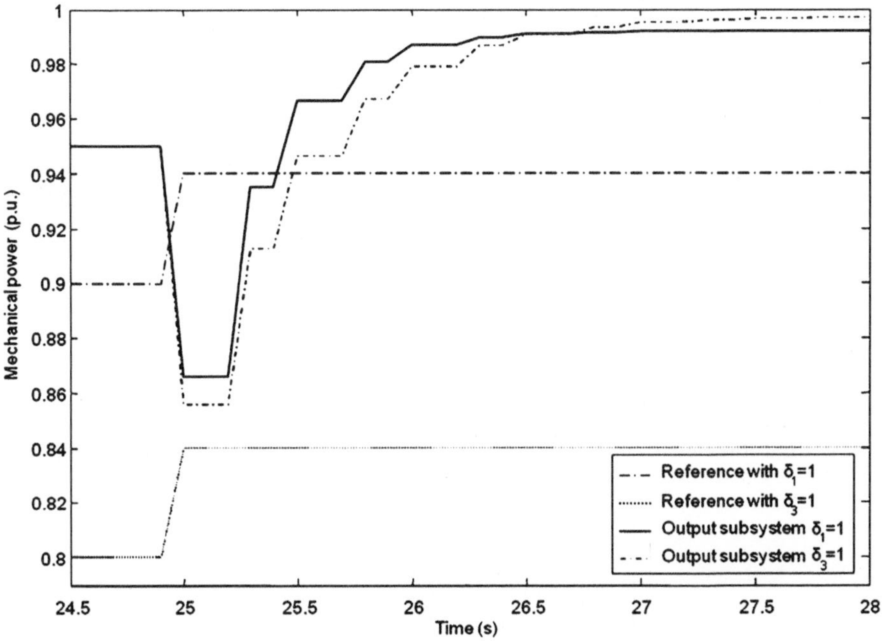

Fig. 12.1 Open loop step response of example for different operational conditions

then
$$y_1(k) = 0.52y\,(k-1) - 2.10u(k) + 2.58u\,(k-1)$$
$$y_2(k) = 0$$

where
$$1 \geq y_1(k) \geq 0.$$

As it is illustrated in Fig. 12.1, where the open loop step responses of unit 1 when $\delta_1 = 1$ and $\delta_3 = 1$ are shown, the description of the system depends on the value of δ_i. Two reference values were used to set the subsystem at the same initial output power, then a small step of 0.04 p.u. was applied at $t = 25$ s. $\delta_1 = 1$ imply that there is only one unit active and the lower hydraulic head causes a lower open loop gain and hence a lower final value. $\delta_3 = 1$ imply that the hydraulic head is higher and both units are active. This increases the water starting time and therefore the system has a longer time constant, so the response is slower. In the following section, a model of a power plant, including the guide vane and hydraulic subsystems, will be assessed for one to six units operational.

12.3 MLD Predictive Model

12.3.1 Description of the MLD Predictive Model

A predictive model to be used in the GPC controller is set up, in this section, using the MLD method. The most important component of the model is the MIMO linear model discussed in early chapters and in other works as [161–163].

By defining:

$$\delta_i = 1 \leftrightarrow U_0 = i \quad \forall i = 1, \ldots, 6, \tag{12.7}$$

where $\bigoplus_{i=1}^{6} [\delta_i = 1]$.

The first three cases of the piecewise linear model of the hydroelectric plant are

$$Y(s) = \begin{cases} G_1(s)U(s) & \text{if} \quad \delta_1 = 1 \\ \begin{bmatrix} G_2(s) & X_2(s) \\ X_2(s) & G_2(s) \end{bmatrix} U(s) & \text{if} \quad \delta_2 = 1 \\ \begin{bmatrix} G_3(s) & X_3(s) & X_3(s) \\ X_3(s) & G_3(s) & X_3(s) \\ X_3(s) & X_3(s) & G_3(s) \end{bmatrix} U(s) & \text{if} \quad \delta_3 = 1 \end{cases} \tag{12.8}$$

where $G_i(s)$ and $X_i(s)$ are, respectively, the direct and cross-coupling transfer functions, which vary according to the operating point (O_p) and the number of units active (U_o), where

$$G_i(s) = f\left(U_\mathrm{o}, O_\mathrm{p}\right) \quad \text{and} \quad X_i(s) = f\left(U_\mathrm{o}, O_\mathrm{p}\right). \tag{12.9}$$

There are computational benefits in maintaining the model in a piecewise form because it needs fewer logical variables than the full MLD model and the logic is less complex [153, 159].

12.3.2 Evaluation of the MLD Predictive Model

The accuracy of the predictions made by the MLD model can be determined by simulation. The basis of comparison was the nonlinear nonelastic discussed in Chap. 4. The output of the linear MIMO model was set at the six-unit operational case. The relationship between the nonlinear nonelastic model and the full nonlinear model has been discussed previously (Chap. 4).

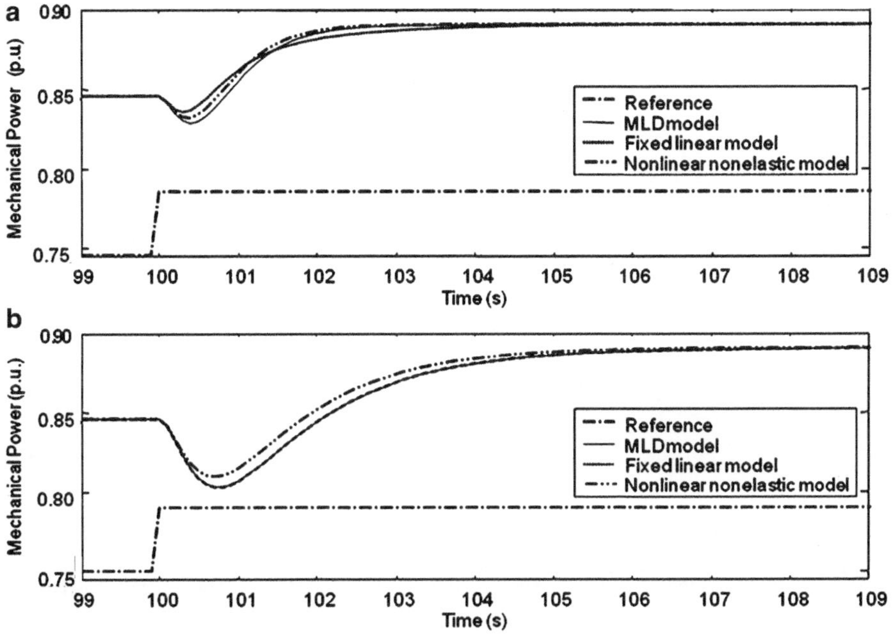

Fig. 12.2 Open loop step response of MLD and nonlinear models with 0.75 as operating point: (**a**) one unit operational and (**b**) six units operational

The open loop responses of the hydraulic models are shown in Fig. 12.2, including guide vanes; for this simulation, an operating point was fixed to 0.75 p.u. for all units was first established, then a step of 0.04 p.u. is applied to unit 1 at simulation time $t = 100$ s. Figure 12.2a, b shows the responses with one unit and six units activate, respectively. In both graphs, the response of the MLD model is a good approximation to the response of the nonlinear model. Note that the number of units determines the behaviour of the model. With six units operational both linear model have exactly the same response, as expected. The nonminimum phase characteristic is more marked in both linear cases; this and other differences between the models are mainly because the linear models are not taking into account other factors, such as friction losses.

As it was expected the MLD model has better or equal performance to the fixed linear model and it follows the nonlinear model accurately, in spite of the number of units active. This adaptive characteristic is really important when a controller is designed to preserve the same performance under different operational conditions.

Figure 12.3 shows the direct and cross-coupling step responses of the MLD and nonlinear models. The lower graph is shown the response of units 2–6 in synchrony and the upper graph shows the response of unit 1. For this simulation first is established an operating point of 0.75 p.u. for all units, then a step of 0.04 is applied to unit one at 100 s of simulation, followed by a step of the same amplitude

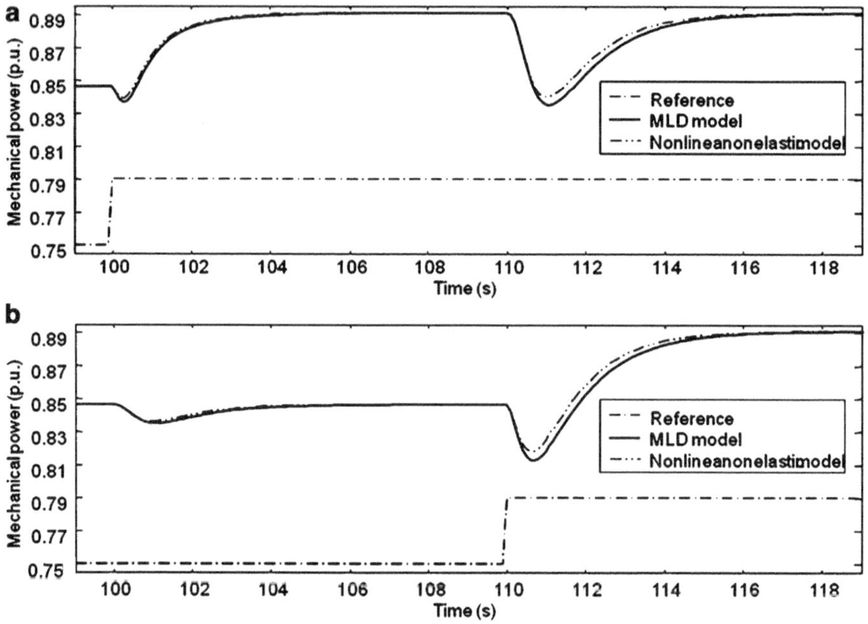

Fig. 12.3 Direct and cross-coupling step response: (**a**) response of unit 1 and (**b**) response of units 2–6

applied to units 2–6 at 110 s. The responses of the models are comparable with best agreement occurring, as expected, for the case of six units operational, which was the case selected for linearization.

To show how the MLD model adapts according to the number of units active, a simulation involving large and small step responses was conducted, Fig. 12.4. Unit 1 is set to 0.75 p.u. at $t = 5$ s, units 2–6 being inactive. At $t = 25$ s a 0.1 p.u. step is applied to unit 1. Units 2 and 3 are turned on at $t = 50$ s and at the same time unit 1 is set back to 0.75. A 0.1 step is then applied to units 1–3. This simulation is then repeated when units 4–6 are turned on at simulation time $t = 100$ s.

Figure 12.5 shows the response of the fixed linear model; it can be seen that this response deviates more from the reference than the MLD model's response. The MLD linear model demonstrates approximately the same accuracy in all cases. The MLD model could also modify its response to take into account changes to the operating point (i.e. the load) of a single turbine.

The cross-coupling response of unit 1, when other units are starting, is shown in Fig. 12.6. Although the undershoot of the MLD model response is bigger, after a few seconds it moves close to the response of the nonlinear model.

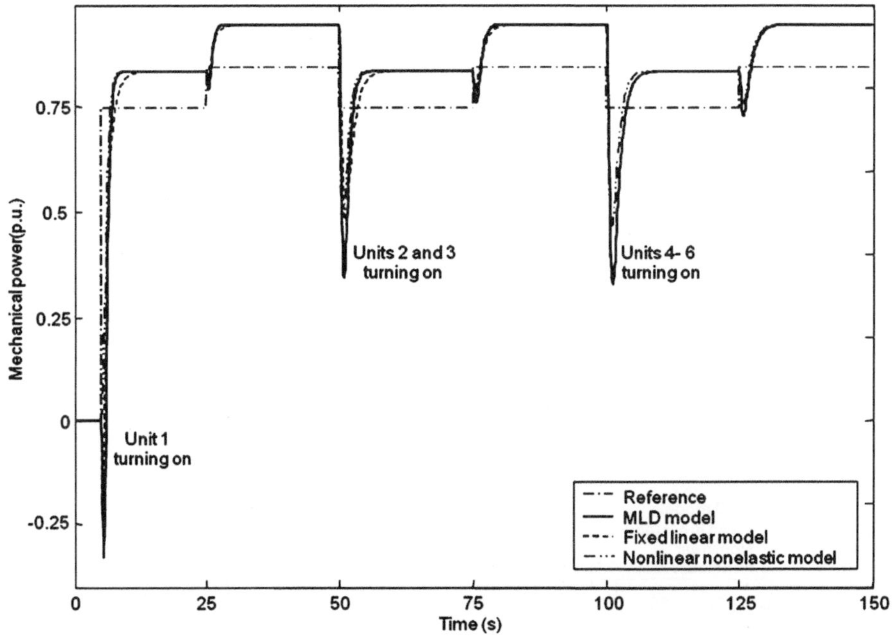

Fig. 12.4 Open loop step response of MLD and nonlinear models with different units operational

12.4 Model Predictive Control Using MLD Prediction Models

12.4.1 Predictive Controllers and MLD Systems

As discussed in Chap. 11, model predictive control (MPC) has to calculate the future output and control signals of the plant at every step of iteration; different MPC approaches use different types of models of the plant to obtain these predictions. For example, generalised predictive control (GPC) uses a CARIMA representation of the plant and dynamic matrix control (DMC) uses the step response to model the plant [136]. Hybrid systems have been used as the model of the plant in some MPC approaches. Branicky et al. [156] proposed the use of a hybrid model to represent dynamic systems together with discrete phenomena. Bemporad and Morari [153] have used MLD models to represent plants and discrete (qualitative and quantitative) inputs and outputs as a part of MPC control. Applications of MLD or equivalent models to controlling or monitoring power systems have recently appeared. Lu [164] has applied hybrid control to a hydro-turbine-generator set. Thomas et al. [165] have used a MLD framework for fault detection in the sensors of a steam generator.

In this chapter, the plant is modelled using the nonlinear elastic representation (Chap. 4) while a MLD model is considered as the prediction model for GPC. The

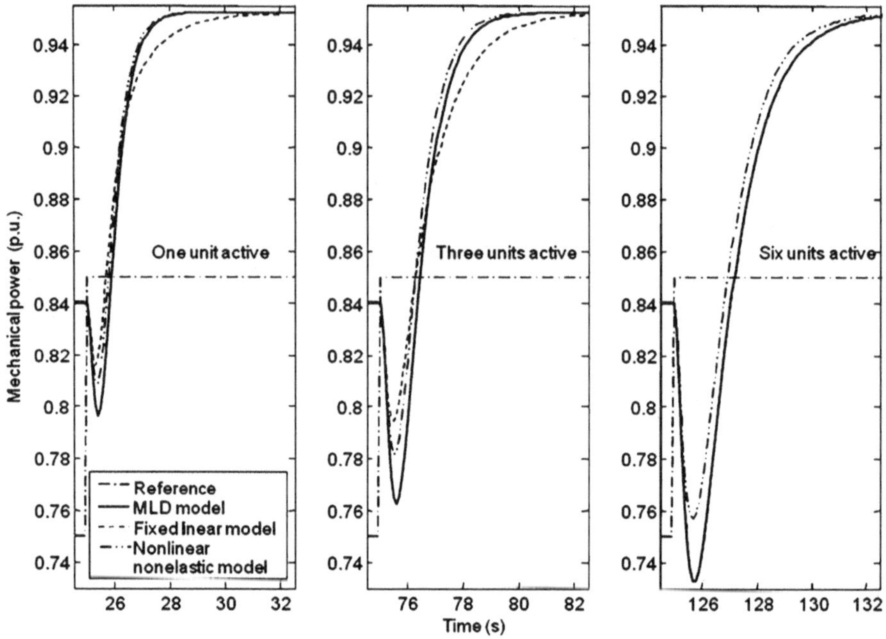

Fig. 12.5 Detail of the open loop small step response of MLD and nonlinear models with different units operational

MLD model is a piecewise CARIMA model, whose structure and value depends on the number of units active. The MLD model also depends on the operating point.

12.4.2 Applying a MLD-GPC to the Hydroelectric Station

Since the MLD model is an accurate representation of the plant, it allows the use of a shorter horizon of prediction (N), thus decreasing the computational effort (see Chap. 11). Also, lower values of the weight control factor (λ) can be selected, which accelerates the response of the system.

In order to demonstrate the functionality of the GPC with a MLD system as the prediction model (MLD-GPC), it was applied to a computational model of Dinorwig; for these simulations, the MLD-GPC was tuned with $N = 10$ (prediction horizon), $N_u = 10$ (control horizon) and $\alpha = 0$ (weight factor on the reference trajectory), see Chap. 11.

The λ value was tuned using the strategy of gain scheduling [166]. $\lambda = 40$ was selected for one unit operational, linearly incrementing by 3, per unit active, up to $\lambda = 55$ for six units operational [161]. The same constraint limits were employed for CGPC as in Chap. 11.

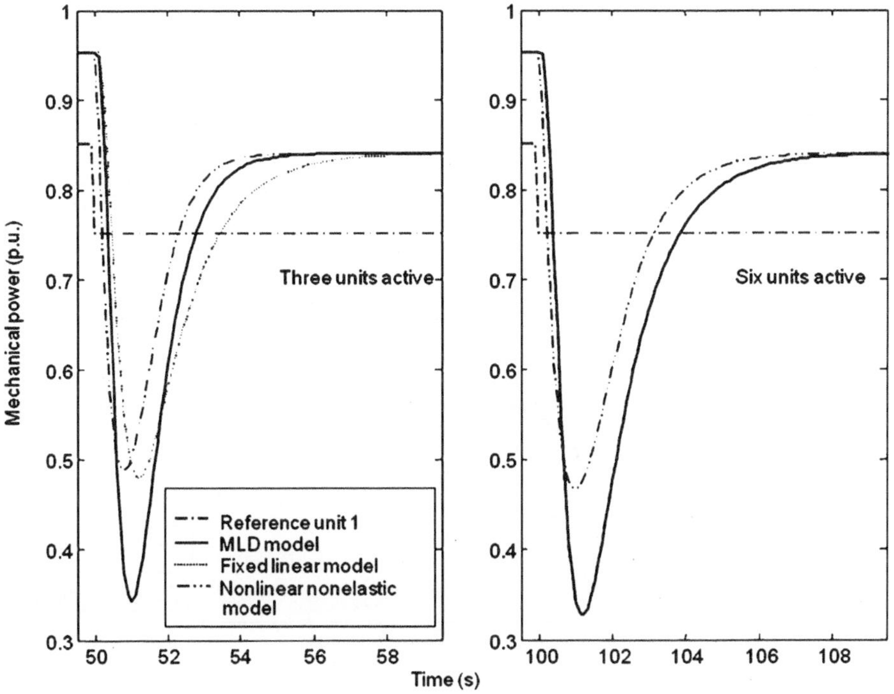

Fig. 12.6 Detail of the cross-coupling response of MLD and nonlinear models with different units operational

The MLD-GPC was compared with the constrained GPC (CGPC), the CGPC has a fixed linear model for prediction and has $N = 40$, $N_u = 10$, $\lambda = 250$ and $\alpha = 0$. Figure 12.7 shows the performance of both controllers when the Dinorwig power plant is represented by a full nonlinear simulation [33] with a large step (0.3 p.u.). The lower graph shows the one-unit operational case and the upper graph shows the six-unit operational case; in both cases MLD-GPC is faster. Lower values of λ can be used to tune the controller because the better prediction of the MLD model. Consequently, the controller can use more gain in the control loop hence a faster response. CGPC was tuned for the case of six units operational and, as can be seen from Figs. 12.7 and 12.8, it is in this operational case that CGPC has its best performance.

The response of the plant to a small step (0.04 p.u.) is shown in Fig. 12.7. It can be seen that the MLD-GPC is faster than CGPC in the one-unit operational (lower graph) case. The responses for six units operational are comparable. The primary response, [131] of the system using MLD-GPC was 5.4 s, while the system using CGPC has a primary response of 8.45 s, a considerable improvement.

The responses of the system when a large ramp signal is applied is shown in Fig. 12.9; once more the MLD-GPC is faster than the CGPC controller.

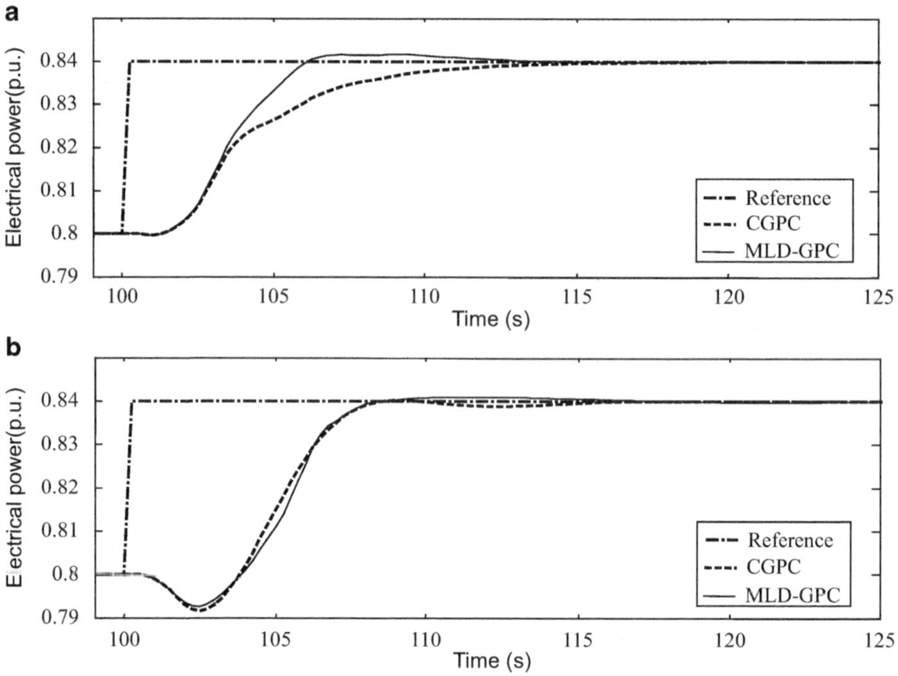

Fig. 12.7 Small step response for six and one unit operational: (**a**) one unit operational and (**b**) six units operational

If the CGPC is re-tuned in order to reach the same performance as the MLD-GPC in one unit operational, then a poorly damped response is produced at six units operational. One strategy to reduce the damping could be increasing or relaxing the constraints, although this leads to a gain scheduled approach similar to the MLD-GPC, see Fig. 12.10. A CGPC controller re-tuned, with a reduced value of lambda and its other parameters tuned at their original values, is compared with the original CGPC and the MLD-GPC controllers. A small step of 0.04 p.u. was applied to the system at 0.8 initial operating point. The re-tuned CGPC produces a decaying oscillatory response, as can be seen from Fig. 12.10a. Figure 12.10b, c shows the control signals. The control signals of the CGPC when $\lambda = 175$ produces rapid changes that could cause premature fatigue to the guide vane.

The frequency-control response of the hydroelectric plant under MLD-GPC and CGPC controllers with six units operational and a power grid model was evaluated [131]. The GPC controllers were retuned according to this sample time of the grid model (1/3 s). Once more, the parameters were chosen to match the step and ramp responses as discussed previously in this section. Figures 12.11 and 12.12 show that the responses of the system under both controllers are very similar. This performance was expected because the small step and ramp responses are similar

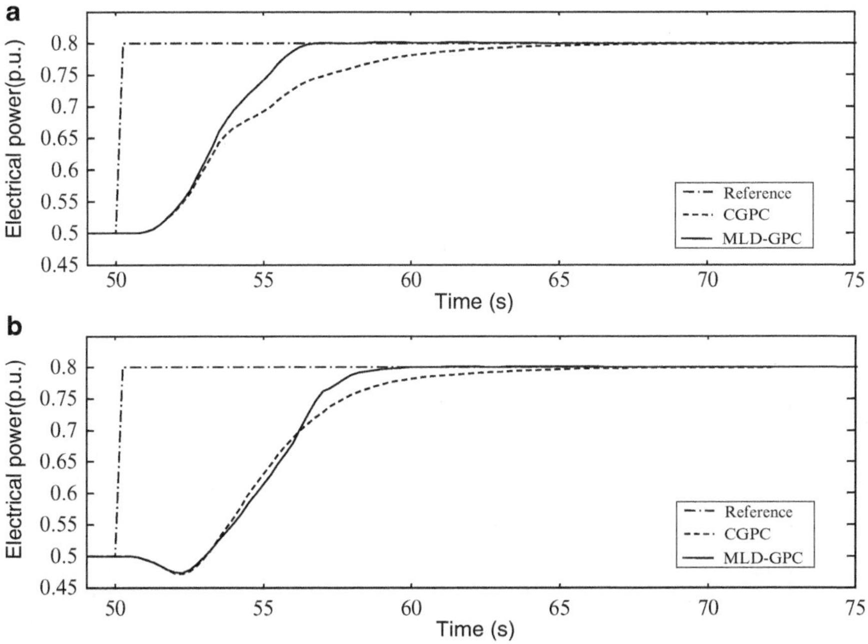

Fig. 12.8 Large step response for six and one unit operational: (**a**) one unit operational and (**b**) six units operational

with six units operational, which was the case evaluated in this simulation. Also, they use the same droop gain (0.01), the parameter that defines the rate at which the unit picks up or sheds load relative to other units [19].

12.5 Modelling High-Level Control Rules with MLD

12.5.1 Hierarchical Control

For many years financial and operational aspects have been taken into account for the optimal control of plant, in any type of industry. Hierarchical control is often used to integrate these aspects at all levels of control [153, 167].

In hierarchical control, Fig. 12.13, the lowest or execution level handles the sequence of actions necessary to manipulate the system. The medium levels constitute a supervisory system with the capacity of alarm monitoring, automatic protection, fault diagnosis, and reconfiguration. The top level is able to plan and make decisions about the strategic operating planning. MLD systems and hybrid systems, in general, can be used to develop a platform to integrate all levels of control [153].

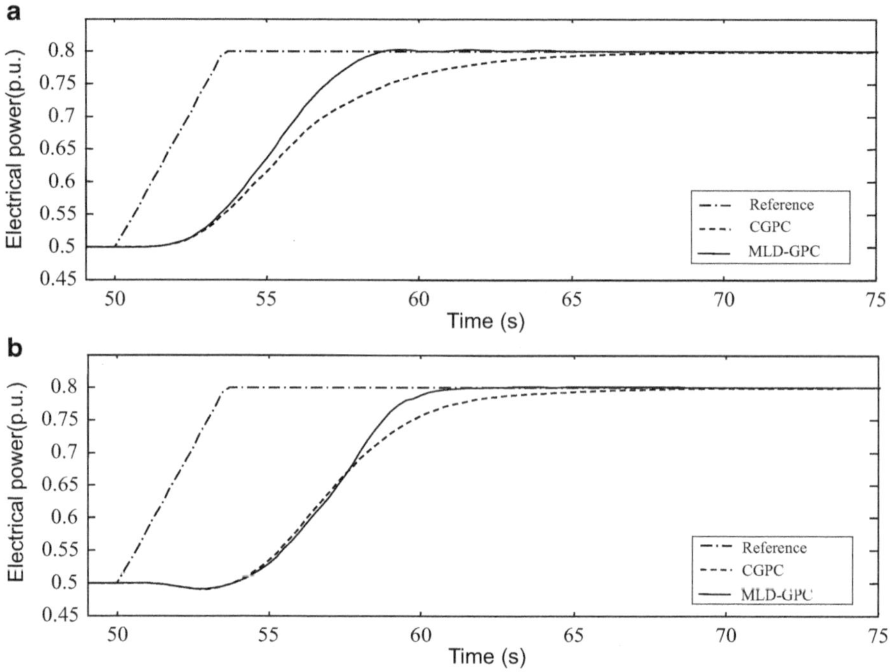

Fig. 12.9 Large ramp response for six and one unit operational: (**a**) one unit operational and (**b**) six units operational

A hierarchical structure is not the only paradigm to deal with higher levels of control; other approaches exist. However, all of them distinguish between higher and lower levels of control. Logic rules and optimal functions are commonly used to represent the high control levels; therefore, MLD is also suited to these paradigms.

In the electrical sector, studies have appeared where the financial aspects are the main component at all levels of day-to-day operational decisions [168, 169]. Different approaches, such as fuzzy logic, have been proposed to optimise the operation of power systems [170]. Fuzzy logic control has been evaluated in Dinorwig by King et al. [171], who have described a study where a fuzzy inference system is assessed as the basis of a governor. MLD systems can be applied to represent control logic that involves the characteristics of both high (economics, lifetime) and low (dynamic response, accuracy) levels of control.

Recently, some studies on hydroelectric power plants have explicitly used hybrid systems techniques to integrate different levels of control. Gallestey et al. [172] have applied MPC and MLD to optimise the operation of a power plant taking into accounts not only immediate profits but also lifetime consumption. Chang et al. [173] have shown how mixed integer linear programming can be applied to short-term scheduling of a hydro system, while Lu et al. [168] have analysed how the market clearing price can be taken into account to establish an operational strategy.

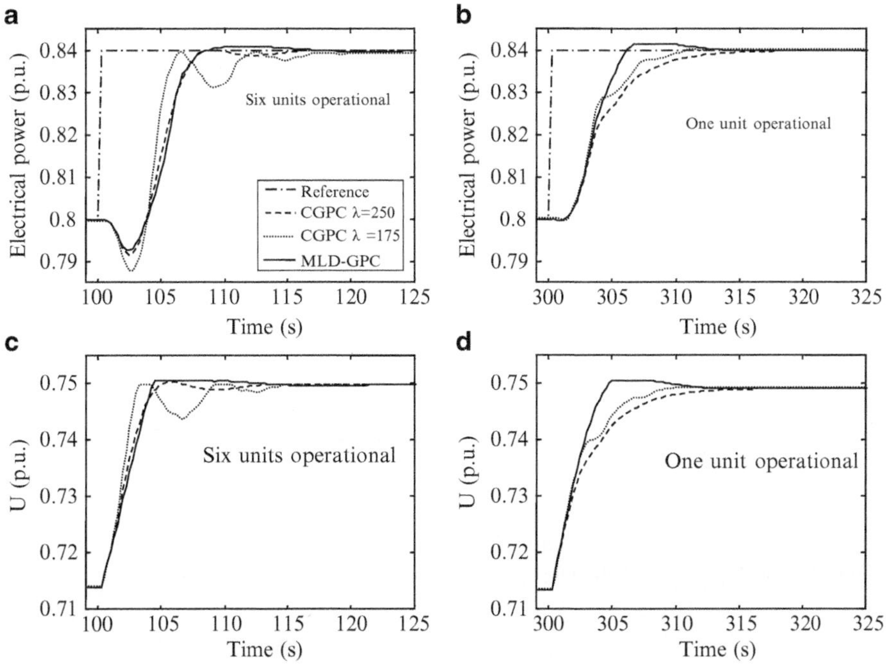

Fig. 12.10 Comparison of the responses and control signals produced by MLD-GPC and CGPC with two different values of lambda

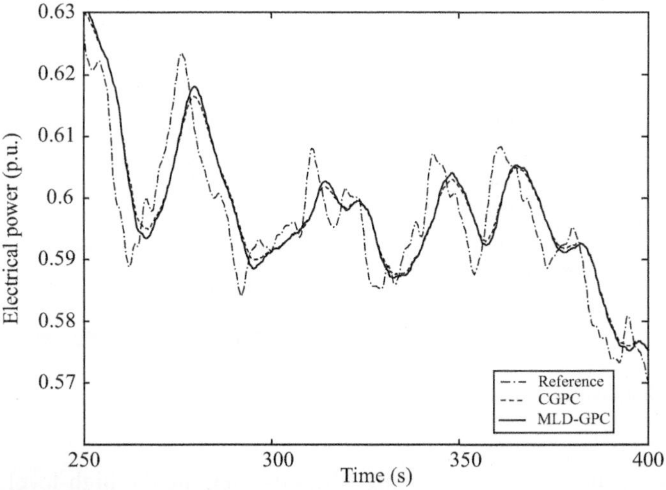

Fig. 12.11 Comparison of the frequency responses produced by MLD-GPC and CGPC controls, six units operational

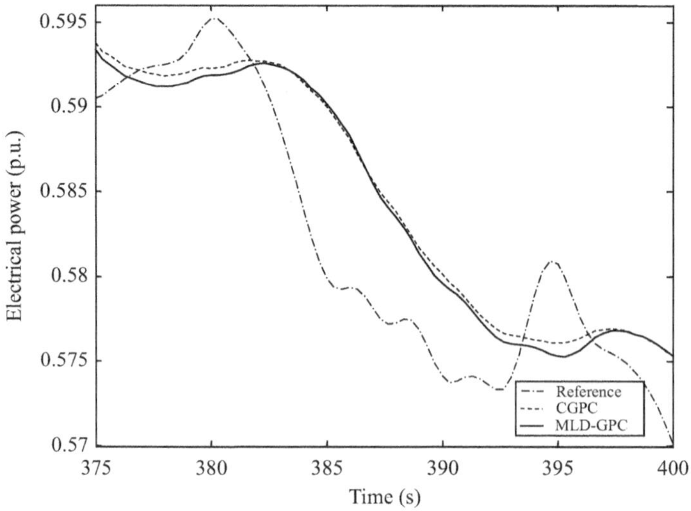

Fig. 12.12 Detail of the comparison of the frequency responses produced by MLD-GPC and CGPC controls, six units operational

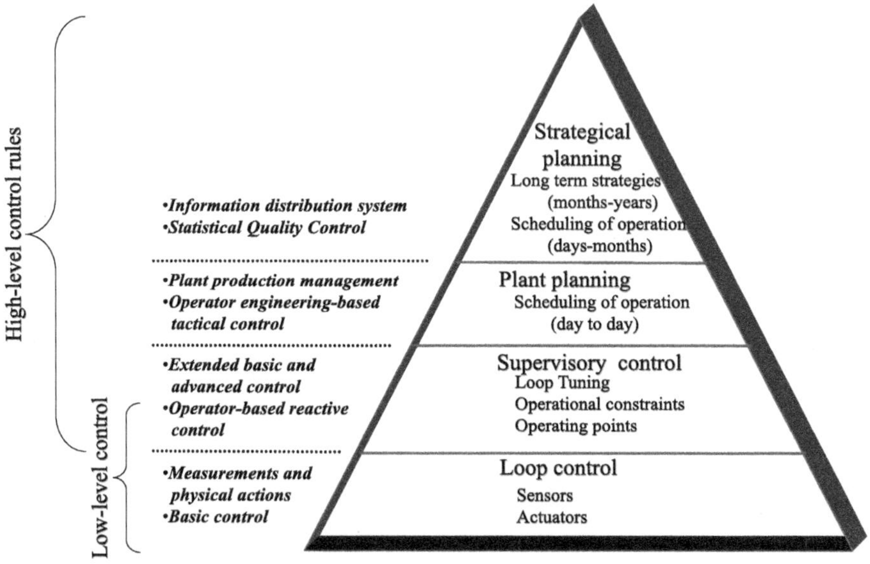

Fig. 12.13 Hierarchical control

In this work, some examples of using MLD to integrate the high-level control are presented. The models to generate high-level parameters and reference signals are not discussed here because these details are not within the scope of the study. Nevertheless, they could be assessed in future research.

12.5.2 Lifetime Consumption

Lifetime consumption models can determine the aging process, providing information about direct relationship between plant load and plant aging, and registering the operating history of the components [172]. The aging model can be used to modify constraint limits or to alter the objective function, for example to include more emphasis on the reduction of control effort than on the speed of the plant response. Predictive control can deal with this situation. For example, when Rossiter et al. [133] applied predictive control to a fossil-fired power station, they were able to reduce the effort of control without detriment to the plant performance.

The problem of operational optimisation is an objective function subject to several constraints. The generic form of this equation is

$$J(u) = \int_{t}^{t+N_{\mathrm{T}}} e\left(\tau, r_{\mathrm{u}}(\tau)\right) + c\left(\tau, r_{\mathrm{u}}(\tau)\right) - q\left(\tau, r_{\mathrm{u}}(\tau)\right) \mathrm{d}\tau, \tag{12.10}$$

where N_{T} is the time-optimisation horizon (few days), r_{u} the set of power plant references and constraints, e the aging rate cost of plant components, c the cost rates and q the revenue rates.

The control constraints and references are then calculated by Eq. 12.10. The process could be seen as a double loop of control, where a fast internal loop, with a sample period of seconds or lower, calculates the control signal and a slow external loop, with a sample period of minutes or higher, updates the constraint limits and fixes the references taking into account not only the performance of the loop of control but also other criteria, such as the lifetime consumption. For example, if a reduction in the effort of control is required, then a constraint that is based on Eq. 12.10 can be applied. Figure 12.14 shows the comparison of the effects of this constraint in the output and control signals of a nonlinear model under a MLD-GPC, which was tuned as in the previous section. As can be seen from the figure, the response with constraints is not only faster but also requires less control effort, since it involves less activity of the actuator.

12.6 MPC Real-Time Applications

A short discussion of real-time systems was presented in Sect. 8.2; however, the features that limit the application of MPC to real plants have been not mentioned. Those aspects could be, among other, restrictions on the computation time and on the storage space. The computation time for MPC can be substantially greater than other solutions because, under certain assumptions on the problem structure, the control solution uses linear or quadratic programming at each sampling instant. On the other hand, the storage space is related to the sample rate; both depend on the complexity of the algorithm and the hardware implementation.

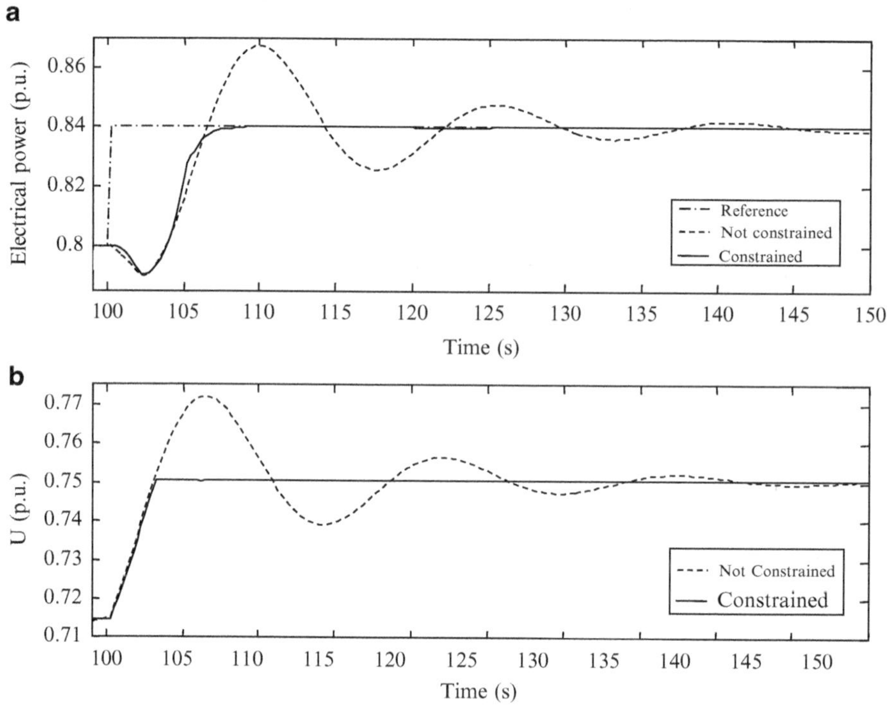

Fig. 12.14 Output and control signals of the response with and without constraints: (**a**) output signals and (**b**) control signals

Many applications of MPC have been reported, where different approximations to solve the optimization problem have been explored. For instance, Qin and Badgwell presented an overview of commercially available MPC technology [174]; for this work, information was collected from vendors starting in mid-1999, reflecting the status of MPC practice just prior to the new millennium, approximately 25 years after the first applications.

A common characteristic in almost all current available MPC methods, industrial and academic, is that they compute the optimal controller online or derive an explicit control law, but there are some applications where due to the systems' dynamic or the limited hardware (where the controller is embedded) that 'optimal' solution is not feasible. This has increased the interest in the development of new methods to either improve online optimization or to approximate explicit solutions [175–178].

Another practical solution to the optimization problem is to use dedicated hardware such as digital signal processors (DSPs), field programmable gate arrays (FPGAs) or, in general, system on-a-chip implementation; these devices could give an order of magnitude reduction of the computing time, although this would lead to increased development costs and require interfacing to the industrial PLC used to implement the governor [179–185].

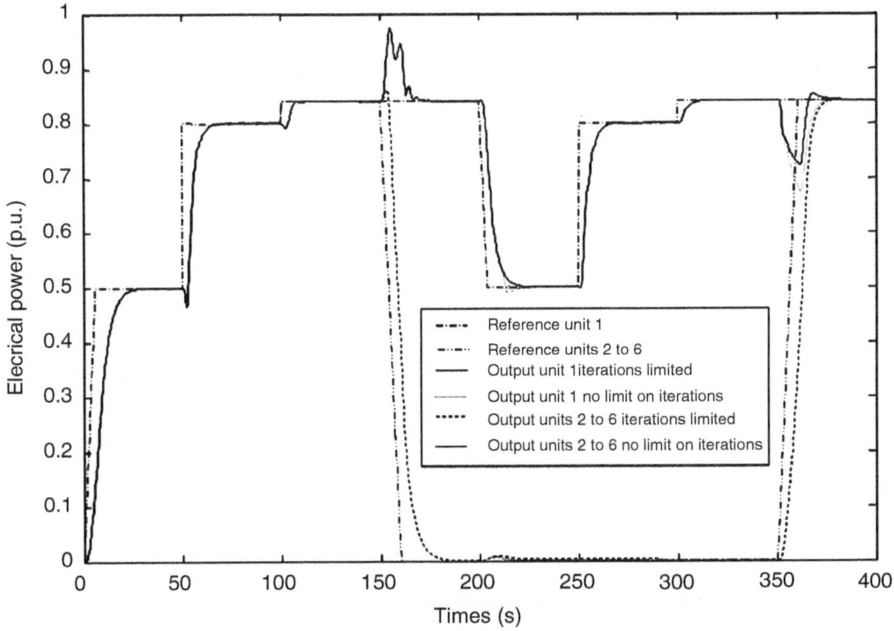

Fig. 12.15 Output responses for different CGPC, with and without limit on the number of iterations

A work, where the computation time for the evaluation of MPC applications on Dinorwig power plant, has been published by Munoz-Hernandez and Jones [135]; this work shows a comparative measure of the computation times of CGPC and GPC executing on the same platform. In this study, a simulation was set up where the power reference was varied in small and large steps and ramps, in order to activate the CGPC control constraints. The plant responses for the two CGPC cases were compared over the simulation period and were found to have only a significant discrepancy only occurring in the extreme case of the six units being ramped from zero to full load together, Fig. 12.15. The authors concluded that only a small number of QP iterations need to be done to obtain almost all the benefits of the CGPC algorithm. As can be seen in Fig. 12.16, the loop time for the GPC controller is small and constant, whereas the loop times for CGPC are much longer. In the case of unlimited iterations, the loop times are variable but when the number of iterations is limited to five, the loop time is reduced although it remains much greater than GPC; the loop time for CGPC appears bounded when the number of iterations is limited. This is summarised in Table 12.1. The maximum loop time of 360 ms in Table 12.1 is an order of magnitude greater than the 40 ms sample rate used on the Dinorwig governors. However, this is a measurement made in SIMULINK®'s offline interpretive simulation mode and is subject to considerable overhead. Speed

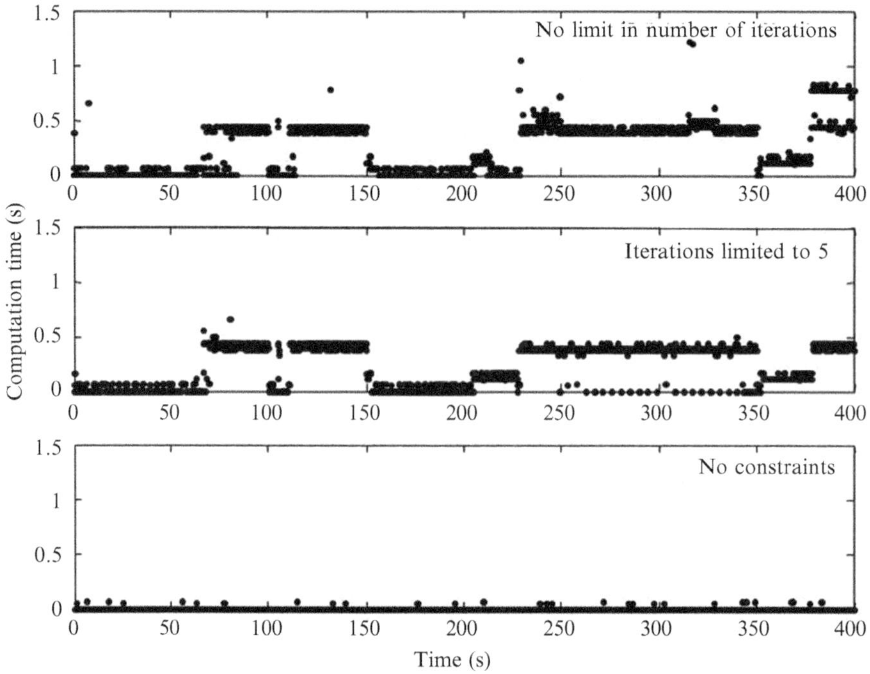

Fig. 12.16 Comparison of computation times for the GPC algorithm, showing the effect of QP search iterations needed in the constrained version

Table 12.1 Comparison of computation times for the GPC algorithms

Algorithm	Average (s)	Maximum (s)
GPC	0.001	0.02
CGPC limited to five iterations	0.115	0.36
CGPC unlimited iterations	0.214	1.56

Source: Munoz-Hernandez and Jones [135], Table 2, p. 50

gains can be achieved by rewriting S-functions in the C language. A further speed gain can be achieved if Real-Time Workshop [94] is used to generate optimised and model-specific code, which can then be compiled.

12.7 Conclusions

The results have shown how MLD-GPC can be applied to a hydroelectric pumped-storage station to improve its fast-response characteristics. The direct transient responses are improved compared with CGPC, when the multivariable nature of the plant is taken explicitly into account. In particular, selecting different values of λ and different predictive models, depending on the number of units active, allows MLD-

GPC to produce a fast response across the operational range. Inclusion of rate and saturation constraints in both GPC controllers yields a fast, well-damped response in the common case when only a single unit is in operation, without compromising stability when multiple units are online.

The brief discussion of the integration of high-level control rules into MLD-GPC and the simple examples presented are intended to help in the elucidation of future research rather than form a comprehensive study. Nevertheless, the versatility of this approach has been described and its potential for application to pumped storage stations has been discussed.

Chapter 13
Outlook and Conclusions

13.1 Outlook

The U.S. Energy Information Administration (EIA) publishes an annual outlook for international energy markets, including electricity [186]. The future prospects for hydroelectricity are linked to the overall consumption of electricity, and worldwide this is projected to rise by an average of 2.3% per year over the period 2007–2035, as shown in Fig. 13.1.

Figure 13.1 differentiates between countries which belong to the Organisation for Economic Cooperation and Development (OECD) and those which do not. OECD countries comprise:

- OECD North America (USA, Canada, Mexico)
- OECD Europe
- OECD Asia (Japan, South Korea, Australia, New Zealand)

It is interesting to note that the projected worldwide average growth of 2.3% exceeds the corresponding figure of 1.9% recorded between 1990 and 2007, and that much of this acceleration is due to countries outside the OECD. This trend is largely due to the extension of networked electricity to areas and populations that currently have little or no service.

Over the same period, it is projected that the average growth rate of renewable electricity generation will be about 3% *per annum* and that its share of the total will increase from 18% to 23%. Of this, 54% is predicted to come from hydroelectricity and 26% from wind power. It is not foreseen that other sources will be competitive with fossil fuels and will make only a modest contribution.

Figure 13.2 shows that a difference in trend is expected between the OECD and non-OECD regions. In the latter, significant numbers of large-scale sites remain that can be exploited; as noted in Chap. 1, almost all current or planned schemes are in non-OECD countries. In contrast, hydroelectricity development in OECD countries is confined mostly to sites that are small or more difficult to exploit and the growth of base load hydroelectric generation is limited. On the other hand, there is a very

G.A. Munoz-Hernandez et al., *Modelling and Controlling Hydropower Plants*, Advances in Industrial Control, DOI 10.1007/978-1-4471-2291-3_13,

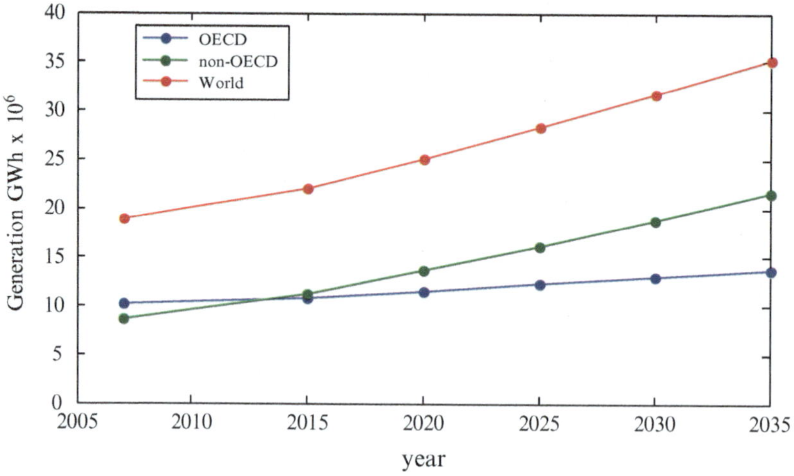

Fig. 13.1 Projected net electricity generation for 2007–2035. (Source: U.S. Energy Information Administration (EIA) [186], Table 11, p. 78)

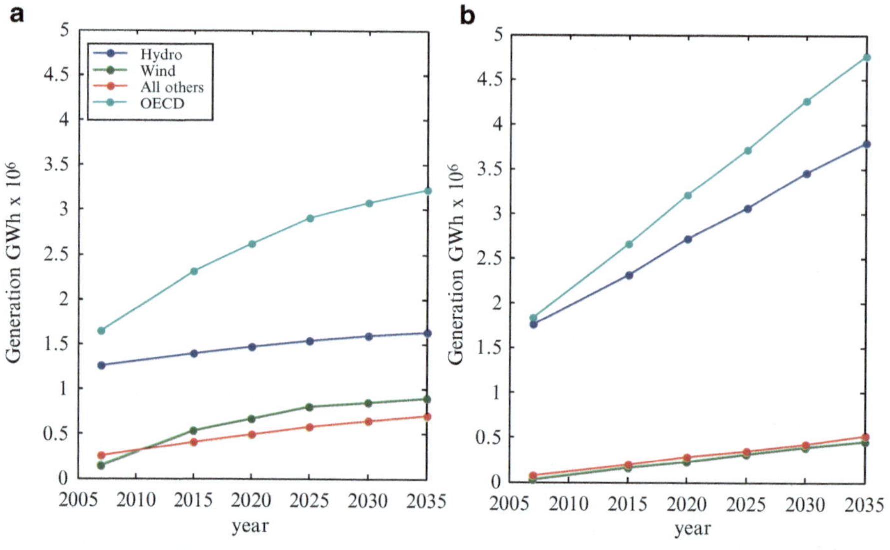

Fig. 13.2 Projected net renewable electricity generation for 2007–2035, (**a**) OECD region (**b**) non-OECD regions (Source: U.S. Energy Information Administration (EIA) [186], Table 12, p. 81)

pronounced trend in OECD countries to expand wind and other renewable sources which are characterised by their intermittent nature. Here, the role of hydropower is more likely to be as an agent for storage and network control rather than raw generation. Summing up, it may be foreseen that:

- In non-OECD countries, hydropower will be an important contributor to base load renewable generation; strong growth can be anticipated in the BRIC countries.
- Within the OECD, continued development of hydropower as a renewable resource is likely to be encouraged by means of policies such as tax incentives and feed-in tariffs. However, it is likely that attention will be focused on its potential for energy storage and as a mechanism for counteracting intermittency of supply.

13.2 Future Role of Pumped Storage

If the indications of the previous section are correct, there is an important future role for pumped storage hydropower, perhaps mostly within the OECD but increasingly worldwide as grid electrification expands. This presents both an opportunity and a challenge. Modelling, control and simulation – the topics of this book – will continue to be one of the tools that helps meet these challenges.

The UK Renewable Energy Strategy [187] outlines a plan to produce 15% of the energy consumed in the UK from renewable sources by 2020, of which 30% or more will be electrical energy. For an average electrical load of 40 GW, renewable sources will therefore need to provide about 12 GW. The dominant supplier (if the target is indeed met, either wholly or partially) is likely to be offshore wind farms which, assuming a capacity factor of 33%, implies that about 36 GW of capacity needs to be installed. Any curtailment of power from wind farms that reduces their capacity factor must be avoided, if for no other reason than to assuage the fears of an often sceptical public that investment in wind farms is ill-founded. That the fluctuating nature of the wind is an impediment to its use as a source of power has long been recognised. In a very readable summary of this well-known problem, MacKay [188] quotes remarks made as far back as 1866 by the British economist and logician Jevons on the deficiencies of windmills as a method of draining mines [189]. MacKay identifies the main problems as lulls (periods of several days with light winds) and slews (rapid changes in either supply or demand) and discusses the potential of pumped storage as a means of compensating. The prescience of Jevons[1] comes to the fore again:

> ... if there is anything which could be cheaply done by wind, it is the raising of large masses of water where occasional irregularities are of no consequence, the rain and wind mostly coming together.

[1]In writing this brief outlook, the authors are acutely aware of the potential embarrassment of attempting to foretell the future because, in the same book, Jevons also comments on the possibilities of manned flight: 'Uncertainty will for ever render aërial conveyance a commercial impossibility'.

A more up-to-date analysis of the impact of pumped storage on power systems affected by high levels of wind power is provided by Tuohy and O'Malley [190]. They conclude that pumped storage reduces wind power curtailment and enables better use to be made of base load plant thus reducing system costs. However, they also show that this is true only for a network with rather high levels of wind penetration (>50%) and that constructing additional pumped storage stations is uneconomic because of their high capital cost.

Denmark, where wind power already supplies over 20% of electricity, in effect has adopted pumped storage as a solution to variability – except that the storage is provided by their European neighbours. Danish wind power is balanced by exporting it across interconnectors to Norway and Sweden who can use their extensive hydro storage to compensate. A more detailed account of the experience of integrating wind power on the Danish network is given by Söder et al. [191]. For many countries, such a strategy is not tenable at present, though a European HVDC 'Super-Grid' is under investigation.

We are reminded by Taulan et al. [192] that the initial premise of pumped storage was simple 'load-shifting' to compensate for the lack of variability of conventional base load plant, including the early nuclear stations. Generally, excess generation during the night was stored and released during the day, allowing the base load plant to operate efficiently at more or less fixed output. The hydroelectric company derived its revenue from the day/night price differential for electricity. For the UK's 'island' grid, the need for a 'peak lopping' function was more acute than in continental Europe. This placed more emphasis on constructing stations such as Cruachan and Dinorwig that have a high slew rate on spinning reserve that can counteract major load events, such as sudden disconnection of a large base generator. The next natural step was automatic control of grid frequency in a closed loop. Note the trend–load shifting on a timescale of hours, peak lopping on a timescale of a few minutes and frequency control on a timescale of a few seconds. It is interesting to reflect that the pumped-hydro concept, originally intended to compensate for the lack of variability of base load plant, is now viewed as a solution for the excessive variability of renewable sources!

Electricity wholesale markets generally work on a day-ahead basis. Generators bid in an auction to supply electricity for slots (usually an hour or half-hour) in the following day. Once the contract price has been fixed, the generator must fulfil this commitment. If they fail to do so then they must make up the difference at the spot market price (generally higher than the day-ahead price) and also pay a penalty. The market is co-ordinated by a transmission system operator, such as National Grid in the UK. So a generator's aspiration, broadly speaking, is to offer all the electricity it can economically deliver and then supply exactly that amount. Clearly, this poses a big problem for renewable sources. One strategy is to 'pair' wind farms with pumped-hydro storage, and optimal policies for unit commitment under such arrangements are a topic of current research [193]. There are many performance studies on local, isolated grids supplied by hybrid renewable sources combined with pumped storage hydro [194, 195]. However, the major challenge will be their integration into national and international power networks.

This discussion gives a clear indication of the future requirements for any system intended to control the grid (meaning active/reactive power balance, frequency regulation and quality of supply) – make it quick, make it accurate and make it flexible. To this may be added the usual overriding provisos that it should be safe and reliable and preferably cheap, or at least not too expensive.

In a recent survey of storage methods for mitigating the variability of renewable energy sources on power networks, Beaudin et al. [8] cite the advantages of pumped storage as:

- High efficiency (65–85%)
- Large power capacity (100–1,000 MW)
- Large storage capacity (up to 20 GWh)
- Long life (30–60 years)
- Low cycle cost (i.e. the store/release cycle causes relatively little wear and tear on the system)

They also note that a high slew rate capability is likely to increase in importance, a trend anticipated by the Kopswerk II pumped storage scheme in Austria [196]. This 820 m head station opened in 2008 and is equipped with three 'ternary groups'. Each group consists of a 180 MW Pelton turbine, a 200 MVA motor/generator, a pump and a torque coupler all mounted on a common shaft. The station does up to 60 load changes a day, with a response time of less than 20 s, supplying up to 525 MW of power to the grid or absorbing up to 450 MW from it. The use of separate pumps and turbines allows the individual components to be optimised for efficiency, thus avoiding the inevitable compromise in a reversible pump/turbine. Crucially, this arrangement can implement a 'hydraulic short-circuit' that allows full control of the power drawn from the grid during pumping as well as very fast transitions between pump and generation modes.

Another technology that is gaining prominence is adjustable-speed pumped storage, whose advantages are [192, 197]:

- *Optimising conversion efficiency*: As head and flow vary, the unit is able to increase or decrease its speed to operate closer to the pump/turbine's best efficiency point. This also makes better use of the reservoir because larger water level variations can be allowed.
- *Adjustable load in pump mode*: This is particularly useful at night to give finer power balance when the station is operating principally in pump mode.
- *Power control*: Independent delivery of both active and reactive power to the grid.
- *Power injection*: Retarding the rotor to a new steady state speed gives a brief but rapid injection of power into the grid, contributing to grid stability.

Kuwabara et al. [198] describe one of the earliest large-scale implementations of adjustable speed pumped storage – a 400 MW unit at Ohkawachi power station in Japan that was developed by the Kansai Electric Power Company and Hitachi. Their evaluation of operational experience at the plant was very favourable and they reported that its instantaneous power capability made a greater contribution to the stabilisation of the power system than expected. This encouraged similar developments in Europe such as the scheme at Goldisthal in Germany, which was commissioned in 2004. This station has two units whose speed can be adjusted from 90% to 104% of synchronous and output power variable in the range 190–290 MW. The variable-speed machines are used for both primary and secondary regulation and are claimed to have an efficiency advantage of about 10% when compared to similar fixed speed units. However, it must not be thought that adjustable speed motor/generators are restricted to large-scale installations because there are several examples of small-scale proposals too [199, 200].

These recent innovations have prompted a revival of pumped storage technology in Europe and the USA and it is tempting to believe that the future of hydropower is assured. However, it must be remembered that technical solutions come at a price. Power network co-ordinators may be grateful that a technical solution exists but this will not prevent them from using it as sparingly as possible and to continually seek cheaper options. In their survey [8] Beaudin et al. list other storage technologies which could in future challenge the position that hydropower currently enjoys:

- Compressed air storage
- Batteries
- Superconducting magnetic energy
- Hydrogen storage
- Flywheels
- Capacitors and super-capacitors

They conclude that pumped storage is good at the load-following and unit-commitment timescales but that flywheels, capacitors and batteries are most suitable for maintaining power quality and grid stability. Indeed, Beacon Power Corp is already regulating the frequency of New York State's grid by means of an 8 MW capacity flywheel, shortly to be upgraded to 20 MW [201]. A plant for commercial scale lithium-ion battery storage is also under construction and may foreshadow the concept of regulating frequency using millions of electric car batteries and demand-side management via a 'Smart-Grid'. Ultimately, this may present the greatest long-term threat to hydropower's pre-eminence in this role [188]. Although these are relatively immature technologies at present, they are a reminder that more R&D and continual improvement of the technology will be required to preserve hydropower's lead.

13.3 Conclusions

In this brief outlook, we have attempted to discern recent and current trends in hydropower and there appear to be several positive indicators towards a bright future. From the perspective of this book, the need for improved simulation models and more sophisticated control methods to ensure system stability and rapid response will continue. It is likely that the power industry will follow the aerospace and automotive industries in making greater use of simulation-based design methods. More use will be made of hardware-in-the-loop simulators for commissioning but it will also be extended to other areas of operation such as maintenance, fault-finding and training. Perhaps the greatest test for an industry which is (quite understandably) rather conservative in nature is how to implement and evaluate some of the advanced methods emerging from the realms of research. It is imperative that system safety and reliability remain the top priority but a greater emphasis on experiment may well be needed as alternative technologies compete for the roles traditionally filled by hydropower.

Appendix A
Dinorwig Simulation Models

This chapter summarises the various mathematical models used to represent the dynamic characteristics of the Dinorwig Hydroelectric Pumped Storage Plant. As described earlier, the models were developed systematically with increasing complexity, each model suitable for a particular system dynamic study. For instance, linear models can be applied to represent the effects of low frequency and guidance in speed control and nonlinear models are required when large changes of speed and power are considered, such as in islanding, load rejection and systems restoration studies. As Dinorwig has six units, the approaches considered both single input single output (SISO) and multivariable (multiple inputs multiple outputs, MIMO) models. The model of the system can be separated into subsystems. Figure A.1 shows a schematic of the Dinorwig power plant model. The full hydroelectric station model is constructed combining the four subsystems: guide vane dynamics, hydraulic subsystem, turbine/generator and sensor filters.

Each block can be selected to represent a diversity of operation modes. For example, there are three models available to simulate the hydraulic subsystem: linear, nonlinear nonelastic and nonlinear elastic. The guide vane dynamics can be selected with or without rate limitation and saturation. The sensor filters block is a fixed block. The grid model can be adjusted to represent different conditions of the grid. Through the governor block classical and advanced controllers can be selected.

The principal constants and parameters used in modelling and simulating are listed in Table A.1.

A.1 Hydraulic Subsystem

The layout of the Dinorwig pumped storage station is shown in Fig. A.2. As was discussed in Sect. 4.1, this scheme consists of the upper reservoir, a low-pressure tunnel, a high-pressure tunnel, a manifold and six individual penstocks.

G.A. Munoz-Hernandez et al., *Modelling and Controlling Hydropower Plants*, Advances in Industrial Control, DOI 10.1007/978-1-4471-2291-3_14,

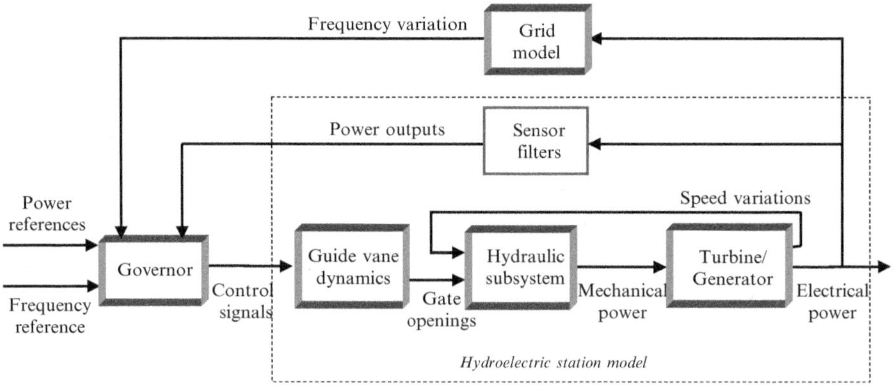

Fig. A.1 Schematic of the SIMULINK® model developed (Fig. 7.12)

The mechanical power (P_m) available from a hydraulic turbine is related to the head (h) and flow by the following equation:

$$P_m = A_t h(q - q_{nl}) - D_n G \Delta n. \tag{A.1}$$

In (A.1) q_{nl}, the no-load flow, is subtracted from the net flow (q) in order to represent the fact that the turbine is not 100% efficient. A_t is the turbine gain; its value depends directly on the turbine MW rating and inversely on the generator MVA rating. A turbine damping effect is also included; this effect is a function of the guide vane opening (G), the speed variation (Δn) and the turbine-damping coefficient (D_n). The models are expressed in the per-unit system, normalised to 300 MW and 50 Hz.

The net flow through the turbine depends on the guide vane opening and the pressure, or head (h), according to

$$q = G\sqrt{h}. \tag{A.2}$$

The nonlinear relationships (A.1) and (A.2) cause the effective gain of the plant to vary significantly with flow and head. They also show that reducing steady state power output requires the guide vanes to be closed. Doing so, it causes a transient increase in the turbine inlet pressure whose amplitude depends on the rate of closure. However, as a consequence of the inertia of the moving water column, there is no change in the instantaneous flow and there is consequently a transient increase in the power output, which is the opposite of the desired effect. This is termed a nonminimum phase (NMP) response and it is an important limiting factor on the dynamic performance of fast-response hydroelectric plant.

Table A.1 Dinorwig parameter used in modelling and simulating

Symbol	Name	Subsystem
Δn	Speed variation	Hydraulic
η	The turbine efficiency	Hydraulic
ρ	Water density	Hydraulic
a	Velocity of sound in water (velocity of the pressure wave moves)	Hydraulic
A	Cross-sectional area of the tunnel	Hydraulic
A_s	Cross-sectional area of the surge tank	Hydraulic
A_t	Turbine gain	Hydraulic
C_s	The storage constant of the surge tank	Hydraulic
d	Penstock diameter	Hydraulic
e	Thickness of the penstock wall	Hydraulic
E	Elastic modulus of the penstock material	Hydraulic
f_p	Head loss coefficient, $m/(m^3/s)^2$	Hydraulic
f_{pn}	Head loss coefficient in penstock	Hydraulic
f_{pt}	Head loss coefficient in main tunnel	Hydraulic
f_r	The friction factor	Hydraulic
F	Friction losses in the penstock	Hydraulic
g_a	Acceleration due to gravity, m^2/s	Hydraulic
G	Guide vane position	Hydraulic
G_{fl}	Guide vane position at full load	Hydraulic
G_{nl}	Guide vane position at no-load	Hydraulic
h	The head at the turbine admission, m	Hydraulic
h_0	The static head of water column, m	Hydraulic
h_f	The head loss due to friction, m	Hydraulic
h_s	The rate of change of the tank level	Hydraulic
k	Bulk modulus of the water	Hydraulic
l	Length of sections in the water passage	Hydraulic
P_m	Turbine output power	Hydraulic
q	Actual turbine flow, m^3/s	Hydraulic
q_{base}	Turbine flow rate	Hydraulic
q_{nl}	No-load flow	Hydraulic
q_p	Flow to turbines	Hydraulic
q_s	Flow into the surge tank	Hydraulic
q_t	Flow down the upper tunnel	Hydraulic
T_e	Wave travel (propagation) time in one penstock	Hydraulic
T_{em}	Wave travel (propagation) time in the main tunnel.	Hydraulic
T_{st}	Surge oscillation period in the surge tank	Hydraulic
T_w	Water time constant	Hydraulic
T_{wn}	Water starting time of a 'n' single penstock	Hydraulic
T_{wt}	Water starting time of the main tunnel.	Hydraulic
v	The water velocity, m/s	Hydraulic
$v^2/2g$	The velocity head	Hydraulic
Z_0	Surge impedance single penstock	Hydraulic

(continued)

Table A.1 (continued)

Symbol	Name	Subsystem
Z_{0T}	Surge impedance main tunnel	Hydraulic
D_n	Turbine-damping coefficient	Hydraulic-Electrical
T_m	Machine starting time	Hydraulic-Electrical
β	The stiffness	Electrical
Δf_{ss}	Frequency deviation	Electrical
δ_m	Mechanical torque angle of the rotor, rad.	Electrical
ω	Angular velocity of the rotor in electric rad/s	Electrical
ω_0	Base rotor electrical speed	Electrical
ω_m	The mechanical angular velocity	Electrical
ω_{m0}	Rated angular velocity in mechanical radians per second	Electrical
D	Load damping factor	Electrical
D_{pf}	The frequency sensitivity parameters of the model for active power	Electrical
D_{qf}	The frequency sensitivity parameters of the model for reactive power	Electrical
f_0	Rated frequency	Electrical
f_{fl}	Steady state frequency at full load	Electrical
f_{nl}	Steady state frequency at no load	Electrical
F_a	Average frequency	Electrical
F_f	Power system frequency	Electrical
H	Per unit inertia kinetic energy constant of the machine at rated speed per machine volt-ampere rating (turbine/generator inertia constant)	Electrical
J	Combined moment of inertia of the generator and turbine, kg.m^2	Electrical
K_D	The per-unit coefficient of damping torque	Electrical
K_s	Synchronising torque coefficient	Electrical
M	The combined inertia constant of the local machine and the effective rotating inertia of all the other machines connected via the power system	Electrical
p_n	The number of generator poles	Electrical
P	Active power	Electrical
P_0	Active power at normal frequency	Electrical
Q	Reactive power	Electrical
Q_0	Reactive power at normal frequency	Electrical
R	Speed droop	Electrical
t	Time, s	Electrical
T_{acc}	Accelerating torque, N m	Electrical
T_{elec}	Electromagnetic torque, N m	Electrical
T_{mech}	Mechanical torque, N m	Electrical
T_s	The synchronising coefficient of the power transmission line	Electrical
T_{sa}	Sampling period	Electrical

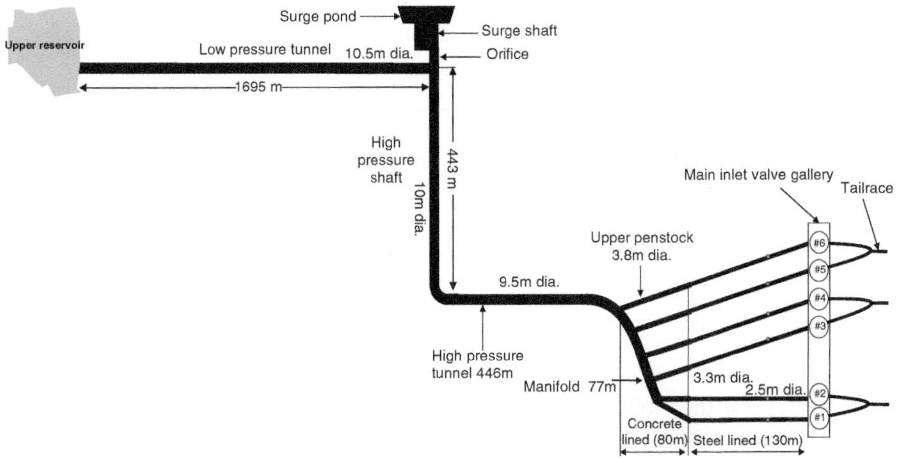

Fig. A.2 Dinorwig hydraulic system (Fig. 4.1)

A.1.1 Linearised Model

A.1.1.1 SISO Linearised Model

For small variations around an operating point, the hydraulic model can be linearised, considering the main dynamic characteristics of the hydraulic subsystem, by Eq. A.3. In this approximation, the hydraulic coupling is not considered. This SISO linear model is appropriate to represent the operation of Dinorwig with a single unit active or when multiple units active are working synchronously. It is also suitable for studies for control tuning using linear analysis tools.

$$G(s) = \frac{P_{\mathrm{m}i}(s)}{G(s)} = \frac{A_{\mathrm{t}}(1 - G_{\mathrm{o}} T_{\mathrm{wt}i} s)}{\left(1 + \frac{G_{\mathrm{o}} T_{\mathrm{wt}i}}{2} s\right)} \tag{A.3}$$

In the transfer function (A.3), $P_{\mathrm{m}i}$ is the mechanical power produced by a single turbine and G_{o} is the operating point. T_{mt} is the *water starting time* of the main tunnel, $T_{\mathrm{w}i}$ is the water starting time of any single penstock and $T_{\mathrm{wt}i}$ is the water starting time of the main tunnel and a single penstock, that is $T_{\mathrm{wt}i} = T_{\mathrm{mt}} + T_{\mathrm{w}i}$. Kundur defines the *water starting time* as the time required for a given head to accelerate the water in the penstock from standstill to a specific velocity [19]. The values of T_{mt}, $T_{\mathrm{w}i}$ and $T_{\mathrm{wt}i}$ depend directly on the constructional dimensions of the main tunnel and penstocks.

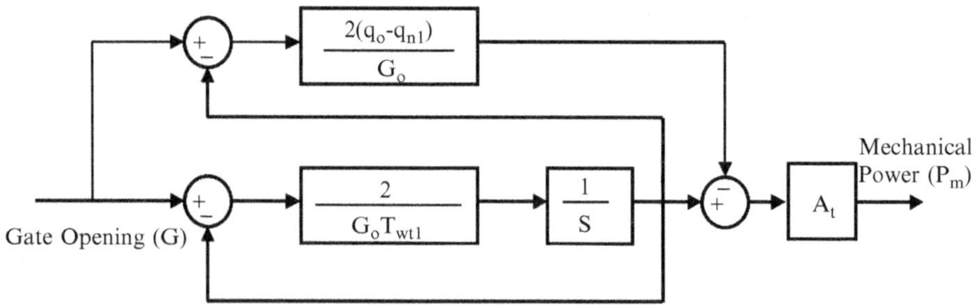

Fig. A.3 Linear model of the hydraulic subsystem

Figure A.3 is the block diagram for Eq. A.3. Friction losses are neglected but the effects of flow variation are considered [19, 28]. q_{o} is the steady state flow rate at the operating point and q_{nl} is the no-load flow, both values in the per-unit system. Note that $G_{\text{o}} = q_{\text{o}}$.

A.1.1.2 MIMO Linearised Model

Like many other stations, the hydraulic subsystem at Dinorwig is inherently multivariable, because the common supply conduit produces significant dynamic coupling between the turbines [28]. This is known to have an adverse effect on the stability margin in closed loop. Units that are online react, via their governors, to pressure and flow (and therefore power) perturbations caused by other units, whereas those that are offline have their guide vanes closed and do not interact. The structure of the plant therefore varies with time, depending on the number of active units. For this, MIMO linearised model with inelastic water column and negligible losses are assumed, also all six units are assumed to be identical (although minor differences due to manufacturing tolerances do occur in practice).

The rate of change of the flow in the penstock can be determined as

$$\frac{\mathrm{d}q}{\mathrm{d}t} = (h_o - h - h_l)\,\frac{gA}{l} \tag{A.4}$$

where A is the cross-sectional area of tunnel, h_o the static head of water column in per unit notation, h the head at the turbine admission in per unit notation, h_l the head loss due to friction, l length of tunnel and g acceleration due to gravity.

The momentum of the water in an individual penstock is

$$h - h_i = \frac{l_i}{A_i g_{\text{a}}}\left(q_{i0}\frac{\mathrm{d}q_i}{\mathrm{d}t}\right). \tag{A.5}$$

The sum of the flows in the individual penstocks must be equal to the total flow in the common tunnel, then using (A.4) the momentum equation for the water at the common tunnel is

$$h_o - h = \frac{l}{Ag_{\mathrm{a}}}\left(\frac{\mathrm{d}q_1}{\mathrm{d}t} + \frac{\mathrm{d}q_2}{\mathrm{d}t} + \cdots + \frac{\mathrm{d}q_n}{\mathrm{d}t}\right). \tag{A.6}$$

Eliminating h in Eq. A.6 using Eq. A.5, the hydraulic MIMO model can be described using a matrix of relations (A.7); this matrix changes its value depending on the number of units active [39]:

$$\begin{bmatrix} \dot{q}_1 \\ \dot{q}_2 \\ \vdots \\ \dot{q}_n \end{bmatrix} = \begin{bmatrix} T_{\mathrm{wt}n} & T_{\mathrm{mt}} & \cdots & T_{\mathrm{mt}} \\ T_{\mathrm{mt}} & T_{\mathrm{wt}n} & \cdots & T_{\mathrm{mt}} \\ \vdots & \vdots & \ddots & \vdots \\ T_{\mathrm{mt}} & T_{\mathrm{mt}} & \cdots & T_{\mathrm{wt}n} \end{bmatrix}^{-1} \begin{bmatrix} h'_1 \\ h'_2 \\ \vdots \\ h'_n \end{bmatrix}. \tag{A.7}$$

The effect of hydraulic coupling can be expressed as an effective increase of the water starting time as the number of units online increases [33, 39]. The total water starting time, $T_{\mathrm{wt}n}$, is given by

$$T_{\mathrm{wt}n} = T_{\mathrm{w}} + T_{\mathrm{mt}}. \tag{A.8}$$

In this model, the variation of the hydraulic coupling will be considered as a discrete function; therefore, the contribution of any active unit, in the starting time, is always considered at its maximum value even if its guide vane is not open at 100%.

Considering the case of two units active:

$$\begin{bmatrix} T_{\mathrm{wt}} & T_{\mathrm{mt}} \\ T_{\mathrm{mt}} & T_{\mathrm{wt}} \end{bmatrix}^{-1} = \begin{bmatrix} \frac{2T_{\mathrm{wt}}}{G_{\mathrm{o}}\left(T_{\mathrm{wt}}^2 - T_{\mathrm{mt}}^2\right)} & \frac{-2T_{\mathrm{mt}}}{G_{\mathrm{o}}\left(T_{\mathrm{wt}}^2 - T_{\mathrm{mt}}^2\right)} \\ \frac{-2T_{\mathrm{mt}}}{G_{\mathrm{o}}\left(T_{\mathrm{wt}}^2 - T_{\mathrm{mt}}^2\right)} & \frac{2T_{\mathrm{wt}}}{G_{\mathrm{o}}\left(T_{\mathrm{wt}}^2 - T_{\mathrm{mt}}^2\right)} \end{bmatrix}. \tag{A.9}$$

In Eq. A.9, the two penstocks are considered equal, that is $T_{\mathrm{wt}} = T_{\mathrm{wt1}} = T_{\mathrm{wt2}}$. Figure A.4 shows this case. In Fig. A.5 the model includes all six machines.

In Fig. A.5, ΔG is the guide vane opening, ΔP_{m} is the turbine's mechanical power and ΔP_{e} is the electrical power. The current governor comprises an individual classical PI controller on each turbine.

The hydrodynamics block of the extended MIMO model has the ability to change the matrix transfer function, which relate mechanical power with gate opening, depending on the number of units active. The elements of the matrix transfer function ($G(s)$) are the direct transfer function (diagonal) $G_i(s)$ and cross-coupling transfer function (symmetric) $X_i(s)$. The rules to determine whether a unit is online or offline are:

- Unit n comes online when $P_{\mathrm{di}} > 0$
- Unit n goes offline when $[(\Delta P_{\mathrm{di}} = 0)$ and $(\Delta P_{\mathrm{mi}}) = 0]$

Table A.2 summarises the elements of the matrix transfer function $G(s)$ of the extended MIMO model. The values are expressed in the per-unit system, normalised to 300 MW and 50 Hz, and assume a grid system with infinite busbars.

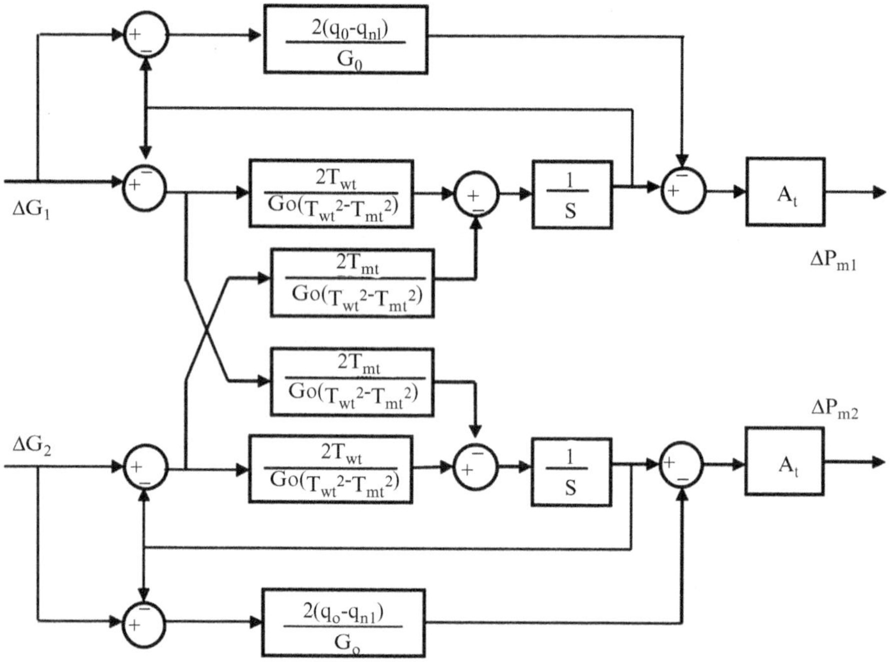

Fig. A.4 MIMO linear model of the hydraulic subsystem (Fig. 11.12)

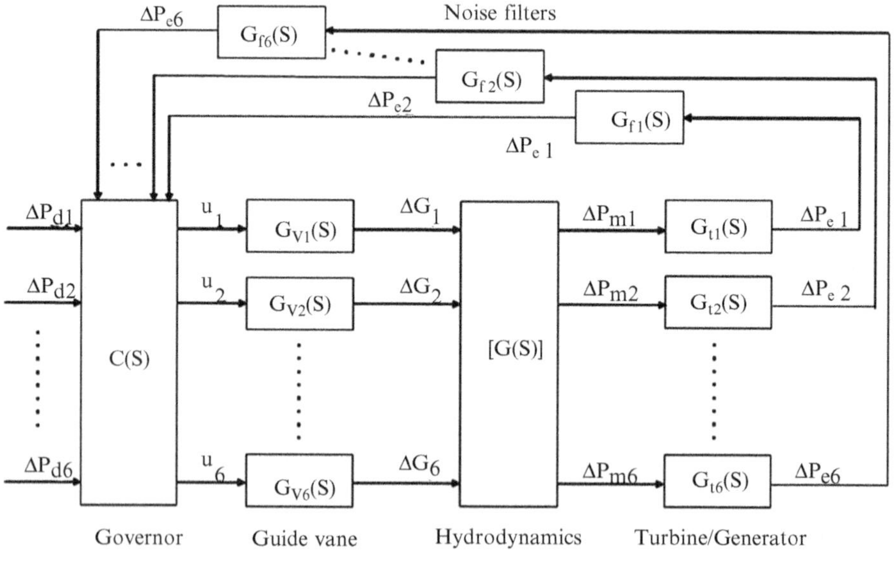

Fig. A.5 Extended MIMO model of the hydroelectric plant (Fig. 11.11)

Table A.2 Variation of transfer function matrix with number of active units (0.95 operating point), based on Table 11.3

U_o	$G_i(s)$	$X_i(s)$
1	$\dfrac{-2.358s + 3.395}{0.076s^3 + 0.8204s^2 + 2.788s + 3.031}$	0
2	$\dfrac{-2.358s^2 - 5.454s + A.96}{0.076s^4 + 1.26s^3 + 7.213s^2 + 16.69s + 13.35}$	$\dfrac{-8.559s}{0.076s^4 + 1.26s^3 + 7.213s^2 + 16.69s + 13.35}$
3	$\dfrac{-2.358s^2 - 1.986s + 11.01}{0.076s^4 + 1.221s^3 + 6.643s^2 + A.1s + 9.83}$	$\dfrac{-6.301s}{0.076s^4 + 1.221s^3 + 6.643s^2 + A.1s + 9.83}$
4	$\dfrac{-2.358s^2 + 0.03428s + 8.711}{0.076s^4 + 1.198s^3 + 6.311s^2 + 12.59s + 7.778}$	$\dfrac{-4.985s}{0.076s^4 + 1.198s^3 + 6.311s^2 + 12.59s + 7.778}$
5	$\dfrac{-2.358s^2 + 1.357s + 7.207}{0.076s^4 + 1.183s^3 + 6.093s^2 + 11.6s + 6.435}$	$\dfrac{-4.124s}{0.076s^4 + 1.183s^3 + 6.093s^2 + 11.6s + 6.435}$
6	$\dfrac{-2.358s^2 + 2.289s + 6.145}{0.076s^4 + 1.173s^3 + 5.94s^2 + 10.9s + 5.487}$	$\dfrac{-3.517s}{0.076s^4 + 1.173s^3 + 5.94s^2 + 10.9s + 5.487}$

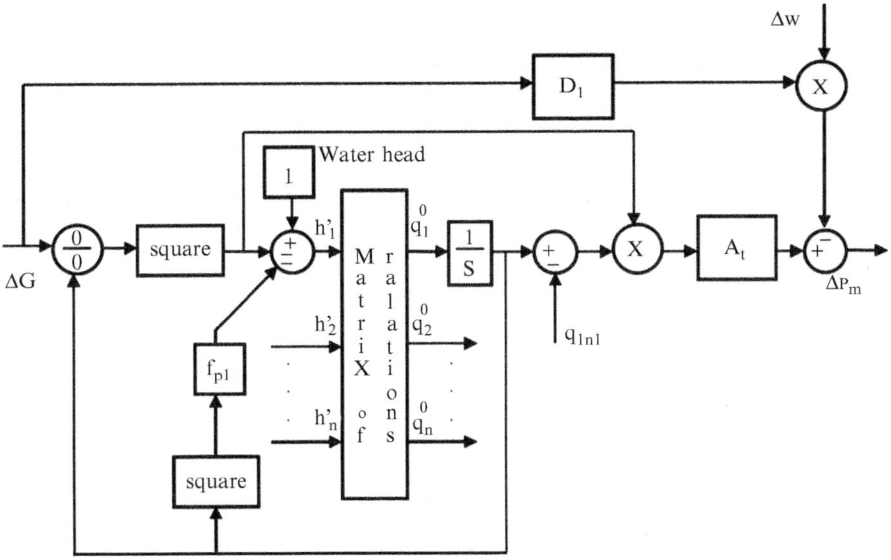

Fig. A.6 Nonlinear nonelastic hydraulic model for multiple penstocks (based on Fig. 4.9)

A.1.2 Nonlinear Nonelastic Model

The parameters of a linearised model vary with the operating point, G_o and q_o, and then the results from simulation can only be accurate near the point selected during the design process. In order to allow the simulation of large changes of speed and power, a nonlinear model, which assumes an incompressible fluid, a rigid conduit and an unrestricted head, was considered [28]. Figure A.6 shows this multivariable

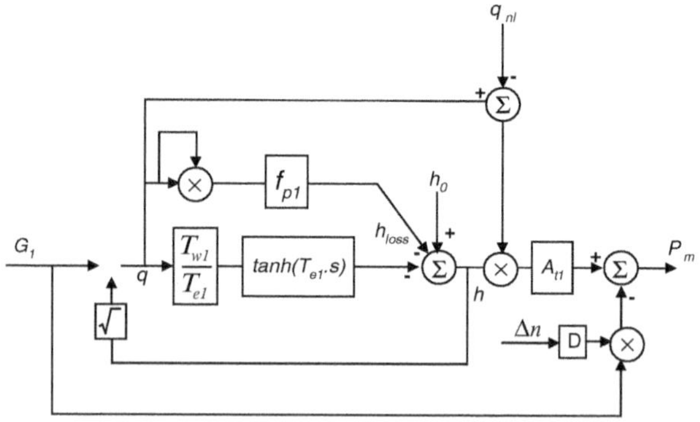

Fig. A.7 Nonlinear elastic model of a single penstock (Fig. 4.10)

nonlinear nonelastic model, Δw is the variation of the generator's speed, D_n is the turbine-damping coefficient and f_{pj} is the head loss coefficient for the jth unit. The matrix of relations utilised in this figure is Eq. A.8.

A.1.3 Nonlinear Elastic Model

Modelling of water columns assuming elastic behaviour is important when the hydraulic subsystem has large penstocks [28]. In this section, a nonlinear model that takes into account the effects of the water column, including water compressibility and pipe wall elasticity, is presented. The coupling effect is represented with the inclusion of the main tunnel, which is modelled with the same form of transfer function as a penstock. Figure A.7 shows the nonlinear elastic model of a single penstock.

A.2 Guide Vanes

The flow in the penstocks is regulated by the guide vane subsystem. The position of the guide vane depends on the control signal from the governor. The guide vane dynamics can be seen as a two-stage system where the input signal is the desired position and the output signal is the actual position of the guide vane [19, 33]. The electrical signals from the governor act as the reference to the guide vane subsystem; these signals are converted to hydraulic force and drive the servomotors that adjust the guide vane's positions. The guide vane subsystem dynamics are represented by the transfer function:

$$\frac{G(s)}{U(s)} = \frac{1}{(0.19s + 1)(0.4s + 1)}, \tag{A.10}$$

where $G(s)$ is the guide vane position and $U(s)$ is the control signal.

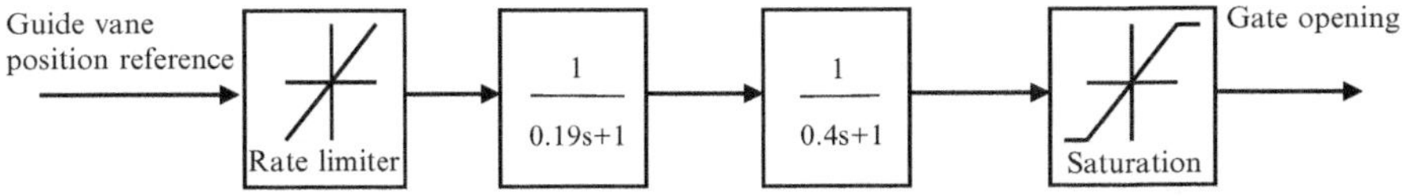

Fig. A.8 The guide vane subsystem

As shown in Fig. A.8, a *saturation constraint* limits the maximum guide vane opening to about 95% of the physical aperture to prevent it from hitting its end-stop. A fixed *rate-limit* at which the guide vane can open or close prevents excessive variation in tunnel pressure (for safety reasons and to minimise fatigue stresses on the wall material). This plays a vital role in alleviating the NMP response, which occurs during the initial part of rapid power transients.

A.3 Electrical Subsystem

The quality of the power supply is strongly dependent on the dynamic behaviour of the generators. Dinorwig has six synchronous generators that are fed with mechanical power from the hydraulic subsystem to produce electrical power at a specific voltage and frequency. The generator model is based on its response to frequency changes because the system frequency depends on the active power balance.

A.3.1 Dinorwig Electrical Subsystem

In Fig. A.9 how the turbine's mechanical power output ΔP_{m} drives the electrical subsystem can be seen. Figure A.9 represents the well-known 'swing' equations [19]. The electrical power is measured and fed back to the governor:

$$\frac{\Delta P_{\mathrm{e}}(s)}{\Delta P_{\mathrm{mech}}(s)} = \frac{K_{\mathrm{s}}\omega_0/2H}{s^2 + (K_{\mathrm{D}}/2H)\,s + K_{\mathrm{s}}\omega_0/2H}, \tag{A.11}$$

where H is the turbine/generator inertia constant, K_{s} the synchronising torque coefficient, K_{D} the damping coefficient and ω_0 the base rotor electrical speed. A first-order filter for noise reduction is included in each power feedback loop, which has the transfer function:

$$\frac{\Delta P_{\mathrm{es}}(s)}{\Delta P_{\mathrm{e}}(s)} = \frac{1}{s+1}. \tag{A.12}$$

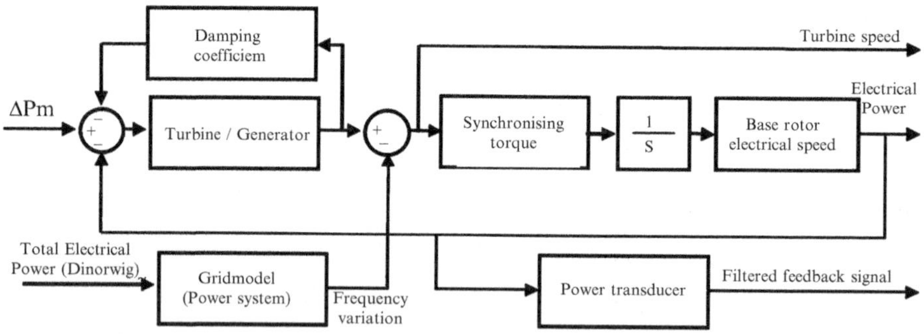

Fig. A.9 The Dinorwig electrical subsystem

The electrical coupling between generators connected in parallel, as in Dinorwig, is considerable [33]. This coupling produces an action reaction effect; for example, if a generator has a disturbance at its output, this changes the balance of the system and all the generators react, as is illustrated by the feedback of the frequency variation from the power system. This variation appears when the power system load changes, its frequency varies and the power supplied will be increased or decreased from the controlled units to attempt to return the system frequency to its base value.

A.3.2 Load Model

The load model can be seen as a second-order transfer function (A.13) [128], where the response is dominated by two time constants, one that is the sum of all inertias of rotating machines associated with the system and another with the collective effect of all regulatory mechanisms (T_R):

$$\Delta f(s) = \frac{\omega_n^2 T_R}{\beta}\left(\frac{s+\frac{1}{T_R}}{s^2+2\zeta\omega_n s+\omega_n^2}\right)(\Delta P_e(s) - \Delta P_L(s)). \tag{A.13}$$

In (A.13), the frequency (Δf) is related to the difference between the power input by Dinorwig (ΔP_e) and the load imbalance power (ΔP_L), ω_n is the natural frequency, ζ is the damping factor and β is the grid stiffness. ω_n, ζ and β depend on the droop settings, regulatory time constants and inertia of all the plant connected to the power system. They are time variant.

This monograph uses an auto regressive moving average and exogenous input (ARMAX) approximation to the load model that was proposed by Jones [131]. The main advantage of this representation is that it can calculate the parameters using operational data and taking into account the input. Then (A.13) can be written in polynomial form as

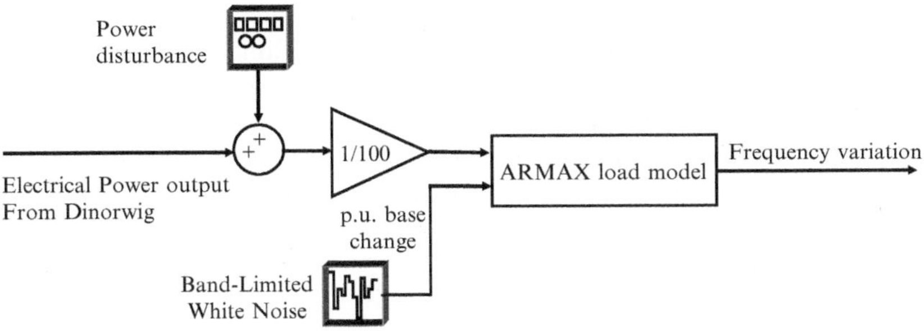

Fig. A.10 Grid model

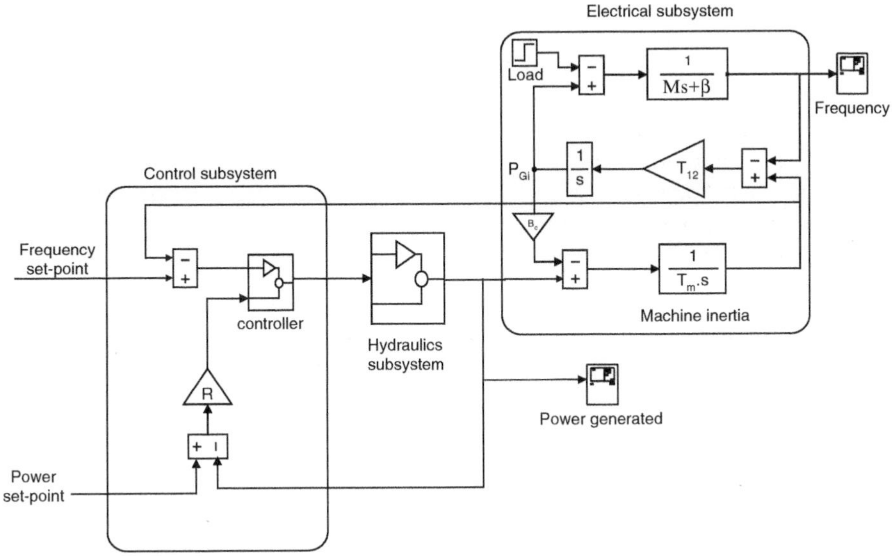

Fig. A.11 Single tunnel hydropower plant connected to the power system (Fig. 7.3)

$$\Delta f(k) = -a_1 f\,(k-1) - a_2 f\,(k-2) - a_3 f\,(k-3) + b_0 \Delta P_e(k) + b_1 \Delta P_e\,(k-1) + c_0 e(k) + c_1 e\,(k-1) + c_2 e\,(k-2) + c_3 e\,(k-3) \tag{A.14}$$

or in a general input–output discrete third-order form, considering the noise equal to zero, as

$$\frac{\Delta f(z)}{\Delta P_e(z)} = \frac{z^2\,(b_0 z + b_1)}{z^3 + a_1 z^2 + a_2 z + a_3}. \tag{A.15}$$

Figure A.10 shows the complete grid model used in this monograph to simulate the frequency variation of the power system. The band-limited white noise is used as a 'driver' for the disturbance component of the ARMAX model. A deterministic power disturbance is used to simulate changes in the power reference target produced by deviations in the national grid frequency. A generic model of single tunnel hydropower plant is assembled as shown in Fig. A.11.

Appendix B
Tuning Guidelines

Classical and modern controllers can satisfy a wide range of specifications when their internal parameters are selected correctly. A good balance between sensitivity, control effort and speed of response is generally the main objective of tuning. Numerous tuning guidelines for MPC have been proposed in the literature; many of them have been converted into suitable tuning rules and investigated with the help of simulations of processes. For instance, Yamuna and Unbehauens have published a study that resumes some of the tuning methods; for DMC and GPC approximations, available at the publication date of their paper [202], they used simulations of two transfer functions, a nonlinear unstable chemical reactor and a real-time laboratory turbo-generator control to evaluate the tuning methods. Garcia et al. have also addressed the tuning problem; in their work [203], they reviewed theorems and some general guidelines about tuning MPC. There are also, in the published literature, some guidelines or tuning discussion for specific MPC approximations, for instance GPC [141], DMC [204, 205] and MPC based on linear programming [176]. The MPC tuning is still an open research area and there are works recently published, for example [206, 207]. The guide presented in this chapter has the objective to resume the strategies used in this monograph to tune the PID and MPC.

B.1 Classical Controllers

B.1.1 PI

Tuning a proportional and integral controller (PI) requires selection of the correct values of K and K_i that allow the controller to achieve a desired plant performance [87, 126]. The values that are currently implemented in the hydroelectric plant are $K = 0.1$ and $K_\mathrm{i} = 0.12$. The PI controller with these parameters has a performance which is a compromise between one and six unit operation. Therefore, to optimise the performance of the plant, different sets of parameters were selected for the

G.A. Munoz-Hernandez et al., *Modelling and Controlling Hydropower Plants*,
Advances in Industrial Control, DOI 10.1007/978-1-4471-2291-3_15,

Table B.1 Parameters of the PI controller (Table 9.2)

	Compromise	Optimised One unit	Optimised Six units
K	0.10	0.175	0.165
K_i	0.12	0.210	0.110

Table B.2 Parameters of the PI controller with anti-windup (Table 9.3)

	Compromise	Optimised One unit	Optimised Six units
K	0.10	0.175	0.165
K_i	1.2	1.2	0.66

extreme cases, one and six units operational. The evaluation accomplished in this monograph shows that suitable ranges for the parameters of the PI are $0.23 \geq K \geq 0.1$ and $0.5 > K_i > 0.1$.

Using classical techniques (Bode and Root Locus) plus a final manual adjustment, different sets of parameters were selected for the PI controller. The goal is to optimise the response of the plant in the one and six units operational modes, reducing the primary response of the system without producing large overshoot. Table B.1 shows the set of values selected for each case.

B.1.2 PI Anti-windup

As was discussed in Sect. 9.5.1.2, although a PI carefully tuned can offer good performance, all processes are subject to constraints, then the plant alters its behaviour and the performance of the PI can be deteriorated significantly [124, 125]. A PI with tracking anti-windup structure was evaluated to be applied on Dinorwig. The saturation limit and the dead zone depend on the constraints fixed by the operator; a value of 0.95 p.u. is commonly used. Table B.2 shows the set of values selected for the cases evaluated for the PI controller.

B.2 MPC

B.2.1 SISO GPC

In GPC a quadratic cost function (11.1) is used to optimise the performance of the system. The parameters to be tuned in the GPC controller are:

Prediction model
N_1, N_2 – minimum and maximum of the prediction horizon
N_u – control horizon
α – future reference trajectory weight factor
λ – cost function control signal weight factor

Table B.3 Parameters of the SISO GPC controller (Table 11.1)

	Compromise	Optimised One unit	Optimised Six units	Constrained
N	5	20	20	20
N_u	10	20	20	20
λ	135	150	600	20
α	0	0	0	0

The prediction model plays a central role in all MPC approaches, then the process of selecting the parameters of the GPC controller utilised in this monograph begins with the selection of an approximated prediction model. After choosing the prediction model with a high value of λ, the maximum horizon of prediction is fixed. N_2 is chosen as low as possible in order to reduce computational time. The horizon of control is normally fixed to the same value as the horizon of prediction, but it can be fixed to a lower value or even $N_u = 1$ if the effort of control is not under consideration. Finally, λ is reduced to the lowest value possible; its optimum value depends on the characteristics of the system and the desired response [136]. α is normally selected equal to zero. λ has a strong effect on the effort of control, low values of this parameter producing an 'active' control while high values produce a 'passive' control [123, 141, 152].

When a SISO GPC is evaluated for Dinorwig, Eq. 11.14 is used as the prediction model; this equation is reproduced as (B.1) with a sample time $T_s = 0.25$ s and a transport delay $T_d = 2.07$ s:

$$G(z) = \frac{0.197}{z^8(z - 0.8245)}. \tag{B.1}$$

Table B.3 resume the set of parameter values selected for the cases evaluated for the SISO GPC (where $N = N_2 - N_1$).

B.2.2 MIMO GPC

MIMO GPC and CGPC controllers were designed using Eqs. (B.2) and (B.3) as the direct and cross-coupling transfer functions in the predictive model. Those functions were obtained empirically from the reaction curve of the nonlinear nonelastic model using a small step (0.04 p.u.) when six units are in operation at 95% output. The predictive model is fixed, regardless of how many units are actually in operation or their loading:

$$G(z) = \frac{0.29044}{z^4 - 0.7422z^3} \tag{B.2}$$

$$X(z) = \frac{-0.04757z^3 - 0.02751z^2 + 0.06817z + 0.006913}{z^4 - 1.802z^3 + 1.093z^2 - 0.2616z + 0.02113}. \tag{B.3}$$

Table B.4 Parameters of the MIMO GPC controllers (Table 11.2)

	MIMO linear		MIMO nonlinear	
	Unconstrained (GPC)	Constrained (CGPC)	Unconstrained (GPC)	Constrained (CGPC)
N	40	40	40	40
N_u	20	10	40	10
λ	425	350	300	250
α	0	0	0	0

Following the same process described before (Sect. B.2.1), the parameters for the MPC approximations were selected; the control rate-limit was fixed at $-0.2 \le \Delta u \le 0.2$ p.u. and the strategy for saturation involving P_d/A_t is followed, see Sect. 11.3. Table B.4 summarises the set of parameter values selected for the cases evaluated for the MIMO GPC and CGPC (again $N = N_2 - N_1$).

References

1. Reynolds, T.S.: Stronger Than a Hundred Men: A History of the Vertical Water Wheel. Studies in the History of Technology. Johns Hopkins University Press, Baltimore (1983)
2. Denny, M.: The efficiency of overshot and undershot waterwheels. Eur. J. Phys. **25**, 193–202 (2004)
3. Layton Jr., E.T.: James B Francis and the rise of scientific technology. In: Pursell, C.W. (ed.) Technology in America: A History of Individuals and Ideas, pp. 92–104. The MIT Press, Cambridge (1990). ISBN 0-262-66067-9
4. Bindon, F.J.L.: Cwm Dyli hydro station. Power Eng. J. **4**, 265–269 (1990)
5. McNeil, I.: An Encyclopaedia of the History of Technology. Routledge, New York (1990)
6. Annual energy review 2009, US Energy Information Administration, Pittsburgh, PA, US (2010)
7. Roberts, B.: Capturing grid power. IEEE Power Energy Mag. **7**(July/August), 32–41 (2009)
8. Beaudin, M., Zareipour, H., Schellenberglabe, A., Rosehart, W.: Energy storage for mitigating the variability of renewable electricity sources: an updated review. Energy Sustain. Dev. **14**, 302–314 (2010)
9. Seoni, R.M., Shadeed, E.N., Simpson, R.J., Warnock, J.G.: Review of trends of large hydroelectric generating equipment. Proc. IEE **123**(10), 1138–1162 (1976)
10. Hiwassee Dam Unit: Hiwassee Dam Unit 2 Reversible Pump Turbine (1956), The American Society of Mechanical Engineers, New York (1981)
11. European Small Hydropower Association: Guide on How to Develop a Small Hydropower Plant, Brussels (2004)
12. Power Engineering Society IEEE: IEEE Standard 1020-1988: Guide for Control of Small Hydroelectric Power Plants IEEE. IEEE, New York (1988)
13. Briand, M.H., Ng, K.: Kinetic energy recovery turbine technology: resource assessment and site development strategy. In: World Energy Conference, Montreal, Canada, 2010
14. Khan, M.J., Bhuyan, G., Iqbal, M.T., Quaicoe, J.E.: Hydrokinetic energy conversion systems and assessment of horizontal and vertical axis turbines for river and tidal applications: a technology status review. Appl. Energy **86**, 1823–1835 (2009)
15. Lago, L.I., Ponta, F.L., Chen, L.: Advances and trends in hydrokinetic turbine systems. Energy Sustain. Dev. **14**, 287–296 (2010)
16. Güney, M.S., Kaygusuz, K.: Hydrokinetic energy conversion systems: a technology status review. Renew. Sustain. Energy Rev. **14**, 2996–3004 (2010)
17. The British Hydropower Association: A Guide to UK Mini-Hydro Developments, Gussage St Michael, UK (2005)
18. Jones, D.I.: Dynamic parameters for the national grid. Proc. IEE Gener. Transm. Distrib. **152**(1), 53–60 (2005)

G.A. Munoz-Hernandez et al., *Modelling and Controlling Hydropower Plants*, Advances in Industrial Control, DOI 10.1007/978-1-4471-2291-3,

19. Kundur, P.: Power System Stability and Control New York. Mc Graw Hill, New York (1994)
20. Erinmez, I.A., Bickers, D.O., Wood, G.F., Hung, W.W.: NGC experience with frequency control in England and Wales – Provision of frequency response by generators. In: IEEE PES 1999 Winter Meeting, vol. 1, pp. 590–593, New York (1998)
21. Vuorinen, A.: Controlling frequency and power balance. Wärtsilä Tech. J., January 10–13 (2009)
22. Koessler, R.J., Feltes, J.W., Willis, J.R.: A methodology for management of spinning reserve requirements. In: IEEE Winter Power Meeting, pp. 584–589, New York (1999)
23. Fasol, K.H.: A short history of hydropower control. IEEE Control Syst. Mag. **22**, 68–76 (2002)
24. Bissell, C.C.: A History of Automatic Control. Springer Handbook of Automation. Springer, London (2009)
25. Leum, M.: The development and field experience of a transistor electric governor for hydro turbines. IEEE Trans. Power Apparat. Syst. **85**, 393–400 (1966)
26. See reference [25]
27. Wylie, E.B., Streeter, V.L.: Fluid Transients. Advanced Book Program. McGraw Hill, New York (1978)
28. IEEE Working group on prime mover energy supply: Hydraulic turbine and turbine control model for system dynamic studies. IEEE Trans. Power Syst. **7**, 167–179 (1992)
29. Kishor, N., Saini, R.P., Singh, S.P.: A review on hydropower plant models and control. Renew. Sustain. Energy Rev. **11**, 776–796 (2007)
30. Thorne, D.H., Hill, E.F.: Extensions of stability boundaries of a hydraulic turbine generating unit. IEEE Trans. Power Apparat. Syst. **PAS-94**, 1401–1409 (1975)
31. Dhaliwal, N.S., Wichert, H.E.: Analysis of P.I.D. governors in multimachine system. IEEE Trans. Power Apparat. Syst. **PAS-97**, 456–463 (1978)
32. Hagihara, S.: Stability of a hydraulic turbine generating unit controlled by P.I.D governor. IEEE Trans. Power Apparat. Syst. **98**, 2294–2298 (1979)
33. Mansoor, S.P.: Behaviour and operation of pumped storage hydro plants. Ph.D. thesis, Bangor University, Bangor (2000)
34. Orelind, G., Wozniak, L., Medanic, J., Whitemore, T.: Optimal PID gain schedule for hydrogenerators – design and application. IEEE Trans. Energy Convers. **4**, 300–307 (1989)
35. Ye, L., Wei, S., Xu, H., Malik, O.P., Hope, G.S.: Variable structure and time-varying parameter control for hydroelectric generating unit. IEEE Trans. Energy Convers. **4**, 293–299 (1989)
36. Mansoor, S.P., Jones, D.I., King, D.J., Bradley, D.A., Aris, F.C., Jones, G.R.: Investigation into a new governor scheme for the Dinorwig pumped-storage plant. In: Proceedings of the International Conference in Hydropower and Dams, pp. 29–38, Kiris, Turkey (2002)
37. Lansberry, J.E., Wozniak, L.: Adaptive hydrogenerator governor tuning with a genetic algorithm. IEEE Trans. Energy Convers. **9**, 179–183 (1994)
38. Eker, I., Tumay, M.: Robust multivariable-cascade governors for hydroturbine controls. Electr. Eng. **84**, 229–237 (2002)
39. Hannett, L.N., Feltes, J.W., Fardanesh, B., Crean, W.: Modeling and control tuning of a hydro station with units sharing a common penstock section. IEEE Trans. Power Syst. **14**, 1407–1414 (1999)
40. Mansoor, S.P., Jones, D.I., Bradley, D.A., Aris, F.C., Jones, G.R.: Reproducing oscillatory behaviour of a hydroelectric power station by computer simulation. Control Eng. Pract. **8**, 1261–1272 (2000)
41. Jones, D.I.: Multivariable control analysis of a hydraulic turbine. Trans. Inst. Meas. Control **2**(3), 122–136 (1999)
42. Gomez-Exposito, A., Conejo, A.J., Canizares, C.: Electric Energy Systems: Analysis and Operation. CRC Press/Taylor & Francis/LLC, Boca Raton (2009)
43. Power Engineering Society IEEE: IEEE Standard 125-2007: Recommended Practice for Preparation of Equipment Specifications for Speed-Governing of Hydraulic Turbines Intended to Drive Electric Generators. IEEE, New York (2007)

44. Power Engineering Society IEEE: IEEE Standard 1010-2006: Guide for Control of Hydroelectric Power Plants. IEEE, New York (2006)
45. Power Engineering Society IEEE: IEEE Standard 1147-2005: Guide for the Rehabilitation of Hydroelectric Power Plants. IEEE, New York (2005)
46. Power Engineering Society IEEE: IEEE Standard 1207-2004: Guide for the Application of Turbine Governing Systems for Hydroelectric Generating Units. IEEE, New York (2004)
47. Power Engineering Society IEEE: IEEE Standard 1248-1998: Guide for the Commissioning of Electrical Systems in Hydroelectric Power Plants. IEEE, New York (1998)
48. Power Engineering Society IEEE: IEEE Standard 1249-1996: Guide for Computer-Based Control for Hydroelectric Power Plant Automation. IEEE, New York (1996)
49. Douglas, J.F., Matthews, R.D.: Solving Problems in Fluid Mechanics, vol. 1, 3rd edn. Addison Wesley, Harlow (1996)
50. Undrill, J., Woodward, J.: Calculation for determining temporary droop. IEEE Trans. Power Apparat. Syst. **PAS-86**, 443–451 (1966)
51. Central Electricity Generating Board: Report on Studies for Hydraulic System. Merz & McLellan, Newcastle, U.K. (1975)
52. Woodward, J.L.: Hydraulic-turbine transfer function for use in governing studies. Proc. IEE **115**, 424–426 (1968)
53. Thorne, D.H., Hill, E.F.: Field testing and simulation of hydraulic turbine governing performance. IEEE Trans. Power Apparat. Syst. **PAS-93**, 1183–1191 (1974)
54. Smith, J.R.: Assessment of hydro-turbine models for power-systems studies. IEE Proc. Part C **130**, 1–6 (1983)
55. Kvaerner Boving Ltd: Commissioning report for unit #1 replacement governors, 55 Montgomery Way, Radcliffe, Manchester, M26 3TG, UK (1995)
56. Schleif, F.R., Wilbor, A.B.: The coordination of hydraulic turbine governors for power system operation. IEEE Trans. Power Apparat. Syst. **PAS-85**, 750–756 (1966)
57. IEEE Committee Report: Dynamic models for hydro turbines in power system studies. IEEE Trans. Power Apparat. Syst. **Pas-92**(6), 1904–1915 (1973)
58. Vournas, C., Zaharakis, A.: Hydro turbine transfer functions with hydraulic coupling. IEEE Trans. Energy Convers. **8**, 527–532 (1993)
59. Ramey, D.G., Skooglund, J.W.: Detailed hydro-governor representation for system stability studies. IEEE Trans. Power Apparat. Syst. **PAS-89**, 106–112 (1970)
60. Warnick, C.C.: Hydropower Engineering. Prentice-Hall, Englewood Cliffs (1984)
61. Wozniak, L., Fett, G.H.: Conduit representation in closed loop simulation of hydraulic systems. ASME Publ. **71**, 1–5 (1971)
62. Hannett, L.N., Feltes, J.W., Fardanesh, B.: Field tests to validate hydro turbine-governor model structure and parameters. IEEE Trans. Power Syst. **9**, 1744–1751 (1994)
63. Munoz-Hernandez, G.A.: Application of model based predictive control to a pumped storage hydroelectric plant. Ph.D. thesis, Bangor University, Bangor (2005)
64. Hannett, L.N.: Field tests to validate hydro turbine-governor model structure and parameters. IEEE Trans. Power Syst. **9**, 1744–1750 (1994)
65. Dutton, K., Thompson, S., Barraclough, B.: The Art of Control Engineering. Addison-Wesley, Menlo Park (1998)
66. Sanathanan, C.K.: Accurate low order model for hydraulic turbine – penstock. IEEE Trans. Energy Convers. **EC-2**, 196–200 (1987)
67. Goodson, R.E.: Distributed system simulation using infinite product expansions. Simulation **15**, 255–263 (1970)
68. Azoury, P., Baasiri, M., Najm, H.: Effects of valve-closure schedule on water hammer. J. Hydraul. Eng. **112**(10), 890–903 (1986)
69. Paynter, H.M.: Methods and results from M.I.T studies in unsteady flow. BSCE J. **39**, 224–227 (1952)
70. Kung, C.S., Yang, X.L.: Energy interpretation of hydraulic transients in power plant with surge tank. J. Hydraul. Res. **31**, 825–839 (1993)

71. Vournas, C.D., Papaioannou, G.: Modelling and stability of a hydro plant with two surge tanks. IEEE Trans. Energy Convers. **10**, 368–375 (1994)
72. Parmakian, J.: Water Hammer Analysis. Dover Publication, New York (1963)
73. Sabersky, R.H., Acosta, A.J.: Fluid Flow. The Macmillan Company, New York (1975)
74. First Hydro Company: FHC Report D21: Head Loss Calculation for Dinorwig System, Llamberis (1995)
75. Nicolet, C., Avellan, F., Allenbach, P., Sapin, A., Simond, J.J.: New tool for the simulation of transient phenomena in Francis turbine power plants. In: Hydraulic Machinery and Systems 21st IAHR Symposium, Lausanne (2002)
76. Nicolet, C., Greiveldinger, B., Herou, J.J., Kawkabani, B., Allenbach, P., Simond, J.J., Avellan, F.: High-order modeling of hydraulic power plant in islanded power network. IEEE Trans. Power Syst. **22**(4), 1870–1880 (2007)
77. Larock, B.E., Jeppson, R.W., Watters, G.Z.: Hydraulics of Pipeline Systems. CRC Press, Boca Raton (1999)
78. de Mello, F.P., Concordia, C.: Concepts of synchronous machine stability as affected by excitation control. IEEE Trans. Power Apparat. Syst. **88**, 316–329 (1969)
79. Concordia, C.: Load representation in power system stability studies. IEEE Trans. Power Apparat. Syst. **PAS-101**, 969–977 (1982)
80. Berg, G.L.: Power system load representation. Proc. IEE **120**, 344–348 (1973)
81. IEEE Task Force: Load representation for dynamic performance analysis. IEEE Trans. Power Syst. **8**, 472–481 (1993)
82. Delfino, B., Denegri, G.B., Bonini, E.C., Marconato, R., Scarpellini, P.: Black-start and restoration of a part of the Italian HV network: modelling and simulation of a field test. IEEE Trans. Power Syst. **11**, 1371–1379 (1996)
83. Basanez, L., Riera, J., Ayza, J.: Modelling and simulation of multiarea power system load-frequency control. Math. Comput. Simul. **26**, 54–62 (1984)
84. Elgerd, O.L., Fosha, C.H.: Optimum megawatt-frequency control of multiarea electric energy systems. IEEE Trans. Power Apparat. Syst. **PAS-89**, 556–563 (1970)
85. Inoue, T., Taniguchi, H., Ikeguchi, Y., Yoshida, K.: Estimation of power system inertia constant and capacity of spinning-reserve support generators using measured frequency transients. IEEE Trans. Power Syst. **12**, 136–143 (1997)
86. Wright, R.M.: Understanding modern governor control. IEEE Trans. Energy Convers. **4**, 453–458 (1989)
87. Astrom, K.J., Hagglund, T.: PID Controllers: Theory, Design and Tuning, 2nd edn. ISA, Research Triangle Park (1995)
88. Franklin, G.F., Powell, J.D., Workman, M.L.: Digital Control of Dynamic Systems, 3rd edn. Addison-Wesley, Reading (2006)
89. ABB: Instruction Manual: ABB Turbine Governor HPC640 PC Diagram. ABB Generation, Sweden (1995)
90. Astrom, K.J., Wittenmark, B.: Adaptive Control. Addison-Wesley, Reading (1989)
91. Wellstead, P.E., Zarrop, M.B.: Self-Tuning Systems. Wiley, New York (1991)
92. Solartron Instruments: 1253 Gain-phase analyser operating manual. Schlumberger Technologies, Houston, TX, USA (1987)
93. Analogue Devices: Instruction Manual: Analogue Devices Voltage-to-Frequency Convertor AD537. Analogue Devices, Norwood (2000)
94. MathWorks: Using Matlab, Version 6. MathWorks, Natick (2000)
95. Undrill, J.M., Strauss, W.: Influence of hydro plant design on regulating and reserve response capacity. IEEE Trans. Power Apparat. Syst. **PAS-93**, 1192–1200 (1974)
96. Franklin, G.F., Powell, J.D.: Feedback Control of Dynamic Systems, 3rd edn. Prentice Hall, Upper Saddle River (2009)
97. Jones, D.I., Mansoor, S.P., Aris, F.C., Jones, G.R., Bradley, D.A., King, D.J.: A standard method for specifying the response of hydroelectric plant in frequency-control mode. Electr. Power Syst. Res. **68**, 19–32 (2004)

98. Munoz-Hernandez, G.A., Jones, D.I.: Applying generalized predictive control to a pumped-storage hydroelectric power station. In: IASTED MIC, pp. 380–385, Grindelwald, Switzerland, 2004
99. Munoz-Hernandez, G.A., Jones, D.I.: MIMO generalized predictive control for a hydroelectric power station. IEEE Trans. Energy Convers. **21**(4), 921–929 (2006)
100. Alpigini, J., Piovoso, M.J.: Visual tuning of controllers. IET Comput. Control Eng. **15**(3), 34–39 (2004)
101. Mansoor, S.P., Jones, D.I., Bradley, D.A., Aris, C., Jones, G.: Modelling of a pumped storage hydro power station for stability studies. Proceedings of 3rd International Conference "Hydropower into the Next Century", Gmunden, Austria, 535–544 (1999)
102. Mansoor, S.P.: Modelling of a multiple pump-storage units connected to a power system. In: Conference Proceeding PREP'99, Manchester, pp. 412–415 (1999)
103. Boldy, A.P.: Dinorwig Power Station Parametric Study, HydroSim Consultants, Wellesbourne, UK (1998)
104. Mansoor, S.P., Jones, D.I., Bradley, D.A., Aris, F.C.: Stability of a pumped storage hydropower station connected to a power system. In: IEEE Power Engineering Society Winter Meeting, New York, pp. 646–650 (1999)
105. Throckmorton, P.J., Wozniak, L.: A generic DSP-based real-time simulator with application to hydro-generator speed controller development. IEEE Trans. Energy Convers. **9**, 238–242 (1993)
106. Mansoor, S.P., Bradley, D.A., Aris, C., Jones, G.; The development of a hardware-in-the-loop simulation of a pump storage hydro power station. In: Proceedings of Mechatronics '98, Skövde, pp. 633–638, Pergamon, Elsevier Science Ltd. Oxford (1998)
107. Mansoor, S.P., Jones, D.I., Bradley, D.A., Aris, F.C., Jones, G.R.: Hardware-in-the-loop simulation of a pumped storage hydro station. Int. J. Power Energy Syst. **23**(2), 127–133 (2003)
108. Fang, H., Shen, Z.: Dynamic real-time simulator for hydraulic turbine generating unit based on programmable computer controller. In: Proceedings of 30th Conference IEEE Industrial Electronics Society, pp. 1502–1507, Busan (2004)
109. Li, X., Wu, C., Hu, X.: Full scope real-time simulation of hydropower plant for a training and research simulator. In. IEEE/PES Transmission and Distribution Conference & Exhibition, Dalian, China (2005)
110. Sharma, J.D., Kumar, A., Singhal, M.K.: Real time digital simulator for small hydropower plant. In: International Conference on Small Hydropower, Kandy, Sri Lanka (2007)
111. Stankovic, J.A.: Real-time computing. Byte **17**, 155–159 (1992)
112. Cooling, J.: Software Engineering for Real-Time Systems. Addison-Wesley, Essex (2003)
113. Peckol, J.K.: Embedded Systems: A Contemporary Design Tool. Wiley, Hoboken (2008)
114. dSPACE: dSPACE User's Guide. dSPACE GmbH. dSPACE, Paderborn (1996)
115. Astrom, K.J., Murray, R.M.: Feedback Systems. Princeton University Press, Princeton (2008)
116. Ogata, K.: Modern Control Engineering, 5th edn. Prentice Hall, Upper Saddle River (2009)
117. Hovey, L.M., Bateman, L.A.: Speed-regulation tests on a hydro station supplying an isolated load. AIEE Trans. **62**, 364–371 (1962)
118. Hovey, L.M.: Optimum adjustment of hydro governors on Manitoba hydro. AIEE Trans. **81**, 585 (1962)
119. Van Ness, J.E.: Root loci of load frequency control systems. IEEE Trans. Power Apparat. Syst. **S82**, 712–724 (1963)
120. Graham, D., Lathrop, R.C.: The synthesis of optimum responses: criteria and standard forms. Trans. Am. Inst. Electr. Eng. **72**, 273–288 (1953)
121. Mason, S.J.: Feedback theory: some properties of signal flow graphs. Proc. IRE **41**, 1144–1156 (1953)
122. Power Engineering Society IEEE: IEEE Standard 122-1991: Recommended Practice for Functional and Performance Characteristics of Control Systems for Steam Turbine-Generator Units. IEEE, New York (1992)
123. Rossiter, J.A.: Model-Based Predictive Control. CRC Press, Boca Raton (2003)

124. Peng, Y., Vrancic, D., Hanus, R.: Anti-windup, bumpless, and conditioned transfer techniques for PID controllers. IEEE Control Syst. Mag. **16**, 48–57 (1996)
125. Bohn, C., Atherton, D.P.: An analysis package comparing PID anti-windup strategies. IEEE Control Syst. Mag. **15**, 34–40 (1995)
126. Goodwin, G.C., Graebe, S.F., Salgado, M.E.: Control System Design. Prentice Hall, Upper Saddle River (2001)
127. Jones, D.I., Mansoor, S.P.: Predictive feedforward control for a hydroelectric plant. IEEE Trans. Control Syst. Technol. **12**, 956–965 (2004)
128. Anderson, P.M., Mirheydar, M.: A low-order frequency response model. IEEE Trans. Power Syst. **5**(3), 720–729 (1990)
129. Ljung, L.: System Identification: Theory for the User, 2nd edn. Prentice Hall, Upper Saddle River (1999)
130. Ljung, L.: Systems Identification Toolbox v5. User's Guide. MathWorks, Natick (2000)
131. Jones, D.I.: Estimation of power system parameters. IEEE Trans. Power Syst. **19**(4), 1980–1989 (2004)
132. Prasad, G., Irwin, G.W., Swidenbank, E., Hogg, B.W.: A hierarchical physical model-based approach to predictive control of a thermal power plant for efficient plant-wide disturbance rejection. Trans. Inst. Meas. Control **24**(2), 107–128 (2002)
133. Rossiter, J.A., Neal, P.W., Yao, L.: Applying predictive control to a fossil-fired power station. Trans. Inst. Meas. Control **3**, 177–194 (2002)
134. Sansevero, G., Pascoli Botura, C.: Model predictive control algorithm for Francis hydro turbo-generators. In: Waterpower XIII, Buffalo, NY (USA) (2003)
135. Munoz-Hernandez, G.A., Jones, D.I.: Simulation studies of a GPC controller for a hydroelectric plant. Trans. Inst. Meas. Control **29**(1), 35–51 (2007)
136. Camacho, E.F., Bordons, C.: Model Predictive Control, 2nd edn. Springer, London (2004)
137. Maciejowski, J.: Predictive Control with Constraints. Prentice Hall, Essex (2001)
138. Richalet, J., Rault, A., Testud, J.L., Papon, J.: Model predictive heuristic control: applications to industrial processes. Automatica **14**, 413–428 (1978)
139. Peterka, V.: Predictor-based self-tuning control. Automatica **20**, 39–50 (1984)
140. De Keyser, R.M.C., Van de Velde, G.A., Dumortier, F.A.G.: A comparative study of self-adaptive long-range control methods. Automatica **24**, 149–163 (1988)
141. Clarke, D.W., Mohtadi, C., Tuffs, P.S.: Generalized predictive control (Part I: The basic algorithm). Automatica **23**, 137–148 (1987)
142. Ramond, G., Dumur, D., Libaux, A., Boucher, P.: Direct adaptive predictive control of an hydro-electric plant. In: IEEE International Conference on Control Applications, pp. 606–611, Mexico (2001)
143. Clarke, D.W., Mohtadi, C., Tuffs, P.S.: Generalized predictive control (Part II: Extensions and interpretations). Automatica **23**, 149–160 (1987)
144. Richalet, J.: Industrial applications of model based predictive control. Automatica **29**, 1251–1274 (1993)
145. Clarke, D.W.: Application of generalized predictive control to industrial processes. IEEE Control Syst. Mag. **8**, 49–55 (1988)
146. Lu, S., Hogg, B.W.: Predictive co-ordinated control for power-plant steam pressure and power output. Control Eng. Prac. **5**, 79–84 (1997)
147. Bordons, C., Camacho, E.F.: A generalized predictive controller for a wide class of industrial processes. IEEE Trans. Control Syst. Technol. **6**, 372–387 (1998)
148. Chow, C., Kuznetsov, A., Clarke, D.W.: Application of multivariable generalised predictive control to the simulink model of a paper machine. Control Appl. **3**, 1675–1691 (1994)
149. Clarke, D.W. (ed.): Advances in Model-Based Predictive Control, pp. 3–21, Oxford Science, Midsomer Norton (1994)
150. Tsang, T.T.C., Clarke, D.W.: Generalised predictive control with input constraints. IEE Proc. Part D **135**, 451–460 (1988)
151. Rossiter, J.A., Kouvaritakis, B.: Constrained stable generalised predictive control. IEE Proc. Part D **140**, 243–254 (1993)

152. Banerjee, P., Shah, S.L.: Tuning guidelines for robust generalized predictive control. In: Conference on Decision and Control, pp. 3233–3234, Tucson, AZ (USA) (1992)
153. Bemporad, A., Morari, M.H.: Control of systems integrating logic, dynamics, and constraints. Automatica **35**, 407–427 (1999)
154. Bemporad, A., Torrisi, F.D., Morari, M.: Performance analysis of piecewise linear systems and model predictive control systems. In: Conference on Decision and Control, pp. 4957–4962, Sydney (2000)
155. Witsenhausen, H.S.: A class of hybrid-state continuous-time dynamic systems. IEEE Trans. Autom. Control **11**, 161–167 (1966)
156. Branicky, M.S., Borkar, V.S., Mitter, S.K.: A unified framework for hybrid control: model and optimal control theory. IEEE Trans. Autom. Control **43**, 31–45 (1998)
157. Gollu, A., Varaiya, P.: Hybrid dynamical systems. In: 28th Conference on Decision and Control, pp. 2708–2712. Omni Press, Tampa (1989)
158. Cavalier, T.M., Pardalos, P.M., Soyster, A.L.: Modelling and integer programming techniques applied to propositional calculus. Comput. Oper. Res. **17**, 561–570 (1990)
159. Williams, H.P.: Logic applied to integer programming and integer programming applied to logic. Eur. J. Oper. Res. **81**, 605–616 (1995)
160. Tyler, M.L., Morari, M.: Propositional logic in control and monitoring problems. Automatica **35**, 565–582 (1999)
161. Munoz-Hernandez, G.A., Jones, D.I., Fuentes-Goiz, S.I.: Modelling and simulation of a hydroelectric power station using MLD. In: Proceedings of the 15th International Conference on Electronics, Communications and Computers (CONIELECOMP), Cholula, Pue, Mexico, 2005
162. Munoz-Hernandez, G.A., Jones, D.I., Mansoor, S.P.: Evaluation of a MLD predictive control in a nonlinear model of a power station. In: IASTED Modelling and Simulation, pp. 422–427, Cancun, Acta Press (2005)
163. Munoz-Hernandez, G.A., Gracios-Marin, C.A., Diaz-Sanchez, A., Mansoor, S.P., Jones, D.I.: Implementing a mixed logical dynamical model of a hydroelectric power station in a FPGA system. In: 2nd International Conference on Electronic Design (ICED), pp. 136–140, Veracruz (2006)
164. Lu, Q.: SDM hybrid control approach for nonlinear systems and application to power systems. In: IEEE International Conference on Power Control, pp. 19–23, China (2002)
165. Thomas, J.J., Dumur, D., Buisson, D., Falinower, C.: Moving horizon state estimation of hybrid systems. Application to fault detection of sensors of a steam generator. In: Conference on Control Applications, pp. 1375–1380 (2003)
166. Mansoor, S.P., Munoz-Hernandez, G.A.: Development of gain scheduling scheme for pumped storage plant. In: IASTED International Conference on Intelligent Systems and Control, pp. 13–18, Honolulu (2006)
167. Ulieru, M.: Approaching intelligent control systems design. In: IEEE International Conference on Intelligent Systems for the 21st Century, pp. 563–568 (1995)
168. Lu, N., Chow, J.H., Desrochers, A.A.: Pumped-Storage hydro-turbine bidding strategies in a competitive electricity market. IEEE Trans. Power Syst. **19**, 834–841 (2004)
169. Pereira, M.V.F., McCoy, M.F., Merrill, H.M.: Managing risk in the new power business. IEEE Comput. Appl. Power **13**, 18–24 (2000)
170. Song, Y., Johns, A.T.: Applications of fuzzy logic in power systems. Power Eng. J. **13**, 97–103 (1999)
171. King, D.J., Bradley, D.A., Mansoor, S.P., Jones, D.I., Aris, F.C., Jones, G.R.: Using a fuzzy inference system to control a pumped storage hydro plant. IEEE Intl. Fuzzy Syst. **3**, 1008–1011 (2001)
172. Gallestey, E., Stothert, A., Antoine, M., Morton, S.: Model predictive control and the optimization of power plant load while considering lifetime consumption. IEEE Trans. Power Syst. **186–191**, 17 (2002)

173. Chang, G.W., Aganagic, M., Waight, J.G., Medina, J., Burton, T., Reeves, S., Christoforidis, M.: Experiences with mixed integer linear programming based approaches on short-term hydro scheduling. IEEE Trans. Power Syst. **16**(4), 743–749 (2001)
174. Qin, S.J., Badgwell, T.A.: A survey of industrial model predictive control technology. Control Eng. Pract. **11**, 733–764 (2003)
175. Bemporad, A., Morari, M., Dua, V., Pistikopoulos, E.: The explicit linear quadratic regulator for constrained systems. Automatica **38**(1), 3–20 (2002)
176. Bemporad, A., Borrelli, F., Morari, M.: Model predictive control based on linear programming-the explicit solution. IEEE Trans. Autom. Control **47**(12), 1974–1985 (2002)
177. Baotic, M., Christophersen, F.J., Morari, M.: Constrained optimal control of hybrid systems with a linear performance index. IEEE Trans. Autom. Control **51**(12), 1903–1919 (2006)
178. Zeilinger, M.N., Jones, C.N., Morari, M.: Real-time suboptimal model predictive control using a combination of explicit MPC and online optimization. IEEE Trans. Autom. Control **56**(7), pp. 1524–1534 (2011)
179. Bleris, L.G., Kothare, M.V., Garcia, J., Arnold, M.G.: Embedded model predictive control for system-on-a-chip applications. In: 7th International Symposium on Dynamics and Control of Process Systems, Boston, July 2004
180. Ling, K.V., Yue, S.P., Maciejowski, J.M.: A FPGA implementation of model predictive control (I). In: American Control Conference, p. Paper WeC15.4, Minneapolis (2006)
181. Bleris, L.G., Vouzis, P.D., Arnold, M.G., Kothare, M.V.: A co-processor FPGA platform for the implementation of real-time model predictive control. In: American Control Conference, p. Paper WeC15.1, Minneapolis (2006)
182. Lin-Shi, X., Morel, F., Llor, A.M., Allard, B., Retif, J.M.: Implementation of hybrid control for motor drives. IEEE Trans. Ind. Electron. **54**(4), 1946–1952 (2007)
183. Naouar, M.W., Naassani, A.A., Monmasson, E., Slama-Belkhodja, I.: FPGA-based predictive current controller for synchronous machine speed drive. IEEE Trans. Power Electron. **23**(4), 2115–2126 (2008)
184. Vouzis, P.D., Bleris, L.G., Arnold, M.G., Kothare, M.V.: A system-on-a-chip implementation for embedded real-time model predictive control. IEEE Trans. Control Syst. Technol. **17**(5), 1006–1017 (2009)
185. Shoukry, Y., El-Kharashi, M.V., Hammad, S.: MPC-on-chip: an embedded GPC coprocessor for automotive active suspension systems. IEEE Embed. Syst. Lett. **2**(2), 31–34 (2010)
186. International Energy Outlook 2010. U.S. Energy Information Administration, Washington, DC (2010)
187. The UK Renewable Energy Strategy. The Stationery Office, London (2009)
188. MacKay, D.J.C.: Sustainable Energy – Without the Hot Air. UIT Cambridge, Cambridge (2009)
189. Jevons, W.S.: The Coal Question: An Enquiry Concerning the Progress of the Nation and the Probable Exhaustion of Our Coal Mines, 2nd edn. Macmillan, London (1866)
190. Tuohy, A., O'Malley, M.J.: Impact of pumped storage on power systems with increasing wind penetration. In: IEEE Power & Energy Society General Meeting, Calgary (2009)
191. Soder, L., Hofmann, L., Orths, A., Holttinen, H., Wan, Y.H., Tuohy, A.: Experience from wind integration in some high penetration areas. IEEE Trans. Energy Convers. **22**(1), 4–12 (2007)
192. Taulan, J.P., Laurier, P., Bourrilhon, M., Bornard, L.: Pump-turbine integration in renewable energy systems. In: Waterpower XVI, Spokane (2009)
193. Schoppe, C.: Wind and Pumped-Hydro Power Storage: Determining Optimal Commitment Policies with Knowledge Gradient Non-Parametric Estimation. Princeton University, Princeton (2010)
194. Bakos, G.C.: Feasibility study of a hybrid wind/hydro power-system for low-cost electricity production. Appl. Energy **72**, 599–608 (2002)
195. Cheng, E.: Economic viability of a wind-powered hydroelectric system. J. Wind. Eng. Ind. Aerodyn. **74–76**, 411–418 (1998)
196. Mitteregger, A.: Austrian pumped storage power stations supply peak demands. World Pumps, 16–21 May (2008)

197. Koutnik, J., Foust, J., Nicolet, C., Saiju, R., Kawkabani, B.: Pump-storage integration with renewables – meeting the needs using various concepts. In: Hydrovision International, Charlotte, 2010
198. Kuwabara, T., Shibuya, A., Furuta, H., Kita, E., Mitsuhashi, K.: Design and dynamic response characteristics of 400MW adjustable speed pumped storage unit for Ohkawachi power station. IEEE Trans. Energy Convers. **11**(2), 376–384 (1996)
199. Ansel, A., Robyns, B.: Modelling and simulation of an autonomous variable speed micro hydropower station. Math. Comput. Simul. **71**, 320–332 (2006)
200. Breban, S., Ansel, A., Nasser, M., Robyns, B., Radulescu, M.M.: Experimental results on a variable-speed small hydro power station feeding isolated loads or connected to power grid. In: IEEE International Aegean Conference on Electrical Machines and Power Electronics, pp. 760–765, Bodrum, Turkey (2007)
201. Fairley, P.: Flywheels keep the grid in tune. IEEE Spectr. **48**, 14–16 (2011)
202. Yamuna Rani, K., Unbehauens, H.: Study of predictive controller tuning methods. Automatica **33**(12), 2243–2248 (1997)
203. Garcia, C.E., Prett, D.M., Morari, M.: Model predictive control: theory and practice a survey. Automatica **25**(3), 335–348 (1989)
204. Shridhar, R., Cooper, D.J.: A tuning strategy for unconstrained SISO model predictive control. Ind. Eng. Chem. Res. **36**, 729–746 (1997)
205. Shridhar, R., Cooper, D.J.: A novel tuning strategy for multivariable model predictive control. ISA Trans. **36**(4), 273–280 (1998)
206. Wibowo, T.C.S., Saad, N., Karsiti, M.N.: A heuristic approach for tuning model predictive controller. Elektrika **11**(1), 8–14 (2009)
207. Di Cairano, S., Bemporad, A.: Model predictive control tuning by controller matching. IEEE Trans. Autom. Control **55**(1), 185–190 (2010)

Index

G.A. Munoz-Hernandez et al., *Modelling and Controlling Hydropower Plants*, Advances in Industrial Control, DOI 10.1007/978-1-4471-2291-3,

W

Z

Printed by Printforce, the Netherlands